Using the Weibull Distribution

Using the Weibull Distribution

Reliability, Modeling, and Inference

JOHN I. McCOOL

A JOHN WILEY & SONS, INC., PUBLICATION

For general information on our other products and services or for technical support, please contact our Customer Care Department within the United States at (800) 762-2974, outside the United States at (317) 572-3993 or fax (317) 572-4002.

Wiley also publishes its books in a variety of electronic formats. Some content that appears in print may not be available in electronic formats. For more information about Wiley products, visit our web site at www.wiley.com.

Library of Congress Cataloging-in-Publication Data:

McCool, John, 1936–
  Using the Weibull distribution : reliability, modeling, and inference / John I. McCool.
      p. cm.
  Includes bibliographical references and index.
  ISBN 978-1-118-21798-6 (cloth)
  1. Weibull distribution—Textbooks.   2. Probabilities—Textbooks.   I. Title.
  QA273.6.M38 2012
  519.2'4–dc23

                                                        2012002909

Printed in the United States of America

10 9 8 7 6 5 4 3 2 1

To Jim, Fran, and Peg
and
To Mary, John, and Jen

Contents

5. Estimation in Single Samples   130

Preface

This book grew out of my experience as a young mechanical engineer working in the research organization of the U.S. subsidiary of SKF, an international ball and roller bearing manufacturer. The ball and roller bearing industry adopted the use of the Weibull distribution to describe the fatigue life of its products back in the 1940s and before the appearance of Weibull's influential article in 1951 that set the stage for its present enormous popularity (Weibull, 1951). I have tried to write the book that I would wish to have if I were today a young engineer asked to become my company's "Weibull Guy."

I began to organize the material in order to teach a number of short courses sponsored by the American Society of Mechanical Engineers (ASME), the U.S. Navy, and later by the American Bearing Manufacturers Association (ABMA). The book benefited from my experience as an adjunct and, since 1988, as a full-time Professor of Systems Engineering at Penn State's School of Graduate Professional Studies in Malvern, Pennsylvania, where I have taught master's level courses in statistics, quality control, and reliability engineering, among others. I have twice used a draft of the book as the text for the Reliability Engineering course in the Master of Systems Engineering curriculum. A sabbatical provided the opportunity to put much of the material into its present form.

The book has also benefited immensely from the opportunity I had while with SKF to develop techniques for inference on the Weibull distribution under the sponsorship of the Aerospace Research Laboratory at Wright Patterson Air Force Base and the Air Force Office of Scientific Research.

The availability of the digital computer is responsible for much of the progress in the development of tools for inference for the Weibull distribution that has taken place since the pioneering work of Lieblein and Zelen (1956). They used a computer to determine the variances and covariances of order statistics needed in the construction of best linear unbiased estimates of the Weibull parameters. Later the computer was essential for making practical the conduct of Monte Carlo determinations of quantities needed for inference and for the computation of estimates for individual samples or sets of samples using the method of maximum likelihood.

The astonishing evolution of personal computing now makes sophisticated inferential techniques accessible from the desktop of the engineer or scientist. Although available computing power is more than equal to the task, many powerful techniques for extracting the most information from expensive data have not yet been as widely adopted as they should because appropriate software has been unavailable. It is my hope that this book and the software that accompanies it will help rectify that situation.

The use of software developed by a succession of talented programmers who have helped me over the years is illustrated in this book. The software may be downloaded for free from my website http://www.personal.psu.edu/mpt. It includes a set of disk operating system (DOS) programs that should serve until such a time as these capabilities become incorporated into commercial software packages. There is also a set of modules written in Mathcad, which both document and perform the calculations associated with parameter estimation in various settings and with other quantitative models such as optimum age replacement and optimum burn-in. Finally it includes some graphical interface software that performs simulations needed for inference in single and multiple samples from the two-parameter Weibull, inference on the location parameter of the three-parameter Weibull, and for determining critical values for a test of goodness of fit.

The reader will find that the exposition is liberally accompanied by numerical examples. All of the chapters conclude with a set of exercises that may be used to test the reader's mastery of the material or as class exercises in a short course.

The Weibull distribution cannot be discussed in isolation from the rest of statistical methodology. A knowledge of elementary probability theory and of distributions such as the binomial, Poisson, exponential, normal, and lognormal is essential for its effective application. The book is intended to be self-contained in this regard, with the inclusion of two chapters treating the fundamentals of probability and statistics needed for what follows.

Chapter 1 introduces the ideas of mutual exclusivity, independence, conditional probability, and the Law of Total probability as they impact reliability-related calculations. The chapter concludes with the application of probability principles to the computation of the reliability of systems in terms of the reliability of its components. Combinations of series, parallel, and crosslinked systems are considered. The idea of the reliability importance of a component is introduced.

Chapter 2 describes singly and jointly distributed discrete and continuous random variables and the concepts of the mean, variance, and covariance. The binomial, Poisson, and geometric discrete random variables are introduced. The binomial distribution is applied to the task of computing the reliability of a system that comprises n identical components and functions as long as at least k ($\leq n$) of the components function successfully. The uniform, normal, and lognormal continuous random variables are discussed and their use is illustrated with examples. The hazard function is introduced and its relationship to the reliability and density functions is explained. Finally the use of the inverse transformation method for simulating samples from a continuous distribution is described and illustrated by an example.

Chapter 3 enumerates the properties of the Weibull distribution. It shows the Weibull to be a generalization of the exponential distribution. It gives the expressions for the mean, variance, mode, skewness, hazard function, and quantiles of the Weibull distribution in terms of its two parameters. The logarithmic transform of a Weibull random variable is shown to follow the distribution of smallest extremes. The power transformation of a Weibull random variable is shown to map it into a different member of the Weibull family. The Weibull distribution conditional on exceeding a specified value is derived and shown to apply to the interpretation of the practice of burn-in used to improve the life of electronic equipment. The computation of the mean residual life is discussed as is the simulation of samples from a Weibull population.

Chapter 4 describes a number of useful applications of the two-parameter Weibull distribution valid when its parameters are known or assumed. This includes the distribution of mixtures of Weibull random variables, the computation of $P(Y < X)$ when X and Y are both Weibull distributed with a common shape parameter, the Weibull distribution of radial error when the location errors in two orthogonal directions are independent and are normally distributed with a common variance. Three types of warranties are discussed (i) a pro rata warranty, (ii) a free replacement warranty, and (iii) a renewing free replacement warranty. Two preventive maintenance strategies are considered: (i) age replacement and (ii) block replacement. Also discussed are optimum bidding in a sealed bid competition and spare parts provisioning when the life distribution is exponential. Weibull renewal theory is discussed and applied to the analysis of the block replacement warranty.

Chapter 5 treats estimation in single samples. The notions of bias and precision of estimation are illustrated by means of the sampling distributions of two competing estimators of the 10th percentile of a Weibull population. Graphical estimation, which for years was the standard means of estimating the Weibull parameters, is explained and various choices of plotting positions are described and compared. Hazard plotting and the Kaplan–Meier method for graphical plotting of randomly censored data are explained. The method of maximum likelihood is introduced and applied to the exponential special case of the Weibull. Complete and type II censoring is discussed. Maximum likelihood estimation for the exponential is applied for the Weibull when the shape parameter is known. Software is illustrated for maximum likelihood estimation of the Weibull parameters in complete and type II censored samples when both parameters are unknown. Exact interval estimation of the shape parameter and percentiles is considered for complete and type II censored samples. Asymptotic results generally applied in the type I censored case are explained and a Mathcad module is given for performing those calculations.

Chapter 6 deals with sample size selection, as well as hypothesis testing. The question most frequently asked by experimenters of their statistical gurus is "what sample size do I need?" Two approaches are offered in this chapter both of which have their parallels in normal distribution theory. One approach is to display for each sample size a precision measure that reflects the tightness of confidence limits on the shape parameter and/or a percentile. The experimenter must express a goal or target value for either of these precision measures, and the needed sample size is determined by a table lookup. The second approach is in terms of the operating characteristic function for a hypothesis test. The experimenter specifies a target value of the shape parameter or a percentile and the desired large probability that the sample will be accepted if the population value equals that target value. The experimenter also must specify an undesirable value of the percentile or shape parameter and the desired small probability that a population having that value will be accepted. Both approaches rely on percentage points of certain pivotal quantities only sparsely available in the current literature. Values are available for just a few of the Weibull percentiles of interest and for limited combinations of sample sizes and numbers of failures. Software described fully in Chapter 7 performs the simulations needed to remove these limitations.

Chapter 6 concludes with a discussion of how to test the hypothesis that a data sample was drawn from a two-parameter Weibull distribution. The discussion distinguishes between the completely specified case in which the Weibull parameters are specified and the more usual case in which the parameters are estimated by the method of maximum likelihood. The computations for the Kolmogorov–Smirnov test are illustrated for an uncensored sample, but the discussion mainly focuses on the Anderson–Darling (AD) test, which has been shown in many studies to be among the more powerful methods of testing goodness of fit to the Weibull. Critical values of the AD statistic may be found in the literature but are not available for many combinations of sample sizes and censoring amounts. A simulation program called ADStat is introduced to overcome this problem. It computes 19 percentage points of the distribution of the AD statistic. It handles complete and type II censored data and will accommodate both the case where the Weibull parameters are completely specified and the case where they need to be estimated. The software will also compute the AD statistic for the user's data sample if needed.

The Weibull and lognormal distribution often compete as models for life test data. Two methods are described and illustrated by examples for testing whether an uncensored sample follows the lognormal or the two-parameter Weibull distribution.

Chapter 7 is devoted to an exposition of the features of the simulation software program Pivotal.exe. Exact inference, that is, confidence limits and hypothesis tests for the Weibull distribution parameters and percentiles based on maximum likelihood estimation in complete or type II censored samples, requires the determination via simulation of percentage points of the distribution of certain pivotal quantities. The program Pivotal.exe described in this chapter performs these calculations for user-selected sample sizes and percentiles. The software is also applicable to inference on series systems of identical Weibull-distributed components and for analyzing the results of sudden death tests.

The software allows the output of 10,000 paired values of the ML estimates of the Weibull shape and scale parameters, which can be post-processed using spreadsheet-based software to provide: (i) operating characteristic curves for hypothesis tests on the Weibull shape parameter or a percentile (ii) confidence intervals on reliability at a specified life and (iii) prediction intervals for a future value.

Chapter 8 is concerned with inference from multiple samples. It discusses how to make best use of the data that result when, not uncommonly, a set of tests is performed differing with respect to the level of some factor such as a design feature, a material, or a lubricant type. Provided that it can be assumed that the shape parameter is the same among the sampled populations, data may be pooled to provide tighter confidence limits for the common shape parameter and for the percentiles of interest among the individual populations. A hypothesis test is provided for assessing whether the common shape parameter assumption is tenable and a simulation software program is described for generating the critical values needed for conducting the test among k sets of samples of size n which are complete or censored at the r-th order statistic. A test for the equality of the scale parameters among the tests is also given which is analogous to the one-way analysis of variance in normal distribution theory as is a multiple comparison test for differences among the scale parameters. Too often in published work reporting on sets of tests, the analysis goes no further than the subjective assessment of a set of Weibull plots.

The chapter also contains a method for setting confidence limits for the value of $P(X < Y)$ when X and Y are Weibull distributed with a common shape parameter based on random samples drawn from the two distributions. $P(X < Y)$ is considered the reliability when X represents a random stress and Y a random strength.

Chapter 9, Weibull Regression, refers to the situation where testing is conducted at several levels of a quantitative factor that might be a stress or load. The Weibull scale parameter is assumed to vary as a power function of the level to which this factor is set, while the shape parameter is assumed not to vary with the level of the factor. Such a model finds use in the analysis of accelerated tests wherein the purpose is to complete testing quickly by running the tests at stresses higher than encountered under use conditions. The fitted model is then used to extrapolate to the lower stress values representative of "use" conditions. The factor need not necessarily be one that degrades performance. The model can account for the scale parameter either increasing or decreasing as a power of the factor level. It is shown that exact confidence intervals may be computed for the power function exponent, the shape parameter, and a percentile of the distribution at any level of the factor. Tables of the critical values are given for two and three levels of the factor. The chapter includes a description of a DOS program for performing the calculations to analyze a set of data, along with a Mathcad module that illustrates the solution of the relevant equations.

The three-parameter Weibull distribution is covered in Chapter 10. It generalizes the two-parameter Weibull to include the location or threshold parameter representing an amount by which the probability density function is offset to the right. In the context of life testing, this offset represents a guarantee time prior to which failure cannot occur. Tables are given for a range of sample sizes, whereby one may (i) test whether the location parameter is zero and (ii) determine a lower confidence limit for the location parameter. A DOS program for performing the computations on a data set is described and a simulation program is given for extending the tabled values to other sample sizes and for exploring the power of the test.

Chapter 11 is entitled Factorial Experiments with Weibull Response. Tests conducted at all combinations of the levels of two or more factors are called factorial experiments. Factorial experiments have been shown to be more efficient in exploring the effects of external factors on a response variable than nonfactorial arrangements of factor levels. In this chapter we present a methodology for the analysis of Weibull-distributed data obtained at all combinations of the levels of two factors. Item life is assumed to follow the two-parameter Weibull distribution with a shape parameter that, although unknown, does not vary with the factor levels. The purpose of the analysis is (i) to compute interval estimates of the common shape parameter and (ii) to assess whether either factor has a multiplicative effect on the Weibull scale parameter and hence on any percentile of the distribution. A DOS program is included for performing the analysis. Tables are given for hypothesis testing for various uncensored sample sizes for the 2×2, 2×3, and 3×3 designs.

I am grateful for support for my work by H.L. Harter, formerly of the Air Force Aerospace Research Laboratory at Wright Patterson Air Force Base, and by the late I. Shimi of the Air Force Office of Scientific Research. I am grateful as well for the help of a number of people in developing the software and tables included in this volume. Mr. John C. Shoemaker, a colleague at SKF, wrote the software used for generating the tables of the distribution of pivotal quantities originally published as an Air Force report. Mr. Ted Staub, when a Penn State student, developed an early version of the Pivotal.exe software, which generates the distribution of pivotal quantities for user-selected sample sizes. Another student, Donny Leung, provided the user interface for Pivotal. exe and extended it produce the Multi-Weibull software. A colleague, Pam Vercellone, helped refine Pivotal.exe and Multi-Weibull. Student Nimit Mehta developed the ADStat software described in Chapter 6. Student Christopher Garrell developed the LocationPivotal software described in Chapter 10.

Finally, I would like to acknowledge the wishes of my grandchildren Jaqueline and Jake to see their names in a book.

JOHN I. McCool

REFERENCES

Lieblein, J. and M. Zelen. 1956. Statistical investigation of the fatigue life of deep-groove ball bearings. *Journal of Research of the National Bureau of Standards* 57(5): 273–316.

Weibull, W. 1951. A statistical distribution function of wide applicability. *Journal of Applied Mechanics* 18(3): 293–297.

CHAPTER 1

Probability

The study of reliability engineering requires an understanding of the fundamentals of probability theory. In this chapter these fundamentals are described and illustrated by examples. They are applied in Sections 1.8 to 1.13 to the computation of the reliability of variously configured systems in terms of the reliability of the system's components and the way in which the components are arranged.

Probability is a numerical measure that expresses, as a number between 0 and 1, the degree of certainty that a specific outcome will occur when some random experiment is conducted. The term random experiment refers to any act whose outcome cannot be predicted. Coin and die tossing are examples. A probability of 0 is taken to mean that the outcome will never occur. A probability of 1.0 means that the outcome is certain to occur. The relative frequency interpretation is that the probability is the limit as the number of trials N grows large, of the ratio of the number of times that the outcome of interest occurs divided by the number of trials, that is,

$$p = \lim_{N \to \infty} \frac{n}{N} \qquad (1.1)$$

where n denotes the number of times that the event in question occurs. As will be seen, it is sometimes possible to deduce p by making assumptions about the relative likelihood of all of the other events that could occur. Often this is not possible, however, and an experimental determination must be made. Since it is impossible to conduct an infinite number of trials, the probability determined from a finite value of N, however large, is considered an estimate of p and is distinguished from the unknown true value by an overstrike, most usually a caret, that is, $\hat{p}$.

Using the Weibull Distribution: Reliability, Modeling, and Inference, First Edition. John I. McCool.
© 2012 John Wiley & Sons, Inc. Published 2012 by John Wiley & Sons, Inc.

1.1 SAMPLE SPACES AND EVENTS

The relationship among probabilities is generally discussed in the language of set theory. The set of outcomes that can possibly occur when the random experiment is conducted is termed the sample space. This set is often referred to by the symbol Ω. As an example, when a single die is tossed with the intent of observing the number of spots on the upward face, the sample space consists of the set of numbers from 1 to 6. This may be noted symbolically as $\Omega = \{1, 2, \dots 6\}$. When a card is drawn from a bridge deck for the purpose of determining its suit, the sample space may be written: $\Omega = \{\text{diamond, heart, club, spade}\}$. On the other hand, if the purpose of the experiment is to determine the value and suit of the card, the sample space will contain the 52 possible combinations of value and suit. The detail needed in a sample space description thus depends on the purpose of the experiment. When a coin is flipped and the upward face is identified, the sample space is $\Omega = \{\text{Head, Tail}\}$. At a more practical level, when a commercial product is put into service and observed for a fixed amount of time such as a predefined mission time or a warranty period, and its functioning state is assessed at the end of that period, the sample space is $\Omega = \{\text{functioning, not functioning}\}$ or more succinctly, $\Omega = \{S, F\}$ for success and failure. This sample space could also be made more elaborate if it were necessary to distinguish among failure modes or to describe levels of partial failure.

Various outcomes of interest associated with the experiment are called *Events* and are subsets of the sample space. For example, in the die tossing experiment, if we agree that an event named A occurs when the number on the upward face of a tossed die is a 1 or a 6, then the corresponding subset is $A = \{1, 6\}$. The individual members of the sample space are known as elementary events. If the event B is defined by the phrase "an even number is tossed," then the set B is $\{2, 4, 6\}$. In the card example, an event C defined by "card suit is red" would define the subset $C = \{\text{diamond, heart}\}$. Notationally, the probability that some event "E" occurs is denoted P(E). Since the sample space comprises all of the possible elementary outcomes, one must have $P(\Omega) = 1.0$.

1.2 MUTUALLY EXCLUSIVE EVENTS

Two events are mutually exclusive if they do not have any elementary events in common. For example, in the die tossing case, the events $A = \{1, 2\}$ and $B = \{3, 4\}$ are mutually exclusive. If the event A occurred, it implies that the event B did not. On the other hand, the same event A and the event $C = \{2, 3, 4\}$ are not mutually exclusive since, if the upward face turned out to be a 2, both A and C will have occurred. The elementary event "2" belongs to the intersection of sets A and C. The set formed by the intersection of sets A and C is written as $A \cap C$. The probability that the outcome will be a member of sets A and C is written as $P(A \cap C)$.

When events are mutually exclusive, the probabilities associated with the events are additive. One can then claim that the probability of the mutually exclusive sets A and B is the sum of $P(A)$ and $P(B)$.

In the notation of set theory, the set that contains the elements of both A and B is called the union of A and B and designated $A \cup B$. Thus, one may compute the probability that either of the mutually exclusive events A or B occurs as:

$$P(A \cup B) = P(A) + P(B) \qquad (1.2)$$

The same result holds for three or more mutually exclusive events; the probability of the union is the sum of the probabilities of the individual events.

The *elementary* events of a sample space are mutually exclusive, so for the die example one must have:

$$P(1) + P(2) + P(3) + P(4) + P(5) + P(6) = P(\Omega) = 1.0. \qquad (1.3)$$

Now reasoning from the uniformity of shape of the die and homogeneity of the die material, one might make a leap of faith and conclude that the probability of the elementary events must all be equal and so,

$$P(1) = P(2) = \ldots = P(6) = p.$$

If that is true then the sum in Equation 1.3 will equal $6p$, and, since $6p = 1$, $p = 1/6$. The same kind of reasoning with respect to coin tossing leads to the conclusion that the probability of a head is the same as the probability of a tail so that $P(H) = P(T) = 1/2$. Dice and coins whose outcomes are equally likely are said to be "fair." In the card selection experiment, if we assume that the card is randomly selected, by which we mean each of the 52 cards has an equal chance of being the one selected, then the probability of selecting a specific card is 1/52. Since there are 13 cards in each suit, the probability of the event "card is a diamond" is $13/52 = 1/4$.

1.3 VENN DIAGRAMS

Event probabilities and their relationship are most commonly displayed by means of a Venn diagram named for the British philosopher and mathematician John Venn, who introduced the Venn diagram in 1881. In the Venn diagram a rectangle symbolically represents the set of outcomes constituting the sample space Ω; that is, it contains all of the elementary events. Other events, comprising subsets of the elementary outcomes, are shown as circles within the rectangle. The Venn diagram in Figure 1.1 shows a single event A.

The region outside of the circle representing the event contains all of the elementary outcomes not encompassed by A. The set outside of A, Ω-A, is

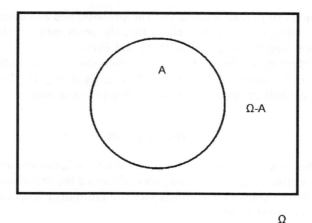

Figure 1.1   Venn diagram showing a single event A.

generally called "not-A" and is indicated by a bar overstrike $\bar{A}$. Since A and $\bar{A}$ are mutually exclusive and sum to the whole sample space, we have:

$$P(A) + P(\bar{A}) = P(\Omega) = 1.0. \tag{1.4}$$

Therefore, the probability of the event not-A may be found simply as:

$$P(\bar{A}) = 1 - P(A). \tag{1.5}$$

Thus, if A is the event that a bearing fails within the next 1000 hours, and $P(A) = 0.2$, the probability that it will survive is $1 - 0.2 = 0.8$. The *odds* of an event occurring is the ratio of the probability that the event occurs to the probability that it does not. The odds that the bearing survives are thus $0.8/0.2 = 4$ or 4 to 1.

Since mutually exclusive events have no elements in common, they appear as nonoverlapping circles on a Venn diagram as shown in Figure 1.2 for the two mutually exclusive events A and B:

1.4   UNIONS OF EVENTS AND JOINT PROBABILITY

The Venn diagram in Figure 1.3 shows two nonmutually exclusive events, A and B, depicted by overlapping circles. The region of overlap represents the set of elementary events shared by events A and B. The probability associated with the region of overlap is sometimes called the joint probability of the two events.

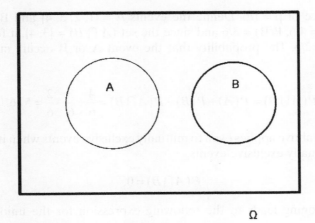

Ω

Figure 1.2 Venn diagram for mutually exclusive events A and B.

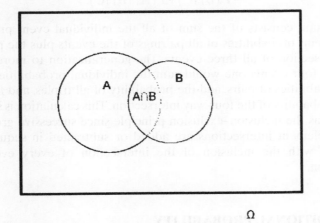

Ω

Figure 1.3 Venn diagram for overlapping events A and B.

In this case, computing the probability of the occurrence of event A or B or both as the sum of P(A) and P(B) will add the probability of the shared events twice. The correct formula is obtained by subtracting the probability of the intersection from the sum of the probabilities to correct for the double inclusion:

$$P(A \cup B) = P(A) + P(B) - P(A \cap B). \qquad (1.6)$$

As an example, consider again the toss of a single die with the assumption that the elementary events are equally likely and thus each have a probability of occurrence of p = 1/6. Define the events A = {1, 2, 3, 4} and B = {3, 4, 5}. Then P(A) = 4/6, P(B) = 3/6 and since the set $(A \cap B) = \{3, 4\}$, it follows that $P(A \cap B) = 2/6$. The probability that the event A or B occurs may now be written:

$$P(A \cup B) = P(A) + P(B) - P(A \cap B) = \frac{4}{6} + \frac{3}{6} - \frac{2}{6} = 5/6.$$

The formula above applies even to mutually exclusive events when it is recalled that for mutually exclusive events,

$$P(A \cap B) = 0.$$

Similar reasoning leads to the following expression for the union of three events:

$$P(A \cup B \cup C) = P(A) + P(B) + P(C) - P(A \cap B) - P(A \cap C) \atop - P(B \cap C) + P(A \cap B \cap C). \tag{1.7}$$

The expression consists of the sum of all the individual event probabilities minus the joint probabilities of all pairings of the events plus the probability of the intersection of all three events. The generalization to more events is similar. For four events one would sum the individual probabilities, subtract all the probabilities of pairs, add the probability of all triples, and finally subtract the probability of the four-way intersection. This calculation is sometimes referred to as the inclusion–exclusion principle since successive groupings of additional element intersections are added or subtracted in sequence until terminating with the inclusion of the intersection of every event under consideration.

1.5   CONDITIONAL PROBABILITY

We know that our assessments of probabilities change as new information becomes available. Consider the event that a randomly selected automobile survives a trip from coast to coast with no major mechanical problems. Whatever the probability of this event may be, we know it will be different (smaller) if we are told that the automobile is 20 years old. This modification of probabilities upon the receipt of additional information can be accommodated within the set theory framework discussed here. Suppose that in the situation above involving overlapping events A and B, we were given the information that event A had indeed occurred. The question is, having learned this, what then is our revised assessment of the probability of the event B? The probability of B conditional on A having occurred is written P(B|A). It is read as "the probability of B given A." Clearly, had the specified events been mutually exclusive instead of overlapping, the knowledge that A occurred would eliminate the possibility of B occurring and so P(B|A) = 0. In general, knowing that A occurred changes the set of elementary events at issue from those in the set Ω to those in the set A. The set A has become the new sample space. Within that new sample space, the points corresponding to the occurrence of B are those contained within the intersection A ∩ B. The probability of B given A is now the proportion of P(A) occupied by the intersection probability P(A ∩ B). Thus:

$$P(B\mid A) = \frac{P(A\cap B)}{P(A)}.\tag{1.8}$$

Similarly, P(A|B) is given by:

$$P(A\mid B) = \frac{P(A\cap B)}{P(B)}.\tag{1.9}$$

The numerator is common to these two expressions and therefore by cross multiplication we see that:

$$P(A\cap B) = P(B)P(A\mid B) = P(A)P(B\mid A).\tag{1.10}$$

One application of this formula is in sampling from finite lots. If a lot of size 25 contains five defects, and two items are drawn randomly from the lot, what is the probability that both are defective? Let A be the event that the first item sampled is defective, and let B be the event that the second item is also defective. Then since every one of the 25 items is equally likely to be selected, P(A) = 5/25. Given that A occurred, the lot now contains 24 items of which four are defective, so P(B|A) = 4/24. The probability that both are defective is then calculated as:

$$\text{Prob(both defective)} = P(A\cap B) = P(A)P(B\mid A) = \frac{5}{25}\cdot\frac{4}{24} = \frac{1}{30}.$$

A similar problem occurs in determining the probability of picking two cards from a deck and finding them both to be diamonds. The result would be (13/52)(12/51) = 0.0588.

When three events are involved, the probability of their intersection could be written as

$$P(A\cap B\cap C) = P(A)P(B\mid A)P(C\mid A\cap B).\tag{1.11}$$

This applies to any ordering of the events A, B, and C. For four or more events the probability of the intersection may be expressed analogously.

1.6 INDEPENDENCE

Two events A and B are said to be *independent* if the conditional probability P(A|B) is equal to P(A). What this says in essence is that knowing that B occurred provides no basis for reassessing the probability that A will occur. The events are unconnected in any way. An example might be if someone tosses a fair coin and the occurrence of a head is termed event A, and perhaps someone else in another country, throws a die and the event B is associated with 1, 2, or 3 spots appearing on the upward face. Knowing that the event B occurred, P(A|B) remains P(A) = 1/2. When two events are independent, the probability of their intersection becomes the product of their individual probabilities:

$$P(A \cap B) = P(B) \cdot P(A \mid B) = P(B) \cdot P(A). \qquad (1.12)$$

This result holds for any number of independent events. The probability of their joint occurrence is the product of the individual event probabilities. Reconsider the previous example of drawing a sample of size 2 from a lot of 25 items containing five defective items but now assume that each item is replaced after it is drawn. In this case the proportion defective remains constant at $5/25 = 0.2$ from draw to draw and the probability of two defects is $0.2^2 = 0.04$. This result would apply approximately if sampling was done without replacement and the lot were very large so that the proportion defective remained essentially constant as successive items are drawn.

When events A and B are independent the probability of either A or B occurring reduces to:

$$P(A \cup B) = P(A) + P(B) - P(A)P(B). \qquad (1.13)$$

Example

A system comprises two components and can function as long as at least one of the components functions. Such a system is referred to as a parallel system and will be discussed further in a later section. Let A be the event that component 1 survives a specified life and let B be the event that component 2 survives that life. If P(A) = 0.8 and P(B) = 0.9, then assuming the events are independent:

$$P[at\ least\ one\ survives] = 0.8 + 0.9 - 0.8 \cdot 0.9 = 0.98.$$

Another useful approach is to compute the probability that both fail. The complement of this event is that at least one survives. The failure probabilities are 0.2 and 0.1 so the system survival probability is:

$$P[at\ least\ one\ survives] = 1 - (0.2)(0.1) = 1 - .02 = 0.98$$

When the system has more than two components in parallel this latter approach has the advantage of simplicity over the method of inclusion and exclusion shown earlier.

The terms independence and mutual exclusivity are sometimes confused. Both carry a connotation of "having nothing to do with each other." However, mutually exclusive events are not independent. In fact they are strongly dependent since $P(A \cap B) = 0$ and not $P(A)P(B)$ as required for independence.

1.7 PARTITIONS AND THE LAW OF TOTAL PROBABILITY

When a number of events are mutually exclusive and collectively contain all the elementary events, they are said to form a partition of the sample space. An example would be the three events $A = \{1, 2\}$, $B = \{3, 4, 5\}$, and $C = \{6\}$. The probability of their union is thus $P(\Omega) = 1.0$. The Venn diagram fails us in representing a partition since circles cannot exhaust the area of a rectangle. Partitions are therefore ordinarily visualized as an irregular division of a rectangle without regard to shape or size as shown in Figure 1.4.

Alternate language to describe a partition is to say that the events are disjoint (no overlap) and exhaustive (they embody all the elementary events). When an event, say D, intersects with a set of events that form a partition, the probability of that event may be expressed as the sum of the intersections of D with the events forming the partition. The Venn diagram in Figure 1.5 shows three events, A, B, C, that form a partition. Superimposed is an event D that intersects each of the partitioning events.

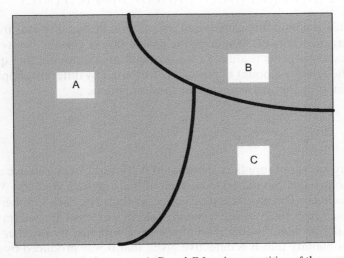

Figure 1.4 Mutually exclusive events A, B, and C forming a partition of the sample space.

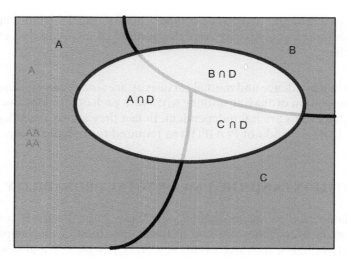

Figure 1.5   An event D superimposed on a partition.

The probability of the event D can be expressed as the sum of the probabilities of the intersections of D with A, B, and C:

$$P(D) = P(A \cap D) + P(B \cap D) + P(C \cap D). \tag{1.14}$$

Using the expression for the joint probability in terms of the probability of D conditioned on each of the other three events, this becomes:

$$P(D) = P(A)P(D \mid A) + P(B)P(D \mid B) + P(C)P(D \mid C). \tag{1.15}$$

This formula is commonly called the Law of Total Probability. It is frequently the only practical way of computing the probability of certain events of interest. One example of its usefulness is in computing overall product quality in terms of the relative amount of product contributed by different suppliers and the associated quality performance of those suppliers.

Example
A company has three suppliers, designated A, B, and C. The relative amounts of a certain product purchased from each of the suppliers are 50%, 35%, and 15%, respectively. The proportion defective produced by each supplier are 1%, 2% and 3%, respectively. If the company selects a product at random from its inventory the probability that it will have been supplied by supplier A is 0.5 and the probability that it is defective given that it was produced by supplier A is 0.01. Let A, B, and C denote the event that a randomly selected part drawn

randomly from the company's inventory was provided by suppliers A, B, and C, we have:

$$P(A) = 0.5, P(B) = 0.35 \text{ and } P(C) = 0.15.$$

The event that an item randomly drawn from inventory is defective is denoted as event D. The following conditional probabilities apply:

$$P(D \mid A) = 0.01, P(D \mid B) = 0.02 \text{ and } P(D \mid C) = 0.03.$$

P(D) then represents the overall proportion defective and may be computed from the Law of Total Probability.

$$P(D) = 0.5 \times 0.01 + 0.35 \times 0.02 + 0.15 \times 0.03 = 0.0165.$$

The company's inventory is thus 1.65% defective.

One creative use of the Law of Total Probability is in the analysis of the randomized response questionnaire (cf. Warner 1965). This questionnaire is aimed at determining the proportion of people who have participated in an activity, such as tax evasion, that they might be loathe to admit if directly asked. Instead two questions are posed, Q1 and Q2. The participant randomly chooses to answer Q1 or Q2 based on a random mechanism such as flipping a coin. Let us say that if the coin is a head, they answer Q1 and otherwise Q2. If the coin is fair, P(Q1) = 1/2 and P(Q2) = 1/2. Now Question Q1 is chosen so that the fraction of affirmative responses is known. For example:

Q1: Is the last digit of your social security number even? Yes/No. The probability of a Yes answer given that Q1 is answered is therefore P(Y|Q1) = 0.5.

Question Q2 is the focus of actual interest and could be something like:

Q2: Have you ever cheated on your taxes? Yes/No.

From the respondent's viewpoint, a Yes answer is not incriminating since it would not be apparent whether that answer was given in response to Q1 or to Q2.

The overall probability of a Yes response may be written as:

$$P(\text{Yes}) = P(Y \mid Q1)P(Q1) + P(Y \mid Q2)P(Q2) = (1/2)(1/2) + 1/2(Y \mid Q2).$$

When the survey results are received the proportion of Yes answers are determined and used as an estimate of P(Yes) in the equation above. For example, suppose that out of 1000 people surveyed, 300 answered Yes. P(Yes) may therefore be estimated as 300/1000 = 0.30.

Substituting this estimate gives:

$$0.30 = 0.25 + 0.5P[Y \mid Q2]$$

So that P[Y| Q2] may be estimated as: (0.30–0.25)/0.5 = 0.10.

1.8 RELIABILITY

One source of probability problems that arise in reliability theory is the computation of the reliability of systems in terms of the reliability of the components comprising the system. These problems use the very same principles as discussed above and are only a context change from the familiar dice, cards, and coins problems typically used to illustrate the laws of probability. We use the term reliability in the narrow sense defined as "the probability that an item will perform a required function under stated conditions for a stated period of time." This definition coincides with what Rausand and Høyland (2004) call survival probability. They use a much more encompassing definition of reliability in compliance with ISO 840 and of which survival probability is only one measure.

Reliability relationships between systems and their components are readily communicated by means of a reliability block diagram. Reliability block diagrams are analogous to circuit diagrams used by electrical engineers. The reliability block diagram in Figure 1.6 identifies a type of system known as a series system. It has the appearance of a series circuit.

1.9 SERIES SYSTEMS

In Figure 1.6, R_i represents the probability that the i-th component ($i = 1...4$) functions for whatever time and conditions are at issue. A series circuit functions if there is an unbroken path through the components that form the system. In the same sense, a series system functions if every one of the components displayed also functions. The reliability of the system is the probability of the intersection of the events that correspond to the functioning of each component:

$$R_{system} = Prob[1\ functions \cap 2\ functions \cap 3\ functions \cap 4\ functions]. \quad (1.16)$$

If the components are assumed to be independent in their functioning, then,

$$R_{system} = R_1 \cdot R_2 \cdot R_3 \cdot R_4 = \prod_{i=1}^{4} R_i. \quad (1.17)$$

It is readily seen that the reliability of a series system is always lower than the reliability of the least reliable component. Suppose that R_3 were lower than

Figure 1.6   A reliability block diagram.

the others, that is, suppose component 3 is the least reliable of the four components in the system. Since R_3 is being multiplied by the product $R_1R_2R_4$, which is necessarily less than or equal to 1.0, the system reliability cannot exceed R_3.

As an example of a series system calculation, if $R_1 = R_2 = 0.9$, $R_3 = 0.8$, and $R_4 = 0.95$, the system reliability is $(0.9)^2(0.8)(0.95) = 0.6156$.

It is clear that a series system comprising a large number of relatively reliable components may nevertheless be quite unreliable. For example, a series system with 10 components each having $R = 0.95$ has a system reliability of only $(0.95)^{10} = 0.599$. One way of improving the system reliability is to provide duplicates of some of the components such that the system will function if any one of these duplicates functions. The practice of designing with duplicates is called redundancy and gives rise to design problems involving optimum tradeoffs of complexity, weight, cost, and reliability.

1.10 PARALLEL SYSTEMS

The reliability of a system that functions as long as at least one of its two components functions may be computed using the rule for the union of two events where the two events are (i) component 1 functions and (ii) component 2 functions. Assuming independence the probability that both function is the product of the probabilities that each do. Thus, the probability that component 1 or component 2 or both function is:

$$R_{system} = R_1 + R_2 - R_1 \cdot R_2. \tag{1.18}$$

Systems of this type are known as parallel systems since there are as many parallel paths through the reliability block diagram as there are components.

The reliability block diagram in Figure 1.7 shows a parallel system having four components. The direct approach shown above for computing system reliability gets more complicated in this case requiring the use of the inclusion–exclusion principle. A simpler but less direct approach is based on the recognition that a parallel system fails only when all of the n components fail.

Assuming independence, the probability that the system functions is most readily computed as 1-Prob[system fails to function]:

$$R_{system} = 1 - \prod_{i=1}^{n}(1 - R_i). \tag{1.19}$$

For $n = 2$, this results in:

$$R_{system} = 1 - (1 - R_1)(1 - R_2) = R_1 + R_2 - R_1 \cdot R_2. \tag{1.20}$$

In agreement with the direct method given in Equation 1.18.

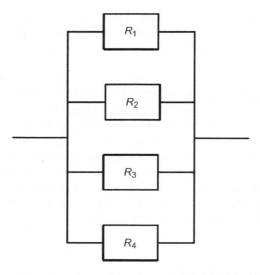

Figure 1.7 A parallel system reliability diagram.

For the component reliabilities considered in the series system depicted in Figure 1.1, letting $R_1 = R_2 = 0.90$, $R_3 = 0.8$, and $R_4 = 0.95$, the reliability of the corresponding parallel system is:

$$R_{system} = 1 - (0.10 \times 0.10 \times 0.20 \times 0.05) = 0.9999.$$

Note that in the parallel case the system reliability is greater than the reliability of the best component. This is generally true. Without loss of generality let R_1 denote the component having the greatest reliability. Subtract R_1 from both sides of Equation 1.19,

$$R_{system} - R_1 = 1 - R_1 - (1 - R_1) \cdot \prod_{i=2}^{n} (1 - R_i). \tag{1.21}$$

Factoring out $(1 - R_1)$, this becomes,

$$R_{system} - R_1 = (1 - R_1) \cdot \left\{ 1 - \prod_{i=2}^{n} (1 - R_i) \right\}. \tag{1.22}$$

Since the values of R_i are all less than or equal to 1.0, the two bracketed terms on the right-hand side are positive and hence $R_{system} \geq R_1$.

We may conclude that the reliability of any system composed of a given set of components is always greater than or equal to the reliability of the series combination and less than or equal to the reliability of the parallel combination of those components.

1.11 COMPLEX SYSTEMS

Systems consisting of combinations of parallel and series arrangements of components can be resolved into a purely parallel or a purely series system. The system depicted in Figure 1.8 is an example:

Replace the series elements on each path by a module whose reliability is equal to that of the series combinations. Multiplying the reliabilities of the three series branches results in the equivalent system of three modules in parallel shown in Figure 1.9.

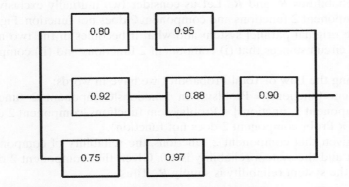

Figure 1.8   Combined series and parallel system reliability block diagram.

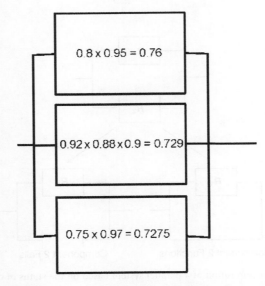

Figure 1.9   System resolved into an equivalent parallel system.

The system reliability may now be computed from the rule for parallel systems:

$$R_{system} = 1 - (1-.76)(1-0.729)(1-.7275) = 0.984.$$

1.12 CROSSLINKED SYSTEMS

The system reliability for more complex systems involving crosslinking can sometimes be found by exploiting the Law of Total Probability. Before examining an example let us consider a simple parallel structure with two components having reliabilities R_1 and R_2. Let us consider two mutually exclusive situations: component 2 functions and component 2 does not function. Figure 1.10 shows the original parallel system and what it becomes in the two mutually exclusive circumstances that (i) component 2 functions and (ii) component 2 fails.

Applying the Law of Total Probability, we have in words:
Prob{system functions} = Prob{system functions|Component 2 functions} × Prob{component 2 functions} + Prob{system functions |component 2 does not function} × Prob(component 2 does not function).

Now, given that component 2 functions, the reliability of component 1 is irrelevant and the system reliability is 1.0. Given that component 2 does not function, the system reliability is simply R_1. Thus:

$$R_{system} = 1 \cdot R_2 + R_1 \cdot (1 - R_2) = R_1 + R_2 - R_1 R_2.$$

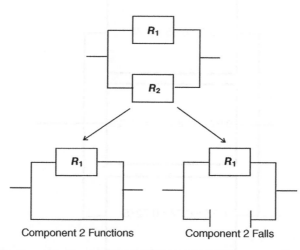

Component 2 Functions   Component 2 Falls

Figure 1.10   Decomposition of a parallel system based on the status of component 2.

We see that the resultant expression is in agreement with the expression previously found for a two-component parallel system. This method of analysis, often called the decomposition method, is always valid, but generally not used for systems that consist of simple combinations of series and parallel subsystems. The power of the decomposition method arises in the analysis of so-called crosslinked systems which cannot be handled by the direct approach used to analyze the system shown in Figure 1.10. Figure 1.11A is the reliability block diagram for such a crosslinked system.

The component labeled 3 causes this block diagram to differ from a pure series/parallel combination. Therefore, component 3 will be chosen as the pivot element in using the decomposition method. Let R^+ denote the system reliability when component 3 functions. The reduced system in this case becomes a series combination of two parallel components as shown in Figure 1.11B.

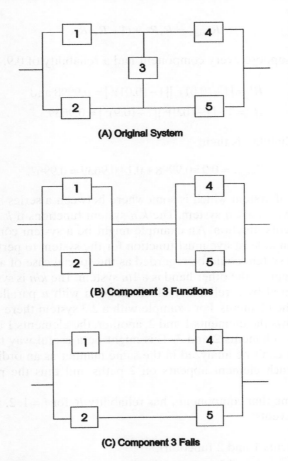

(A) Original System

(B) Component 3 Functions

(C) Component 3 Fails

Figure 1.11 Decomposition of a complex system based on status of component 3.

The reliability R^+ is then the product of the reliabilities of the two parallel modules:

$$R^+ = [1-(1-R_1)(1-R_2)]\cdot[1-(1-R_4)(1-R_5)]. \tag{1.23}$$

When component 3 is in the failed state the reduced system is the parallel combination of two series modules as shown in Figure 1.11C.

The system reliability with component 3 failed is denoted R^- and may be expressed as:

$$R^- = 1-(1-R_1R_4)(1-R_2R_5). \tag{1.24}$$

Using the Law of Total Probability, the system reliability is then expressible as:

$$R_{system} = R_3R^+ + (1-R_3)R^-. \tag{1.25}$$

For example, suppose every component had a reliability of 0.9. In that case:

$$R^+ = [1-(0.01)^2][1-(0.01)^2] = 0.9998 \text{ and}$$
$$R^- = 1-[1-(0.9)^2][1-(0.9)^2] = 0.9639.$$

The system reliability is then:

$$R_{system} = 0.9*0.9998 + 0.1*0.9639 = 0.9962.$$

Another type of system which is somewhere between a series and a parallel system is known as a k/n system. The k/n system functions if k ($\leq n$) or more of its components function. An example might be a system containing eight pumps of which at least five must function for the system to perform satisfactorily. A series system could be regarded as the special case of an n/n system. A parallel system on the other hand is a $1/n$ system. The k/n is system is sometimes represented by a reliability block diagram with n parallel paths each showing k of the elements. For example with a 2/3 system there are 3 parallel paths. One shows the elements 1 and 2, another, the elements 1 and 3 and the third shows the elements 2 and 3. This might be a useful way to convey the situation but it can't be analyzed in the same manner as an ordinary parallel system since each element appears on 2 paths and thus the paths are not independent.

Let us assume that component i has reliability R_i for $i = 1, 2,$ and 3. Define the following events:

A: (components 1 and 2 function),
B: (components 1 and 3 function), and
C: (components 2 and 3 function).

Using the inclusion–exclusion principle, the system reliability is:

$$R_{system} = P(A \cup B \cup C) = P(A) + P(B) + P(C) - P(A \cap B) \\ - P(A \cap C) - P(B \cap C) + P(A \cap B \cap C). \quad (1.26)$$

Now,

$$P(A) = R_1 R_2.$$

Likewise,

$$P(B) = R_1 R_3 \text{ and } P(C) = R_2 R_3.$$

The paired terms and the triple term are all equal to the product $R_1 R_2 R_3$. The final result is therefore:

$$R_{system} = R_1 R_2 + R_1 R_3 + R_2 R_3 - 2 R_1 R_2 R_3. \quad (1.27)$$

This calculation grows quite tedious for larger values of k and n.

For the case where all components have the same reliability the system reliability may easily be computed using the binomial distribution as shown in Section 2.5 of Chapter 2.

1.13 RELIABILITY IMPORTANCE

It is of interest to assess the relative impact that each component has on the reliability of the system in which it is employed, as a basis for allocating effort and resources aimed at improving system reliability. A measure of a component's reliability importance due to Birnbaum (1969) is the partial derivative of the system reliability with respect to the reliability of the component under consideration. For example, the system reliability for the series system shown in Figure 1.6 is:

$$R_s = R_1 R_2 R_3 R_4. \quad (1.28)$$

The importance of component 1 is,

$$I = \frac{\partial R_s}{\partial R_1} = R_2 R_3 R_4 \quad (1.29)$$

and similarly for the other components. Suppose the component reliabilities were 0.95, 0.98, 0.9, and 0.85, respectively, for R_1 to R_4. The computed importance for each component is shown in the table below:

Component	Reliability	Importance
1	0.95	$R_2 R_3 R_4 = 0.7056$
2	0.98	$R_1 R_3 R_4 = 0.6840$
3	0.90	$R_1 R_2 R_4 = 0.7448$
4	0.80	$R_1 R_2 R_3 = 0.8379$

We see that the most important component, the one most deserving of attention in an attempt to improve system reliability is component 4, the least reliable component.

An alternate way of computing the importance of a component comes from the decomposition method. Suppose we seek the importance of component i in some system. We know that the system reliability can be expressed as:

$$R_s = R^+ R_i + R^- (1 - R_i).$$

Differentiating with respect to R_i shows that the importance of component I may be computed as:

$$I = R^+ - R^-. \tag{1.30}$$

Thus, the importance is the difference in the system reliabilities computed when component i functions and when it does not. Referring to the crosslinked Figure 1.11 and the associated computations, the importance of component 3 is the difference:

$$I = 0.9998 - 0.9639 = 0.0359.$$

There is an extensive literature on system reliability, and many other methods, approximations, and software are available for systems with large numbers of components. The book by Rausand and Høyland (2004) contains a good exposition of other computational methods and is a good guide to the published literature on systems reliability calculations. Another good source is the recent text by Modarres et al. (2010).

REFERENCES

Birnbaum, Z. W. 1969. On the importance of different components in a multi-component system. In P.R. Krishnaiah, ed., *Multivariate Analysis*, pp. 581–592. Academic Press, San Diego, CA.

Modarres, M., M. Kaminskiy, and V. Krivtsov. 2010. *Reliability Engineering and Risk Analysis*. CRC Press, Boca Raton, FL.

Rausand, M. and A. Høyland. 2004. *System Reliability Theory: Models, Statistical Methods, and Applications*. 2nd ed. Wiley-Interscience, Hoboken, NJ.

Warner, S. 1965. Randomized response: a survey technique for eliminating evasive answer bias. *Journal of the American Statistical Association* 60(309): 63–69.

EXERCISES

1. Employees at a particular plant were classified according to gender and political party affiliation. The results follow:

Gender	Democrat	Political Affiliation Republican	Independent
Male	40	50	5
Female	18	8	4

If an employee is chosen at random, find the probability that the employee is:

a. Male

b. Republican

c. A female Democrat

d. Republican given that she is a female

e. Male given that he is a Republican

2. Three components, a, b, and c, have reliabilities 0.9, 0.95, and 0.99, respectively. One of these components is required for a certain system to function. Which of the following two options results in a higher system reliability?

   a. Create two modules with a, b, and c in series. The system then consists of a parallel arrangement of two of these modules. This is called high-level redundancy.

   b. The system consists of a parallel combination of two components of type a in series with similar parallel combinations of b and c. This is called low-level redundancy.

   c. If in the low-level redundancy arrangement it were possible to add a third component of either type a or b or c, which would you choose? Why? Show work.

3. In the reliability diagram below, the reliability of each component is constant and independent. Assuming that each has the same reliability R, compute the system reliability as a function of R using the following methods:

   a. Decomposition using B as the keystone element.

   b. The reduction method.

c. Compute the importance of each component if $R_A = 0.8$, $R_B = 0.9$, $R_C = 0.95$, and $R_D = 0.98$.

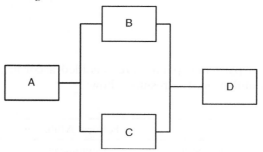

4. A message center has three incoming lines designated A, B, and C which handle 40%, 35%, and 25% of the traffic, respectively. The probability of a message over 100 characters in length is 5% on line A, 15% on line B, and 20% on line C. Compute the probability that a message, randomly selected at the message center, exceeds 100 characters in length.

CHAPTER 2

Discrete and Continuous Random Variables

A random variable is a numerical quantity which is determined in some pre-scribed way from the outcome of an experiment and which can have in any one experiment, any one of a countable or infinite number of possible values. Random variables are classified as discrete or continuous depending upon whether the possible values of the random variable are confined to designated points along the real line or whether they comprise all points within one or more intervals along the real line.

The number of spots on the upward face of a die is an example of a discrete random variable; it can assume the integer values 1 through 6. Another dis-crete random variable not restricted to integer values is the hat size of a randomly selected individual. Most generally, however, discrete random vari-ables arise from counting something. Some examples are the number of mis-prints on a page of text, the number of items on life test that fail prior to a life of 1000 hours, the number of times a bearing must be replaced during a fixed period of operation, and the number of people waiting at any time in a barbershop.

Continuous random variables will be our main concern. An example of a continuous random variable is the time t to failure of a ball bearing operating under a prescribed set of external conditions. Here t assumes any value along the positive real line $(0, \infty)$. Another continuous random variable confined to the positive real line is the breaking strength, that is, the stress at fracture, of a ceramic part. Another example might be the departure, x, of a manufac-tured part from its specified dimension. Here x is confined to be on the real line $(-\infty, \infty)$.

Using the Weibull Distribution: Reliability, Modeling, and Inference, First Edition. John I. McCool.
© 2012 John Wiley & Sons, Inc. Published 2012 by John Wiley & Sons, Inc.

2.1 PROBABILITY DISTRIBUTIONS

A discrete random variable x is described by a list of the values that it may assume, for example, x: {1, 2, 3, 4} and the associated probabilities $p(x)$ with which it will assume those values, for example, $p(x)$: {0.15, 0.25, 0.4, 0.2}. Displayed as a discrete function $p(x)$ against x as shown in Figure 2.1, it is known as a probability distribution function or a probability mass function (pmf). Discrete distributions like this must be obtained by observing the random variable's value in numerous trials and recording the frequency with which each outcome occurs. If the number of trials is large this will converge to the true probability distribution. In other cases the probabilities of each outcome can be computed based on a simple random mechanism and certain assumptions. A case in point is the distribution of the resultant number of heads when a fair coin is tossed four times. Here x varies from 0 to 4 and the probabilities depend on just the probability p of a head in a single toss and the assumption of independence of outcomes. This is an example of the so-called binomial distribution discussed later in this chapter. Another, related random variable whose distribution may be deduced based on a set of assumptions is the number X of coin tosses until the first head occurs. In this example the values of X range from 1 to ∞.

The probability that when observed the value of the random variable depicted above is 2 is written:

$$P[X = 2] = p(2) = 0.25. \tag{2.1}$$

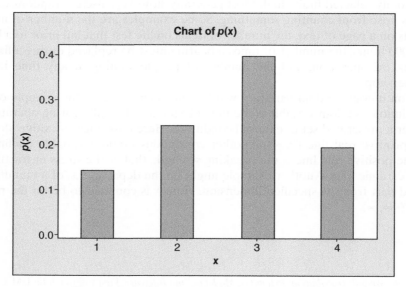

Figure 2.1   A discrete probability distribution function.

The convention is to use an upper case letter to refer to the random variable and a numerical value or lowercase letter to denote the value of the random variable. $P[X = 6]$ is an example, and $P[X = x]$ is the general case.

The values that X may assume are mutually exclusive so that, in the present example, the probability of X assuming either the value 2 or 3 is the sum of the probabilities $p(2) + p(3) = 0.25 + 0.40 = 0.65$. Since, by definition, the random variable *must* assume at least one of its specified values, it follows that the probabilities must sum to unity. That is, if there are k values of X, and distinguishing each value using a subscript, for example, x_i, it follows that:

$$\sum_{i=1}^{k} p(x_i) = 1.0. \qquad (2.2)$$

Here and elsewhere we write, for simplicity, $p(x_i)$ for the more formal $P[X = x_i]$. The cumulative distribution function (CDF) is denoted $F(x)$ and is denoted by:

$$F(x) = P[X \leq x] = \sum_{\min x_i}^{x} p(x_i). \qquad (2.3)$$

Thus, in the example above, $F(3) = 0.8$. The function $F(x)$ increases stepwise at each x value and reaches a maximum of 1.0 at the last value as shown in the tabulation below for the present example.

x	1	2	3	4
$F(x)$	0.15	0.40	0.80	1.00

The probability that X is either 2 or 3 can be computed by adding the associated probabilities as above, or, in terms of the cumulative distribution, by subtraction:

$$P[X \leq 3] - P[X \leq 1] = F(3) - F(1) = 0.80 - 0.15 = 0.65.$$

Correspondingly, the probability that X is 3 or more may be expressed:

$$P[X \geq 3] = 1 - P[X \leq 2] = 1 - F(2).$$

In general,

$$P[X \geq x] = 1 - F[x - 1]. \qquad (2.4)$$

If the random variable X were to represent the lifetime of a product or an organism, rounded to selected discrete values, the function $P[X \geq x]$ would be referred to as the *reliability function* $R(x)$, or, in the biomedical literature, as the survivor function $S(x)$. For the example case, the reliability function is tabulated below:

x	1	2	3	4
$R(x) = P[X \geq x]$	1.0	0.85	0.60	0.2

The mean or expected value of a discrete random variable is:

$$\mu = E(X) = \sum_{i=1}^{k} x_i p(x_i). \tag{2.5}$$

where k is the number of discrete values that the random variable assumes. If x is observed an indefinite number of times, the arithmetic average of those observations approaches $E(X)$. For the example above the mean is $1 \times 0.15 + 2 \times 0.25 + 3 \times 0.4 + 4 \times 0.2 = 2.65$.

When, as in the present example, the values of the random variable consist of consecutive integers starting at $X = 1$, an alternative method of calculating the mean consists of summing the reliability function over each value of the random variable. For the values tabulated above, this gives:

$$\mu = \sum_{i=1}^{k} R(x_i) = 1.0 + 0.85 + 0.60 + 0.2 = 2.65. \tag{2.6}$$

This alternative method of computing the expected value is seldom discussed in expository writing about discrete random variables. The continuous analog in which the summation is replaced by an integral is extensively used in the reliability literature as the preferred way of calculating mean lifetimes.

2.2 FUNCTIONS OF A RANDOM VARIABLE

Sometimes interest is centered on a random variable Y functionally related to the random variable X through an equation of the form $Y = g(X)$. It is generally easy to find the distribution of Y in the discrete case, particularly if the function is single valued, that is, each x value yields only one y value. Suppose for the example above we are interested in the distribution of $Y = X^2$. Since there are no negative x values in this example, there is a unique y for each x. The probability that $Y = 4$ is then the same as the probability that $X = 2$, and we may tabulate the distribution $p(y)$ as follows:

$y = x^2$	1	4	9	16
$p(y)$	0.15	0.25	0.40	0.20

Had the distribution of X included both $x = -1$ and $x = +1$, then $P[Y = 1]$ would have been computed as the sum of $P[X = -1]$ and $P[X = +1]$. Without specifically enumerating the distribution of Y, one may find its mean by the following:

$$E(Y) \equiv \mu_y = \sum_{i=1}^{k} g(x_i) \cdot p(x_i).$$

For the special case where Y is a linear function of X of the form $Y = a + bX$, the mean is:

$$\mu_{a+bx} = \sum_{i=1}^{k} [a + bx_i] \cdot p(x_i) = a \cdot \sum_{i=1}^{k} p(x_i) + b \cdot \sum_{i=1}^{k} x_i p(x_i) = a + b\mu_x. \quad (2.7)$$

For the example, the expected value of $2 + 3X$ would be:

$$E[2 + 3X] = 2 + 3 \cdot 2.65 = 9.95.$$

The variance σ^2 characterizes the scatter in the random variable. It is defined as:

$$\sigma^2 = \sum_{i=1}^{k} (x_i - \mu)^2 p(x_i). \quad (2.8)$$

When expanded and simplified an equivalent expression is:

$$\sigma^2 = \sum_{i=1}^{k} x_i^2 p(x_i) - \mu^2 = E(X^2) - E^2(X). \quad (2.9)$$

For the example above, $E(X^2)$ is computed as: $1^2 \times 0.15 + 2^2 \times 0.25 + 3^2 \times 0.4 + 4^2 \times 0.2 = 7.95$. The variance is then $7.95 - 2.65^2 = 0.9275$. If the random variable has units of say, hours, the variance is in hours^2. The extent of the variability is more readily grasped if scatter is expressed in the same units as the random variable itself, so the square root of the variance, called the standard deviation, is more frequently used to convey the magnitude of the scatter that the random variable exhibits.

The variance of a linear function of x may be computed as:

$$\sigma^2_{a+bx} = E(a + bx)^2 - (a + b\mu_x)^2.$$

Expanding and simplifying leads to the result:

$$\sigma^2_{a+bx} = b^2 \sigma^2_x. \quad (2.10)$$

Note that the additive constant does not appear in the result. This is because adding a constant to a random variable translates the whole distribution to the left or right, depending on the sign of the additive constant, but does not change the scatter. The variance is therefore unaffected. The standard deviation of $a + bX$ is equal to the standard deviation of X multiplied by the constant b.

2.3   JOINTLY DISTRIBUTED DISCRETE RANDOM VARIABLES

Sometimes more than one random outcome is observed in the same random experiment. When a manufactured part is inspected, several dimensions might be measured such as, length, diameter, and weight. In a clinical trial one might measure or record a human subject's weight, age, blood pressure, and so on.

The following example involving two discrete random variables was given by Mosteller et al. (1961).

A coin is tossed three times and the sequence of Heads (H) and Tails (T) is observed. For each toss, one observes two random variables, X, the number of heads, and Y the number of runs. A run is the number of sequences of the same letter H or T. The sequence HHT has two runs. Assuming the coin is fair every sequence has a probability of 1/8 of occurring. The outcomes and the associated values of X and Y are given in Table 2.1.

Table 2.2 gives the probability associated with each $x - y$ pair. For each combination of x and y the entry in the table is the joint probability of that combination noted $P(X = x, Y = y)$ or simply, $p(x, y)$ by an obvious extension of our notation in the univariate case.

Table 2.1 X and Y Values for Each Possible Outcome

Sequence	X = No. of Heads	Y = No. of Runs
HHH	3	1
THH	2	2
HTH	2	3
TTH	1	2
HHT	2	2
THT	1	3
HTT	1	2
TTT	0	1

Table 2.2   Joint and Marginal Distributions of X and Y

$y\downarrow x\rightarrow$	0	1	2	3	Total = $g(y)$
1	0.125	0	0	0.125	0.25
2	0	0.25	0.25	0	0.50
3	0	0.125	0.125	0	0.25
Total = $f(x)$	0.125	0.375	0.375	0.125	

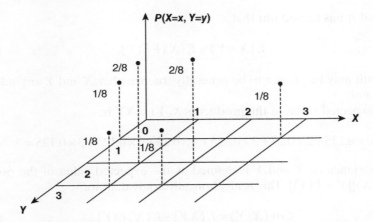

Figure 2.2 Two-dimensional probability mass function. From Mosteller, *Probability with Statistical Applications*, 1st Edition, copyright 1961. Reprinted by permission of Pearson Education, Inc., Upper Saddle River, NJ.

As in the univariate case, the pmf may be shown as a two-dimensional plot with the probabilities of each x, y combination shown as bars drawn in the z direction as in Figure 2.2.

As in the univariate case, the probabilities must sum to 1.0 over all of the possible combinations

$$\sum_{x}\sum_{y} p(x, y) = 1.0. \tag{2.11}$$

The row totals in Table 2.2 give $g(y)$, the probability distribution of Y that would result if X was not recorded. Similarly the column totals give $f(x)$, the distribution of x that would result if Y was not recorded. These are called the marginal distributions presumably because they are the totals recorded in the right and bottom margins of the table. These distributions may be used in the usual way to compute the means of X and Y:

$$E(X) = 0 \times 0.125 + 1 \times 0.375 + 2 \times 0.375 + 3 \times 0.125 = 1.5$$

$$E(Y) = 1 \times 0.25 + 2 \times 0.50 + 3 \times 0.25 = 2.0.$$

The expected value of functions $h(X, Y)$ of the two variables are computed by analogy to the univariate case by evaluating the function at each combination of x and y, multiplying by $p(x, y)$, and summing over all combinations. The sum $h(X, Y) = X + Y$ is one such function of interest. Its expected value is calculated for the example as follows, omitting x, y combinations for which $p(x, y) = 0$:

$$1 \times 0.125 + 3 \times 0.25 + 4 \times 0.125 + 4 \times 0.25 + 5 \times 0.125 + 4 \times 0.125 = 3.5.$$

Note that it has turned out that:

$$E(X+Y) = E(X) + E(Y).\qquad(2.12)$$

This result may be proven to be generally true whether X and Y are independent or not.

The expected value of the product $h(X, Y) = XY$ is:

$$0 \times 0.125 + 2 \times 0.25 + 3 \times 0.125 + 4 \times 0.25 + 6 \times 0.125 + 3 \times 0.125 = 3.$$

The covariance of X and Y is defined as the expected value of the product $[X - E(X)][Y - E(Y)]$. This is algebraically equivalent to:

$$Cov(X, Y) = E(XY) - E(X)E(Y).\qquad(2.13)$$

Note that the covariance of X with itself, $cov(X, X)$, is simply the variance of X.

In the present example,

$$Cov(X, Y) = 3 - 2 \times 1.5 = 0.$$

In this case the covariance is zero and X and Y are said to be uncorrelated. They are not, however, independent. If two random variables are independent, then $p(x, y)$ is the product of the marginal probabilities $p(x)p(y)$. This is clearly not the case in the present example since $p(X = 0, Y = 2) = 0$ and not 0.125×0.50. When two random variables are independent, then one can show that $E(XY) = E(X)E(Y)$ so that the covariance will be zero. Thus, if X and Y are independent, the covariance is 0, but if the covariance is 0, X and Y are not necessarily independent. However, as noted, the expected value of the sum is always the sum of the expected values whether or not the variables are independent.

The covariance has units equal to the product of the units of the two variables such as, for example, volt-inches. The covariance can be positive or negative. When it is positive Y tends to increase with X; when negative Y tends to decrease with X. It is difficult, however, to assess whether a numerical value of the covariance implies a strong relationship because of the mixed units. A more useful, dimensionless, measure of association is the correlation coefficient ρ, defined as:

$$\rho = \frac{cov(X, Y)}{\sigma_X \cdot \sigma_Y}.\qquad(2.14)$$

The magnitude of the correlation coefficient is always between ±1. When it is at either extreme there is a perfect linear relationship between the two variables. So if $Y = 2X$, $\rho = +1$, and if $Y = -2X$, $\rho = -1$.

Recall that in the univariate case the variance, defined as $E[X - E(X)]^2$, could be calculated as $E(X^2) - E^2(X)$. The variance of the sum $X + Y$ follows in the same way, that is,

$$\begin{aligned} Var(X+Y) &= E\big[(X+Y)^2\big] - E^2[X+Y] \\ &= E[X^2+Y^2+2XY] - [E(X)+E(Y)]^2. \end{aligned} \tag{2.15}$$

Expanding and collecting terms results in

$$Var(X+Y) = \sigma_X^2 + \sigma_Y^2 + 2cov(X, Y). \tag{2.16}$$

For three or more variables, the variance of the sum is the sum of the variances plus 2 times all of the pairwise covariance terms. When X and Y are independent the variance of the sum is just the sum of the variances since the covariance is zero. The variance of a linear combination $aX + bY$ is easily shown to be:

$$Var(aX+bY) = a^2\sigma_X^2 + b^2\sigma_Y^2 + 2ab\,cov(X, Y). \tag{2.17}$$

Random variables which are linear combinations of a number of independent random variables $X_1, X_2, \ldots X_n$ are frequently encountered and have the form:

$$Z = c_1 X_1 + c_2 X_2 + \cdots c_n X_n. \tag{2.18}$$

The expected value of Z is:

$$E(Z) = c_1 E(X_1) + c_2 E(X_2) + \cdots c_n E(X_n). \tag{2.19}$$

This equation is true whether or not the variables X_1 to X_n are independent. The variance of Z is:

$$var(Z) = c_1^2 var(X_1) + c_2^2 var(X_2) + \cdots c_n^2 var(X_n). \tag{2.20}$$

And holds only if the variables are independent. If any pair of variables X_i and X_j are dependent the term $2c_i c_j Cov(X_i, X_j)$ should be added to the expression for $var(Z)$.

Example

Imagine blocks of random height X with a mean $E(X) = 1$ inch and a variance of 0.2 in^2.

If we randomly select five such blocks and stack them, the height of the stack will be:

$$H = X_1 + X_2 + X_3 + X_4 + X_5.$$

The expected value of the height is:

$$E(H) = E(X_1) + E(X_2) + E(X_3) + E(X_4) + E(X_5) = 5 \text{ in.}$$

Random selection assures that the block heights are independent so the variance of H is:

$$var(H) = var(X_1) + var(X_2) + var(X_3) + var(X_4) + var(X_5) = 1.0 \text{ in.}^2$$

Now suppose we build another stack but this time we select just one block and, using ultra precise equipment, make four exact duplicates and stack them on the first randomly selected block. This time the height of the stack will be:

$$H = 5X_1.$$

The expected value is:

$$E(H) = 5E(X_1) = 5 \times 1 = 5 \text{ in.}$$

which is the same as the expected value of the sum. The variance however is:

$$var(H) = 5^2 \times var(X_1) = 25 \times 0.2 = 5 \text{ in.}^2$$

Why are the heights of the stacks less variable in the first case than in the second? In the first case larger than average blocks can compensate for smaller than average blocks. In the second case a larger than average block results in five such larger than average blocks and leads to greater variability from stack to stack. It is important not to mistake the sum of n identical random variables for the product of n times the random variable.

2.4   CONDITIONAL EXPECTATION

The distribution of X given a specific value of Y is easily obtained using the laws of conditional probability. So, for example, using the joint distribution from Table 2.2, we have:

$$\text{Prob}[X = 0 \mid Y = 1] = \frac{\text{Prob}[X = 0 \cap Y = 1]}{P(Y = 1)} = \frac{0.125}{0.250} = 0.50.$$

Similarly,

$$\text{Prob}[X = 3 \mid Y = 1] = \frac{\text{Prob}[X = 3 \cap Y = 1]}{P(Y = 1)} = \frac{0.125}{0.250} = 0.50.$$

The distribution of X given $Y = 1$ written as $f(X \mid Y = 1)$ may be set down in a table:

$x\lvert Y=1$	0	3
$f(x\lvert Y=1)$	0.50	0.50

The expected value of $X\lvert Y=1$ is computed in the usual way:

$$E(X\lvert Y=1)=0\times0.50+3\times0.5=1.5.$$

Table 2.3 gives the distribution of $X\lvert Y$ for all three values of Y along with the expected value of X given each Y.

Table 2.3 Distribution of X Conditional on Y and $E(X\lvert Y)$

$x\lvert y\rightarrow$	0	1	2	3	$E(X\lvert Y)$
$Y=1$	0.50	0	0	0.50	1.5
$Y=2$	0	0.50	0.50	0	1.5
$Y=3$	0	0.50	0.50	0	1.5

The expected value happens to be the same for each Y in this example.

The Law of Total Expectation, which is comparable to the Law of Total Probability, states that the unconditional expectation is the sum of the conditional expectations weighted by the probabilities of each Y value:

$$E(X)=E[X\lvert Y=1]P(Y=1)+E[X\lvert Y=2]P(Y=2)+E[X\lvert Y=3]P(Y=3)$$

$$E(X)=1.5\times0.25+1.5\times0.50+1.5\times0.25=1.5.$$

Example

Consider the following joint distribution. Calculate $E(X)$, $E(Y)$, $Var(X)$, $Var(Y)$, $Cov(X, Y)$, $\rho(X, Y)$, and $Var(X+Y)$, and verify $E(X)$ using the Law of Total Expectation.

$y\downarrow x\rightarrow$	0	1	$g(y)$
0	0.1	0.3	0.4
1	0.4	0.2	0.6
$f(x)\rightarrow$	0.5	0.5	

From the marginal distributions of X and Y,

$$E(X)=0\times0.5+1\times0.5=0.5$$

$$E(X^2)=0^2\times0.5+1^2\times0.5=0.5$$

$$Var(X)=E(X^2)-E^2(X)=0.5-0.5^2=0.25$$

$$E(Y)=0.6;\; E(Y^2)=0.6\;\; Var(Y)=0.24$$

$$E(XY)=0\times0.1+0\times0.3+0\times0.4+1\times0.2=0.2$$

$$Cov(X,Y) = E(XY) - E(X)E(Y) = 0.2 - 0.5 \times 0.6 = -0.10$$

$$\rho(X,Y) = \frac{Cov(X,Y)}{\sigma_X \sigma_Y} = \frac{-0.10}{\sqrt{0.25} \cdot \sqrt{0.24}} = -0.4$$

$$Var(X+Y) = 0.25 + 0.24 - 2 \times 0.10 = 0.29.$$

The distribution of $X|Y$ is shown below for both Y values along with their expectation.

$x|Y\rightarrow$	0	1	$E(X|Y)$
$Y = 0$	0.25	0.75	0.75
$Y = 1$	0.667	0.333	0.333

Using the Law of Total Expectation:

$$E(X) = E(X\mid Y = 0)P(Y = 0) + E(X\mid Y = 1)P(Y = 1)$$

$$E(X) = 0.75 \times 0.4 + 0.333 \times 0.6 = 0.5.$$

The Law of Total Expectation works when a random variable is conditioned on any set of mutually exclusive and exhaustive events and not necessarily on another random variable. Suppose the expected value of a certain quality characteristic of a product depended upon which of three suppliers produced it. The table below shows the expected value of that characteristic for each vendor and the proportion that each vendor supplies.

Vendor	$E(X|\text{Vendor})$	Vendor's Share
Vendor 1	1000	0.2
Vendor 2	1200	0.3
Vendor 3	900	0.5

The overall expected value of the quality characteristic is:

$$E(X) = 1000 \times 0.2 + 1200 \times 0.3 + 900 \times 0.5 = 1010.$$

We make use of the Law of Total Expectation in computing the mean of a mixture of Weibull distributions in Chapter 4.

2.5 THE BINOMIAL DISTRIBUTION

A discrete random variable of importance in life testing and many other situations is the one that describes the number of successes X in a series of n independent trials where the probability of a success on any trial is constant from trial to trial and designated p. The distribution of this random variable is called the binomial distribution. The term "success" may not coincide with its positive connotation in ordinary parlance. In the context of applications of the binomial distribution success means only that an event of interest occurred. The event could be the appearance of a defective item, in which case the production of a good item is regarded as a failure, at odds with a production manager's perspective. It could mean that a human member of a cohort in a biomedical study has been stricken with a disease. Quite often successes may be redefined as failures for computational simplicity or convenience. The number of failures is always $n - x$; the probability of a failure is $1 - p$.

The random variable X ranges from 0 (no successes) to n (every trial a success). As an example let the number of trials be $n = 3$ and observe that the following sequential outcomes will result in $X = 1$ success: SFF, FSF, or FFS. These three mutually exclusive outcomes each have the same probability, of occurring, namely, $p(1 - p)^2$. Thus the probability that $X = 1$ is $3p(1 - p)^2$. In the general case, the probability of a sequence that results in x successes and $n - x$ failures is $p^x(1 - p)^{n-x}$. The number of sequences of the n letters S and F that will result in x successes is the number of ways of selecting x positions from a total of n positions to which to assign the letter S. The remaining $n - x$ positions will then be assigned an F. Symbolically this number is denoted $\binom{n}{x}$, and it is evaluated as follows:

$$\binom{n}{x} = \frac{n!}{x!(n-x)!}. \tag{2.21}$$

Evaluating this expression for $x = 1$ and $n = 3$ gives $\binom{3}{1} = 3$, consistent with the example above. Thus, in general, the binomial probability distribution is expressible as:

$$p(x) = \binom{n}{x} p^x (1-p)^{n-x}; x = 0, 1, \ldots n. \tag{2.22}$$

In algebra, this same expression occurs as the coefficient of p^x in the binomial expansion of $(p + q)^n$ when q is taken to be $1 - p$. This fact accounts for the name of this distribution. The expression $\binom{n}{x}$ is often referred to as the number of combinations of n things taken x at a time. Another more informal name is "n choose x."

The mean and variance of the binomial distribution may be shown to be expressible as:

$$\mu = E(X) = np. \tag{2.23}$$

$$\sigma^2 = np(1-p). \tag{2.24}$$

The expression for the binomial mean is intuitively understood by most people. They will correctly answer "50" to the question "how many heads will I expect to see if I flip a fair coin 100 times." The variance of x depends on p. For a large p or a small p the variance is small. The variance is maximum if $p = 0.50$. If the probability that a specimen will withstand a certain stress is $p = 0.99$ and you subject 10 specimens to that stress, the number of successes, that is, specimens surviving, will be 10 or occasionally 9, rarely fewer. On the other hand, if the survival probability is $p = 0.5$ the number of survivors in a test of $n = 10$ specimens will vary widely, with the same chance that $X = 2$ as $X = 8$.

Example

It is known for a certain population of ball bearings that the probability of failing prior to 1000 hours is $p = 0.50$. If $n = 4$ identical bearings are randomly selected from this population and tested for 1000 hours, find the distribution of the number of failures. As an example the probability that $X = 3$ is calculated below:

$$P[X = 3] = \frac{4!}{3!(4-3)!} 0.5^3 0.5^{4-3} = 0.25.$$

The results for all the possible values of X computed from Equation 2.22 are tabled below:

x	0	1	2	3	4
$p(X = x)$	0.0625	0.2500	0.3750	0.2500	0.0625

Figure 2.3 is a plot of the pmf.

The mean and variance may be computed from the tabled probability values as before but in this case we may equivalently use the formulae:

$$\mu = np = 4 \times 0.50 = 2$$
$$\sigma^2 = np(1-p) = 4 \times .5 \times (1-0.5) = 1.0.$$

Individual and cumulative binomial probabilities are readily computed using Excel or any statistical software package. When np and $np(1 - p)$ both exceed 5 the normal distribution with mean np and variance $np(1 - p)$ is a reasonably good continuous approximation to the binomial distribution. An example of the use of the normal approximation with a correction to compensate for the fact that the normal distribution is continuous and the binomial is discrete is given in Section 2.11. For large values of n and correspondingly small values of p such that $np < 0.1$, the Poisson distribution is considered a good approximation for the binomial as shown in the next section.

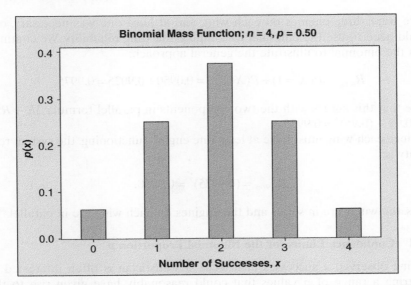

Figure 2.3   Probability mass function (pmf) of the binomial distribution; $n = 4, p = 0.50$.

As mentioned in Chapter 1, the binomial distribution is used to compute the reliability of a system known as a *k/n* system when its components are identical. A *k/n* or "*k* out of *n*" system contains n independently functioning components of which at least k must function satisfactorily for the system to survive a given mission. The probability that a component will survive that mission is the component's reliability *R*. A series system is a special case of the *k/n* system in which *k = n*. A parallel system represents the other extreme special case in which k = 1.

If all n components have the same reliability and letting X denote the number of components that survive the mission, the system reliability is $P[X \geq k]$. Using the binomial with $p = R$, we have:

$$R_{system} = \sum_{i=k}^{n} \binom{n}{i} R^i (1-R)^{n-i}. \qquad (2.25)$$

As an example, suppose a four-engine plane requires two or more engines to function for it to survive. With a mission success probability $R = 0.95$ for each engine, the system reliability is:

$$R_{system} = 0.013538 + 0.171475 + 0.814506 = 0.999519.$$

A twist on this problem is to impose the condition that at least one engine on each wing must function for mission success. In this case each wing becomes a separate two-component parallel system and the ordinary reliability calculation for a parallel system could be used to find the reliability of each wing. If there were three engines on each wing and at least one was necessary, one would need to use the binomial to compute the wing reliability. We continue with the binomial to illustrate the general approach:

$$R_{wing} = P(X = 1) + P(X = 2) = 0.0950 + 0.9025 = 0.9975.$$

Note that this agrees with the two components in parallel formula $2R - R^2 = 2 \times 0.95 - (0.95)^2 = 0.9975$.

Since each wing must have at least one engine functioning, the system reliability is:

$$R_{system} = (0.9975)^2 = 0.9950.$$

Thus, the wings are in series and the engines on each wing are in parallel.

2.5.1   Confidence Limits for the Binomial Proportion p

Having observed x successes in n trials, a statistician is often interested in inferring a range of p values that could reasonably have given rise to the observed proportion. The upper and lower limits on p are known as confidence limits and their values depend on the level of confidence desired. Various approximations for these confidence limits are in common use. Exact limits involve percentage points of the incomplete beta distribution. Most commercial statistical software packages are capable of computing exact or approximate confidence limits on a binomial proportion. For $x = 4$ successes in n = 10 trials Minitab gives the following exact 95% limits for the binomial proportion p: (0.121552, 0.737622).

A useful special case occurs when the observed value is $X = 0$. The exact $100(1 - \alpha)\%$ upper limit on p may then be computed as:

$$p < 1 - (\alpha)^{1/n}. \tag{2.26}$$

So, for a 95% upper limit ($\alpha = 0.05$) based on $X = 0$ successes in n = 10 trials, we have:

$$p < 1 - (0.05)^{1/n} = 0.2589.$$

This expression has useful application in certain reliability demonstration tests. If p is the probability that an item will fail under a given set of test conditions and n items are tested and none fail, a *lower* $100(1 - \alpha)\%$ confidence limit on the reliability $R = 1 - p$ is:

$$R > (\alpha)^{1/n}. \tag{2.27}$$

Suppose it must be demonstrated that with 90% confidence the reliability R under a given set of test conditions must exceed $R_1 = 0.98$. How many items must all survive such a test to justify that claim?

Setting $\alpha = 0.10$ and $R_1 = 0.98$ and solving for n results in:

$$n = \frac{\ln(0.10)}{\ln(0.98)} = 113.9 \approx 114.$$

2.6 THE POISSON DISTRIBUTION

Another popular discrete distribution is the Poisson distribution. The Poisson distribution has a single non-negative parameter designated α. The pmf is given below:

$$P[X = x] = \frac{e^{-\alpha}\alpha^x}{x!}; x = 0, 1 \cdots \infty. \qquad (2.28)$$

The expected value of X may be shown to be equal to α. The variance is also equal to α and so the standard deviation is $\sqrt{\alpha}$. Poisson random variables always represent the number of some kind of occurrences per unit of the region of opportunity for those occurrences. Examples include the number of arrivals per unit time at a message center, blemishes per unit area of cloth, and inclusions per unit volume of steel. Implicit in the applicability of the Poisson distribution is that the occurrence takes up a negligible amount of the region of opportunity. Thus, the arrivals take a negligible amount of time to happen, the blemishes have a negligibly small total area and the inclusions occupy a negligible volume. Unlike the binomial distribution, there is no consideration of non-occurrences in applications of the Poisson distribution. The parameter α adjusts to the units that describe the region of opportunity. Thus, if the mean number of arrivals is 2/minute it can equivalently be quoted as 120/hour.

Example

The most famous application of the Poisson distribution is due to von Bortkiewicz, who used it to explain a celebrated set of data on deaths due to kicks of a horse in the Prussian Cavalry. Ten army corps were observed over a 20-year period and the number of deaths was recorded for each of the 200 corps-years. The number of corps-years with x deaths is given in Table 2.4. The maximum number of deaths observed in any one corps-year was 4, although there is no theoretical upper limit. The average number of deaths per year is computed as $[0 \times 109 + 1 \times 65 + 2 \times 22 + 3 \times 3 + 4 \times 1]/200 = 0.61$. Using $\alpha = 0.61$ (although this is only an estimate of the true value of α) and substituting in the Poisson formula gives the values of $p(x)$ shown in row three of Table 2.4. Finally multiplying the Poisson probabilities by $n = 200$ gives the number of times out of 200 that one would expect each value of x to occur

Table 2.4 Observed Number of Deaths and Expected Number under the Poisson Model

No. of Deaths x	0	1	2	3	4
No. of corps-years	109	65	22	3	1
$p(x)$	0.54335	0.33145	0.10110	0.02055	0.00315
$E(x) = 200 \times p(x)$	108.7	66.3	20.2	4.1	0.61

under the Poisson model. These expected values, shown as row four of the table, are seen to be quite close to the observed numbers, indicating that the Poisson provides an excellent model for this (unfortunate) phenomenon. Formal tests are available to assess whether the expected and observed values are in sufficient agreement to accept the distribution (in this case the Poisson) that has been proposed as an explanation. In this case, it is clear even without such a test that the Poisson model is quite a good fit to the data. The same goodness-of-fit test can indicate whether a model fit is too good, that is, that the agreement is better than chance would dictate. Instances of likely data "fudging" have been detected in this way.

It is clear in this example that non-occurrences are hard to define, let alone record. Would it be a non-occurrence on those occasions when a horse kicked but no soldier was nearby, or when a soldier was kicked but only injured or when a horse thought about kicking a soldier but decided not to? Fortunately data on non-occurrences are not needed to implement the Poisson distribution.

As with the binomial, individual and cumulative Poisson probabilities may be computed using Excel or a statistical software package.

The Poisson is applied in acceptance sampling in quality applications. The problem is to find the probability that a lot of a manufactured product will be accepted if the proportion of defects in the lot is p and the lot is judged acceptable if the number of defects in a sample of size n is less than some prescribed acceptance number c. For large lots, or continuous production, the acceptance probability is computed using the binomial distribution. But if p is small and n is large, the Poisson is a good approximation when the parameter α is taken equal to the product np. For example, the probability that a sample of size $n = 100$ taken from a population that is 1% defective ($p = 0.01$) contains $X = 1$ defects may be computed using the binomial or the Poisson approximation with $\alpha = 100 \times 0.01 = 1$. The results are:

$$\text{Binomial: } P(X = 1) = \binom{100}{1}[0.01]^1[0.99]^{99} = 0.36973$$

$$\text{Poisson: } P(X = 1) = \frac{e^{-1}1^1}{1!} = 0.36788.$$

These results are generally considered to be in sufficient agreement for calculating the behavior of an inspection sampling plan.

A Poisson *process* refers to a random phenomenon that follows a Poisson distribution at every specific value of some indexing quantity such as time or length, and for which the Poisson parameter is proportional to that quantity. Under some circumstances, discussed later, the number of times $N(t)$ that an item fails and is replaced prior to time t, is Poisson distributed with $\alpha = \lambda t$. The mean number of replacements at time t will therefore be $E(N(t)) = \alpha = \lambda t$. The proportionality constant λ thus has the meaning of failures per unit time, that is, the failure rate.

2.7 THE GEOMETRIC DISTRIBUTION

The binomial random variable represents the number of successes in n independent trials when the probability of success remains constant from trial to trial. The geometric random variable is related. It is the number of such trials, X, until the first success is achieved. The probability that $X = 1$ is simply p. The outcome $X = 2$ implies that the first trial resulted in a failure and the second in a success, so $P\{X = 2\} = (1 - p)p$. It follows that $X = k$ signifies that the first $k - 1$ trials were failures and the k-th a success, so that:

$$\text{Prob}[X = x] = (1 - p)^{x-1} \cdot p. \tag{2.29}$$

As x increases each term of the pmf is equal to the preceding term multiplied by $(1 - p)$. The terms thus decrease as a geometric series giving rise to the name. The geometric distribution is known as a waiting time distribution since the random variable represents the wait, measured in trials, until the first success.

The expected value of X is:

$$E(X) = \frac{1}{p}. \tag{2.30}$$

Like the mean of the binomial this result seems to make intuitive sense to people. Most will answer "10" when asked how many "at bats" a 0.100 hitter can expect to have until his/her first hit. This result may be found by summing the series:

$$E(X) = \sum_{k=1}^{\infty} x \cdot (1 - p)^{x-1} \cdot p. \tag{2.31}$$

The expected value may be found more directly and surprisingly, by invoking the Law of Total Expectation and conditioning on the outcome of the first trial. The outcomes, probabilities, and expected value are tabled as follows:

First Trial Result	Probability	$E(X|\text{result})$
Success	p	1.0
Failure	$1 - p$	$E(X) + 1$

If the first trial is a success the expected number of trials to the first success is obviously 1.0. If the first trial is a failure, then the process begins anew, except the expected number of trials increases by 1. Thus, the expectation may be written:

$$E(X) = 1 \cdot p + [E(X) + 1] \cdot (1 - p).$$

Simplifying and solving for $E(X)$ gives the desired result.

The CDF for the geometric distribution may be shown to be:

$$F(x) = P[X \leq x] = 1 - (1 - p)^x. \qquad (2.32)$$

Suppose the probability that an item fails a test of some type is $p = 0.1$. If a number of items are tested sequentially what is the probability that the first failure occurs after the fifth test?

$$P[X > 5] = 1 - F(5) = (1 - 0.1)^5 = 0.590.$$

We apply the geometric distribution in Chapter 4 when considering the cost of a renewing free replacement warranty.

2.8   CONTINUOUS RANDOM VARIABLES

It can be argued that all random variables are discrete because measurements of any type have a finite precision. A reported temperature of 68°F might in fact represent all of the temperatures in the range 67.5° to 68.5° if our ability to measure temperature has a precision of ±0.5°. However, rather than use a discretized distribution having a very large number of possible values it is convenient, and generally not overly inaccurate, to regard measurement data as continuous.

Just as the pmf contains all of the information available about a discrete random variable, all of the information about a continuous random variable is embodied in its probability density function (pdf). The pdf $f(x)$ of a continuous random variable x is a positive continuous function defined over the range of the random variable. The range of the random variable may be finite, as it is for the uniform random variable discussed further below; it may extend over the entire real line from minus to plus infinity as it does for the normal random variable; or it may extend over just the positive real line $(0, \infty)$ as it does for most distributions that represent strength or life length such as the exponential, lognormal, and Weibull. The pdf has the property that its integral between two points, say x_1 and x_2, is the probability that an observed X will have a value between x_1 and x_2. That is,

$$\text{Prob}[x_1 < X < x_2] = \int_{x_1}^{x_2} f(x)dx. \tag{2.33}$$

The shaded area in Figure 2.4 displays for the continuous random variable X defined over the range $(0, \infty)$ the probability $P[1.5 < X < 2.0]$.

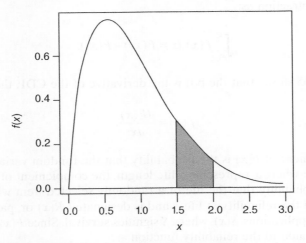

Figure 2.4 Probability density function $f(x)$ of a continuous random variable.

Again, we follow the convention of using the upper case when referring to the random variable and the lower case when referring to a hypothetical observed value or values. Unlike the discrete case, the probability that a continuous random variable is exactly equal to any specific value, for example, $P[X = 2]$, is zero. For this reason it is unnecessary to write *less than or equal* in Equation 2.33 since the probability that a continuous random variable is *equal* to any specific value is zero. If $x_2 - x_1 \equiv \Delta x$ is small, then as an approximation,

$$\text{Prob}[x_1 < X < x_2] = \int_{x_1}^{x_2} f(x)dx \approx f(x)\Delta x.$$

Since, by definition, x *must* lie within its stated range, between 0 and ∞ in this case, one has

$$\text{Prob}[X < \infty] = \int_0^\infty f(x)dx = 1.0. \tag{2.34}$$

The CDF of a positive continuous random variable is defined as the probability $F(x)$ that the random variable X is less than a general value x. That is,

$$F(x) = \text{Prob}[X < x] = \int_0^x f(x)\,dx. \tag{2.35}$$

The argument x is the upper limit of the integral. The x in the integrand is a dummy variable of integration and could be replaced by another symbol such as x' or y as some writers choose to do.

In view of Equation 2.34, one must have $F(\infty) = 1.0$.

Given the CDF $F(x)$, the area between two values x_1 and x_2 may be computed by subtraction as:

$$\int_{x_1}^{x_2} f(x)\,dx = F(x_2) - F(x_1). \tag{2.36}$$

Equation 2.35 shows that the pdf is the derivative of the CDF, that is,

$$f(x) = \frac{dF(x)}{dx}. \tag{2.37}$$

The complement of $F(x)$ is the probability that the random variable exceeds x. In the case where x represents a life length, the complement of $F(x)$ represents the probability that the life of a randomly selected item will exceed x. This is called the reliability at life x and is designated $R(x)$ or, particularly in biomedical applications, $S(x)$, where S signifies survival. Since $F(x) = 1 - R(x)$, the pdf is related to the reliability function as,

$$f(x) = -\frac{dR(x)}{dx}. \tag{2.38}$$

The distribution of a continuous random variable may contain one or more constants known as parameters that characterize its appearance. Denoting these by the letters $\alpha_1, \alpha_2, \ldots, \alpha_n$ the density function can be written more generally as $f(x; \alpha_1, \alpha_2, \ldots, \alpha_n)$.

The mean μ or expected value, $E(X)$, of a random variable is defined as:

$$\mu = E(X) = \int_{-\infty}^{+\infty} x \cdot f(x; \alpha_1, \alpha_2, \ldots \alpha_n)\,dx. \tag{2.39}$$

$E(X)$ will be a function of the parameters $\alpha_1, \alpha_2, \ldots, \alpha_n$. As in the discrete case the average of a sample of size n approaches $E(X)$ as n increases. (More precisely, the probability that the sample mean deviates from $E(X)$ by more than a prescribed amount ε diminishes with n.)

Example

Find $E(X)$ for the right triangular density:

$$f(x, \alpha) = \frac{2x}{\alpha^2}; 0 < x < \alpha. \tag{2.40}$$

The density of Equation 2.40 is depicted in Figure 2.5.

Equation 2.34 is satisfied since:

$$\int_0^\alpha \frac{2x}{\alpha^2} dx = \frac{2x^2}{2\alpha^2}\bigg|_0^\alpha = 1.0.$$

The expected value $E(X)$ is calculated from Equation 2.39 as follows:

$$E(X) = \int_0^\alpha x \cdot \frac{2x}{\alpha^2} dx = \frac{2x^3}{3\alpha^2}\bigg|_0^\alpha = \frac{2\alpha}{3}.$$

For this right triangular distribution, $E(X)$ is a linear function of the parameter α. Physically a line parallel to the ordinate that intersects the x axis at the point where $x = E(x)$ is the centroid of the density function. If a density function were cut out of homogeneous sheet metal it could be balanced on a knife edge placed perpendicular to the x axis at $x = E(X)$.

When a random variable is positive, for example, life length or strength, the mean may alternately be computed as the integral of the reliability function $R(x) = 1 - F(x)$ over the range of variation of the random variable, that is,

$$\mu = E(X) = \int_0^\infty R(x) dx. \tag{2.41}$$

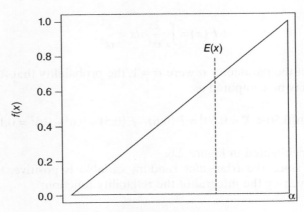

Figure 2.5 The triangular density.

This may be shown by integrating Equation 2.39 by parts after substituting the negative derivative of $R(x)$ for $f(x)$. This expression for the mean is quite convenient for computing the expected value of system life in terms of the system reliability function.

The expected value of a function $Y = g(X)$ of a random variable X is, analogously to the discrete case,

$$E(Y) = \int_{-\infty}^{\infty} g(x) f(x) dx. \tag{2.42}$$

In particular the expected value of the linear function $Y = g(X) = a + bX$ is:

$$E(Y) = \int_{-\infty}^{\infty} (a + bx) f(x) = a \int_{-\infty}^{\infty} f(x) dx + b \int_{-\infty}^{\infty} x f(x) dx$$

or, by Equations 2.34 and 2.39,

$$E(Y) = a + b E(X). \tag{2.43}$$

Thus, as in the discrete case, the expected value of a linear function of X is that same linear function of the expected value $E(X)$.

Example
Determine the CDF of the right triangular distribution whose pdf is given by Equation 2.40.

$$F(x) = \int_0^x \frac{2x}{\alpha^2} dx = \frac{x^2}{\alpha^2}.$$

If the value of the parameter α were $\alpha = 1$, the probability that X is between 0.5 and 0.6 may be computed as:

$$\text{Prob}[0.50 < X < 0.60] = F(0.6) - F(0.5) = 0.6^2 - 0.5^2 = 0.11.$$

This area is highlighted in Figure 2.6.

As noted, since the triangular random variable is positive, the mean is alternately given by the integral of the reliability function:

$$R(x) = 1 - F(x) = 1 - \frac{x^2}{\alpha^2}.$$

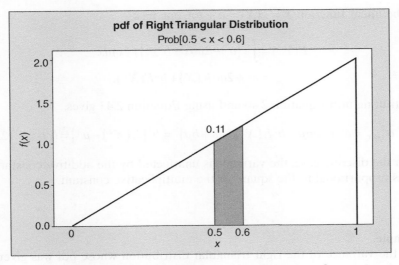

Figure 2.6   Graphical depiction of Prob[0.5 < *X* < 0.6].

So $E(X)$ may be computed as:

$$\mu = E(X) = \int_0^\alpha \left[1 - \frac{x^2}{\alpha^2}\right]dx = x - \frac{x^3}{3\alpha^2}\bigg|_0^\alpha = \frac{2\alpha}{3}.$$

This agrees with the value previously found directly from the definition of the expected value.

The variance σ^2 of a random variable is defined to be the expected value of the square of the difference between the random variable and its mean. As in the discrete case it characterizes the spread in the distribution.

That is,

$$\sigma^2 = E(X - \mu)^2. \tag{2.44}$$

Expanding $(x - \mu)^2$ and integrating its product with the density function $f(x)$:

$$\sigma^2 = \int_{-\infty}^\infty \left[x^2 - 2x\mu + \mu^2\right]f(x)\,dx = \int_{-\infty}^\infty x^2 f(x)\,dx - 2\mu\int_{-\infty}^\infty xf(x)\,dx + \mu^2.$$

Since $\int_{-\infty}^\infty xf(x) = \mu$, the result is the same as in the discrete case, namely,

$$\sigma^2 = E(X^2) - \mu^2. \tag{2.45}$$

By analogy, the variance of a function $Y = g(X)$ of X may be expressed as:

$$var(Y) = E(Y^2) - E^2(Y). \tag{2.46}$$

for the linear function $Y = g(X) = a + bX$ one has:

$$E(Y^2) = \int_{-\infty}^{\infty} [a^2 + 2abx + b^2x^2] f(x) dx$$
$$= a^2 + 2abE(X) + b^2 E(X^2).$$

Substituting into Equation 2.46 and using Equation 2.43 gives,

$$\sigma_{a+bx}^2 = a^2 + 2ab\mu + b^2 E(X^2) - [a + b\mu]^2 = b^2 [E(X^2) - \mu^2] = b^2 \sigma^2. \quad (2.47)$$

As in the discrete case, the variance is unaffected by the additive constant a, and is proportional to the square of the multiplicative constant.

Example
Find the variance of the right triangular distribution whose pdf was given in Equation 2.40.

$$E(X^2) = \int_0^\alpha x^2 \frac{2x}{\alpha^2} dx = \frac{2x^4}{4\alpha^2}.$$

Using $\mu = E(X) = \dfrac{2\alpha}{3}$ in Equation 2.45 results in:

$$\sigma^2 = \frac{\alpha^2}{2} - \left(\frac{2\alpha}{3}\right)^2 = \frac{\alpha^2}{18}.$$

For the same distribution, what is the variance of of $2 + 3X$?

$$\sigma_{2+3X}^2 = 3^2 \sigma^2 = 9 \frac{\alpha^2}{18} = \frac{\alpha^2}{2}.$$

The p-th quantile of a distribution is the value, designated x_p, for which

$$F(x_p) = p. \quad (2.48)$$

That is, x_p is the point on the x axis for which the area under the pdf to the left of x_p is p. The p-th quantile is frequently called the $100p$-th percentile. Thus, for example, the value $x_{0.10}$ denotes the 10th percentile of the distribution. The 50th percentile is called the median. Figure 2.7 shows a pdf having a median of 78.3. The area below the median is by definition equal to 0.50 and is shown shaded.

The median of the triangular distribution is the value $x_{0.50}$ found by equating $F(x)$ to 0.50 and solving for x:

$$F(x_{0.50}) = \frac{x_{0.50}^2}{\alpha^2} = 0.50; \ x_{0.50} = 0.707\alpha.$$

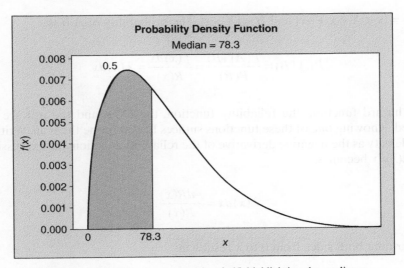

Figure 2.7 Probability density function (pdf) highlighting the median $x_{0.50}$.

2.8.1 The Hazard Function

The hazard function $\lambda(x)$ is a concept specific to the study of life length. The probability of failing in the differential interval $(x, x + dx)$ conditioned on having survived up to life x is given by $\lambda(x)dx$. The hazard function is also known as the instantaneous failure rate, and, in actuarial studies, the force of mortality. If $\lambda(x)$ increases with x it indicates that vulnerability increases with age due to wearout. If $\lambda(x)$ decreases with x it means that specimens that survive have diminished vulnerability because weaker specimens fail early.

One intuitive method of deriving the expression for the hazard function is to consider that if a very large number N of items are run to failure the expected number that will fail in the interval $(x, x + dx)$ is $Nf(x)dx$. The number expected to survive life x is $NR(x)$. For large enough N the ratio of the expected number failing in $(x, x + dx)$ to the number that were still functioning at time x will converge to the probability that a survivor of life x will fail in that interval. Thus,

$$\text{Prob}[x < X < x + dx \mid X > x] = \frac{Nf(x)dx}{NR(x)} = \frac{f(x)}{R(x)}dx.$$

The hazard function is thus given by:

$$\lambda(x) = \frac{f(x)}{R(x)}. \tag{2.49}$$

A more formal derivation follows when the laws of conditional probability are applied to events A and B defined as follows:

$A = \{x < X < x + dx\}$ and $B = \{X > x\}$. The probability of $A|B$ is:

$$P(A \mid B) = \frac{P(A \cap B)}{P(B)} = \frac{f(x)\,dx}{R(x)} = \lambda(x)\,dx.$$

The hazard function, the reliability function, the CDF, and the pdf are all linked. Knowing one of these functions suffices to determine them all. Writing the density as the negative derivative of the reliability function, the expression for $\lambda(x)\,dx$ becomes:

$$\lambda(x)\,dx = \frac{-dR(x)}{R(x)}.$$

Integrating both sides from 0 to x results in:

$$-\int_0^x \lambda(x)\,dx = \ln[R(x)]\big|_0^x = \ln[R(x)].$$

Since $R(0) = 1$, $\ln[R(0)] = 0$.

Exponentiating both sides then expresses $R(x)$ in terms of $\lambda(x)$ as:

$$R(x) = \exp\left[-\int_0^x \lambda(x)\,dx\right]. \tag{2.50}$$

$F(x)$ then follows from $1 - R(x)$ and $f(x)$ can be found by differentiating $F(x)$. The integral of the hazard function is called the cumulative hazard $\Lambda(x)$:

$$\Lambda(x) = \int_0^x \lambda(x)\,dx. \tag{2.51}$$

The average failure rate over the interval (x_1, x_2) is:

$$\bar{\lambda} = \frac{\int_{x_1}^{x_2} \lambda(x)\,dx}{(x_2 - x_1)} = \frac{\Lambda(x_2) - \Lambda(x_1)}{(x_2 - x_1)}. \tag{2.52}$$

Using Equations 2.50 and 2.51 the average failure rate may alternately be written as:

$$\bar{\lambda} = \frac{\ln[R(x_1)] - \ln[R(x_2)]}{(x_2 - x_1)}. \tag{2.53}$$

As an example of the derivation of a life distribution in terms of a specified hazard function, consider the case where the hazard function is constant, that is, $\lambda(x) = \lambda$.

Then $R(x)$ is:

$$R(x) = \exp\left[-\lambda \int_0^x dx\right] = e^{-\lambda x}. \tag{2.54}$$

$F(x)$ is then $1 - e^{-\lambda x}$, and $f(x)$ is $\lambda e^{-\lambda x}$. This distribution is known as the exponential distribution. Its mean is equal to $1/\lambda$. As we shall see in the next chapter, it is a special case of the Weibull distribution. The constant hazard function means that failure is not due to wearout since aged specimens are as good as new. It is characteristic of a failure mode such as an accident or shock that terminates a specimen's life regardless of its present age. We will discuss the hazard function further when we begin our exploration of the Weibull distribution.

2.9 JOINTLY DISTRIBUTED CONTINUOUS RANDOM VARIABLES

When two continuous random variables are being studied the density function is replaced by the joint density function $f(x, y)$. The probability that $x_1 < X < x_2$ and simultaneously, $y_1 < Y < y_2$ is given by the integral of $f(x, y)$ over that region:

$$\text{Prob}[x_1 < X < x_2 \cap y_1 < Y < y_2] = \int_{x_1}^{x_2} \int_{y_1}^{y_2} f(x, y)\,dy\,dx. \tag{2.55}$$

When x and y are independent $f(x, y) = f(x)f(y)$.

The other results are as for the discrete case but with summations replaced by integration. The marginal distribution of X is found by integrating $f(x, y)$ over the range of y. The marginal distribution of y is found correspondingly. The covariance is:

$$cov(X, Y) = E(XY) - E(X)E(Y). \tag{2.56}$$

With $E(XY)$ given by:

$$E(XY) = \int_{-\infty}^{\infty} \int_{-\infty}^{\infty} xyf(x, y)\,dx\,dy. \tag{2.57}$$

The correlation coefficient and variance of a sum is the same as in the discrete case.

2.10 SIMULATING SAMPLES FROM CONTINUOUS DISTRIBUTIONS

A distribution of interest in conjunction with simulation studies is the uniform or rectangular distribution. Figure 2.8 shows a uniform distribution over the interval (1.0, 3.0).

That a random variable X follows the uniform distribution over an interval (a, b) is sometimes communicated by the notation $X \sim U(a, b)$. The area of the rectangle representing the probability that $a < X < b$ must be unity so $f(x)$, the height of the rectangle must be $1/(b - a)$. In the pdf depicted in Figure 2.8, the value of $f(x)$ is seen to be $\frac{1}{2} = 0.50$. One can see by inspection and confirm by integration that the mean of the uniform distribution is:

$$\mu = E(X) = (a+b)/2. \tag{2.58}$$

The variance can be shown to be:

$$\sigma^2 = \frac{(b-a)^2}{12}. \tag{2.59}$$

The CDF is:

$$F(x) = \frac{x}{(b-a)}; a < X < b. \tag{2.60}$$

The uniform distribution having $a = 0$ and $b = 1$ is of special interest. Consider the transformation of a random variable by means of a function equal to its own CDF; that is, let $Y = F(X)$. This is a nondecreasing function of X. The CDF

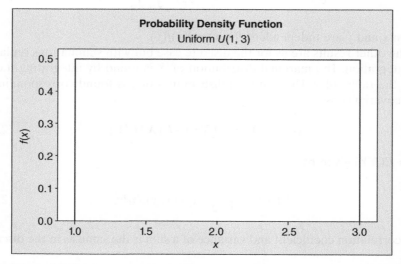

Figure 2.8 Probability density function (pdf) of the uniform distribution over the interval (1, 3).

of Y is denoted $G(y) = \text{Prob}(Y < y)$. Now $Y < y$ whenever $X < F^{-1}(y)$ where $F^{-1}(y)$ denotes the inverse of $F(x)$ evaluated at the value y. Thus,

$$G(y) = \text{Prob}(Y < y) = \text{Prob}(X < [F^{-1}(y)]) = F(F^{-1}(y)) = y.$$

So $G(y)$ is of the form of the uniform CDF with $(b - a) = 1$. Since $y = 1$ when x is at its largest and $y = 0$ when x is at its smallest, $a = 0$ and $b = 1$. Thus, $y = F(x)$ follows the uniform distribution with $a = 0$ and $b = 1$; that is, $F(x)$ is $U(0, 1)$. This fact can be exploited to generate simulated observations from a density function whose CDF is expressible in closed form. The method is to generate a sample of values from $U(0, 1)$ and equate them to $F(x)$. The associated values of x may then be found as the inverse $x = F^{-1}(u)$. For the right triangular pdf given earlier, the CDF is:

$$F(x) = \frac{x^2}{\alpha^2}.$$

Choosing the parameter $\alpha = 2$ for illustration, equating the CDF to a uniform variable u, and solving for x gives:

$$x = \sqrt{4u}. \tag{2.61}$$

Selecting a random sample of values from the distribution $U(0, 1)$ and using Equation 2.61 will result in a random sample of values of x. Methods of computing values of u that behave as if they were random samples have been studied extensively (cf. Marsaglia and McLaren 1965). Because they are computed and hence actually deterministic they are referred to as pseudorandom. Figure 2.9 is a histogram of 1000 values of x computed by substituting 1000 random values of u into Equation 2.61. The right triangular shape is evident in the histogram of the simulated observations.

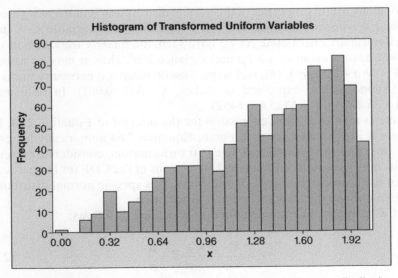

Figure 2.9 Histogram of 1000 simulated observations from triangular distribution.

2.11 THE NORMAL DISTRIBUTION

The most important distribution in applied statistics is the normal or gaussian distribution.

It is a two-parameter distribution having the density function:

$$f(x; \alpha_1, \alpha_2) = \frac{1}{\sqrt{2\pi\alpha_2}} \cdot \exp\left(-\frac{[x - \alpha_1]^2}{2\alpha_2^2}\right). \tag{2.62}$$

As noted previously, the expected value and standard deviation will be functions of the parameters. For the normal distribution, the expected value $\mu = E(X)$ can be shown to be equal to just α_1 and the custom is to use μ in lieu of α_1 when writing the normal density function. Likewise the standard deviation may be shown to equal α_2 and so σ is customarily used in lieu of α_2 in the expression for the normal density.

The fact that a random variable X is normally distributed with expected value μ and variance σ^2 is compactly communicated by the following notation.

$$X \sim N(\mu, \sigma^2). \tag{2.63}$$

The cumulative form of the normal distribution is given by :

$$F(x) = \frac{1}{\sqrt{2\pi}\sigma} \int_{-\infty}^{x} \exp(-[x - \mu]^2 / 2\sigma^2) dx. \tag{2.64}$$

One of many useful properties of the normal distribution is that a new variable obtained as a linear combination $Y = a + bX$ of a normal variable $X \sim N(\mu, \sigma^2)$ is itself normally distributed. As we have seen, the linearly transformed variable will have a mean of $a + b\mu$ and variance $b^2\sigma^2$. Thus, it may be asserted that $Y \sim N(a + b\mu, b^2\sigma^2)$. This fact is quite useful when it is necessary to change units. Suppose that, expressed in inches, $X \sim N(1, 0.001)$. In millimeters, $Y = 0 + 25.4X$, so $Y \sim N(25.4, 0.6452)$.

There is no closed form expression for the integral of Equation 2.64. Fortunately it is not necessary to integrate Equation 2.64 numerically for every value of μ and σ of concern. A linear transformation, considered presently, permits the evaluation of Equation 2.64 in terms of the CDF for the particular normal distribution having $\mu = 0$ and $\sigma = 1$. This specific normal distribution has been dubbed the standard normal distribution.

Let Z be the linear transformation of X defined as follows:

$$Z = \frac{X - \mu}{\sigma} = \frac{1}{\sigma}X - \frac{\mu}{\sigma}. \tag{2.65}$$

Equation 2.65 is of the form $a + bX$ wherein $a = -\dfrac{\mu}{\sigma}$ and $b = \dfrac{1}{\sigma}$. The expected value of Z is therefore,

$$E(Z) = -\frac{\mu}{\sigma} + \frac{1}{\sigma} E(X) = -\frac{\mu}{\sigma} + \frac{1}{\sigma}\mu = 0.$$

and the variance of Z is:

$$\sigma_z^2 = \frac{1}{\sigma^2} var(X) = \frac{\sigma^2}{\sigma^2} = 1.$$

Thus $Z = \dfrac{X - \mu}{\sigma}$ is $N(0, 1)$. The CDF for the standard normal distribution is customarily denoted $\Phi(z)$ and is expressed by the integral:

$$\Phi(z) = \frac{1}{\sqrt{2\pi}} \int_{-\infty}^{z} \exp\left[-\frac{x^2}{2}\right] dx. \tag{2.66}$$

The standard normal density is generally written as $\varphi(z)$. The standard normal CDF $\Phi(z)$ has been evaluated numerically and is extensively tabulated. A short table sufficient for illustrative purposes is given below. In practice Excel or a statistical package can be used to compute normal probabilities directly without the use of the standard normal. Since the use of the standard normal is entrenched in the literature we cover its use here. The standard normal is frequently invoked in applied statistics in conjunction with discussions of random variables that are asymptotically normal.

Given that X is $N(\mu, \sigma^2)$ one may find the probability that X is less than some specified value A, in terms of the distribution of $Z = \dfrac{X - \mu}{\sigma}$ as follows:

$$\text{Prob}[X < A] = \text{Prob}\left[\frac{X - \mu}{\sigma} < \frac{A - \mu}{\sigma}\right] = \text{Prob}\left[Z < \frac{A - \mu}{\sigma}\right] = \Phi\left(\frac{A - \mu}{\sigma}\right). \tag{2.67}$$

Standard Normal CDF $\Phi(z)$

z	$\Phi(z)$	z	$\Phi(z)$
−3	0.0013499	0	0.5
−2.9	0.0018658	0.1	0.5398278
−2.8	0.0025551	0.2	0.5792597
−2.7	0.003467	0.3	0.6179114
−2.6	0.0046612	0.4	0.6554217
−2.5	0.0062097	0.5	0.6914625

(Continued)

z	$\Phi(z)$	z	$\Phi(z)$
−2.4	0.0081975	0.6	0.725746,9
−2.3	0.0107241	0.7	0.7580363
−2.2	0.0139034	0.8	0.7881446
−2.1	0.0178644	0.9	0.8159399
−2	0.0227501	1	0.8413447
−1.9	0.0287166	1.1	0.8643339
−1.8	0.0359303	1.2	0.8849303
−1.7	0.0445655	1.3	0.9031995
−1.6	0.0547993	1.4	0.9192433
−1.5	0.0668072	1.5	0.9331928
−1.4	0.0807567	1.6	0.9452007
−1.3	0.0968005	1.7	0.9554345
−1.2	0.1150697	1.8	0.9640697
−1.1	0.1356661	1.9	0.9712834
−1	0.1586553	2	0.9772499
−0.9	0.1840601	2.1	0.9821356
−0.8	0.2118554	2.2	0.9860966
−0.7	0.2419637	2.3	0.9892759
−0.6	0.2742531	2.4	0.9918025
−0.5	0.3085375	2.5	0.9937903
−0.4	0.3445783	2.6	0.9953388
−0.3	0.3820886	2.7	0.996533
−0.2	0.4207403	2.8	0.9974449
−0.1	0.4601722	2.9	0.9981342
		3	0.9986501

Example

A random variable is distributed as $N(20, 4)$. Find the probability that an observed value of this random variable has a value less than 23.

$$\text{Prob}[X < 23] = \text{Prob}\left[Z < \frac{23 - 20}{2}\right] = \Phi(1.5) = 0.9332$$

For the same random variable find the probability that $19 < X < 22$.

$$\text{Prob}[19 < X < 22] = \text{Prob}[X < 22] - \text{Prob}[X < 19]$$

$$\text{Prob}\left[Z < \frac{22 - 20}{2}\right] - \text{Prob}\left[Z < \frac{19 - 20}{2}\right]$$

$$= \Phi(1) - \Phi(0.5) = 0.8413 - 0.3085 = 0.5328.$$

Example

If X is $N(\mu, \sigma^2)$ find the probability that X is within $a = \pm 1\ \sigma$ interval surrounding its mean:

$$\text{Prob}\left[\mu-\sigma<X<\mu+\sigma\right]=\text{Prob}\left[Z<\frac{\mu+\sigma-\mu}{\sigma}\right]-\text{Prob}\left[Z<\frac{\mu-\sigma-\mu}{\sigma}\right]$$

$$=\text{Prob}\left[Z<1\right]-\text{Prob}\left[Z<-1\right]$$

$$\Phi(1)-\Phi(-1)=0.8413-0.1587=0.6826.$$

Repeating this computation for a $\pm2\sigma$ interval establishes the well-known fact that a two-sigma interval contains roughly 95% of the area under the density function. If exactness is preferred to simplicity, a $\pm1.96\sigma$ interval will be closer to the 95% coverage. Figure 2.10 shows the standard normal density with the area between ±1.0 shown shaded.

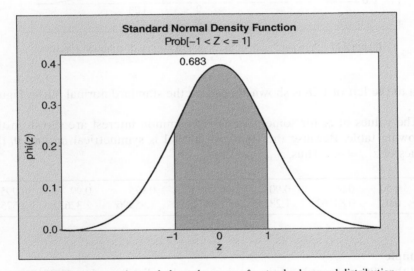

Figure 2.10 A one σ interval about the mean of a standard normal distribution.

Example

Find the 90th percentile for the standard normal variable $z\sim N(0,1)$. By definition:

$$\Phi(z_{0.90})=0.90=\int_{0}^{z_{0.90}}\varphi(z)\,dz.$$

The preceding table is insufficiently detailed to determine a precise value of z for which $\Phi(z)=0.90$. It suffices only to show that $z_{0.90}$ is less than 1.3 and greater than 1.2. The actual value to three significant figures is $z_{0.90}=1.28$. The

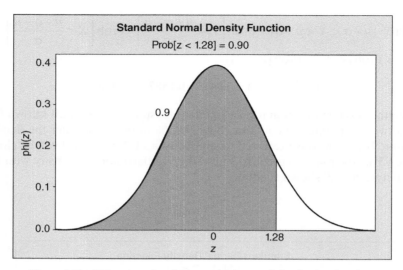

Figure 2.11 90th percentile of the standard normal distribution function.

area to the left of 1.28 is shown shaded on the standard normal pdf in Figure 2.11.

The values of z_p for some p values of common interest are shown in the following table. Because the standard normal is symmetrical around 0, the values of $z_{1-p} = -z_p$. Thus, $z_{0.10} = -1.28155$.

p	0.50	0.80	0.90	0.95	0.975	0.99	0.995
z_p	0.0	0.84162	1.28155	1.644485	1.95996	2.32635	2.57583

Example

Find the 90th percentile of the random variable distributed as $N(25, 9)$:

$$0.90 = \text{Prob}[X < x_{0.90}] = \text{Prob}\left[Z < \frac{x_{0.90} - 25}{3}\right].$$

But by the last example,

$$\text{Prob}[Z < z_{0.90}] = 0.90,$$

so that

$$\frac{x_{0.90} - 25}{3} = z_{0.90} = 1.282.$$

and hence $x_{0.90} = 1.282 \times 3 + 25 = 28.84$.

In general,

$$x_p = z_p \sigma + \mu. \tag{2.68}$$

The normal is only occasionally used as a model for life length. Its reliability function is:

$$R(x) = 1 - F(x) = 1 - \Phi\left(\frac{(x-\mu)}{\sigma}\right). \tag{2.69}$$

The hazard function is:

$$\lambda(x) = \frac{f(x)}{R(x)} = \frac{\varphi\left[\dfrac{x-\mu}{\sigma}\right]}{\sigma}\{1 - \Phi[(x-\mu)/\sigma]\}^{-1}. \tag{2.70}$$

Figure 2.12 is a plot of $\sigma\lambda(x)$ and shows that a normal model for life length implies an increasing hazard function.

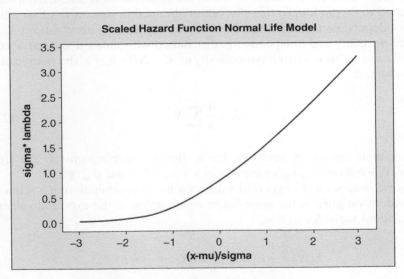

Figure 2.12 The normal hazard function.

It was stated in Section 2.5 that, provided, np and $n(1 - p)$ both exceed 5, the normal distribution with mean np and variance $np(1 - p)$ may be used to approximate the binomial. As an example, let $n = 20$ and $p = 0.4$ and suppose we wish to compute $P[X = 10]$. The mean is $E(X) = 20 \times 0.4 = 8$ and the variance is $var(X) = 20 \times 0.4 \times 0.6 = 4.8$. Now the probability that a continuous random variable is exactly equal to any specific value is 0. To compute $P[X = 10]$ we must determine the area under the approximating normal distribution between 9.5 and 10.5. Thus,

$$P[X = 8] = \Phi\left(\frac{(10.5 - 8)}{\sqrt{4.8}}\right) - \Phi\left(\frac{9.5 - 8}{\sqrt{4.8}}\right) = 0.119863.$$

The exact value of $P[X = 10]$ is 0.117142.

To compute a value of the cumulative binomial using the normal, a continuity correction of 0.5 should be added to the upper limit. In the present example $P[X \leq 10] \approx \Phi\left(\frac{10.5 - 8}{\sqrt{4.8}}\right) = 0.873083.$ The exact value using the binomial distribution is 0.872479.

2.12   DISTRIBUTION OF THE SAMPLE MEAN

Let n observations be made on the random variable following the distribution $N(\mu, \sigma^2)$. Denote these observations by $X_1, X_2, X_3 \ldots X_n$. The subscript on each observation denotes the order in which the observation is taken. Sampling is assumed to be random so that the values obtained in successive observations do not depend on the order number. We may say that the observations X_i and X_j are normally and independently distributed with mean μ and variance σ^2 for all i and j. This is written symbolically as $X_i \sim NID(\mu, \sigma^2)$. The sample mean $\bar{X}$ is defined:

$$\bar{X} = \frac{1}{n}\sum_{i=1}^{n} X_i. \tag{2.71}$$

The sample mean is seen to be a linear combination of the form $Y = c_1X_1 + c_2X_2 + \cdots c_nX_n$, where $c_1 = c_2 = \cdots c_n = 1/n$ and $Y = \bar{X}$.

As we have seen the expected value of a linear combination of a sequence of random variables is the same linear combination of the expected values of the random variables so that,

$$E(\bar{X}) = \frac{1}{n}E(X_1) + \frac{1}{n}E(X_2) + \cdots \frac{1}{n}E(X_n).$$

but $E(X_i) = \mu$ for all i and so,

$$E(\bar{X}) = \frac{1}{n}\mu + \frac{1}{n}\mu + \cdots \frac{1}{n}\mu = \mu. \tag{2.72}$$

It is important to realize that the sample mean is a random variable just as X itself is. By taking repeated samples of, say, size 5 and calculating successive values of the sample mean $\bar{X}_{(5)}$ one is sampling from the distribution of the random variable $\bar{X}_{(5)}$.

Intuitively we would expect that this distribution would have the same expected value as the parent distribution, that is, that $E[\bar{X}_{(5)}] = E(X) = \mu$. We would expect it to scatter less about its expected value however. Moreover, if we were considering the mean $\bar{X}_{(10)}$ of samples of size 10 we would expect still less scatter about μ. The variance of $\bar{X}$ must therefore be a decreasing function of sample size n.

It can be proved that $\bar{X}$ tends to normality when n is large irrespective of the distribution of X. The mathematical support for this behavior is called the Central Limit Theorem. How large n must be for this to occur depends on how nearly normal the original distribution is.

Since $\bar{X}$ is normally distributed, one need only find its mean and variance to describe its distribution completely.

The variance of a linear combination of random variables is, when the variables are independent, given by:

$$\sigma_Y^2 = c_1^2 \sigma_1^2 + c_2^2 \sigma_2^2 + \cdots c_n^2 \sigma_n^2. \tag{2.73}$$

For $Y = \bar{X}_{(n)}, \sigma_1^2 = \sigma_2^2 = \cdots = \sigma_n^2 = \sigma^2, \quad c_1 = c_2 = c_n = \dfrac{1}{n}.$

Hence

$$\sigma_{\bar{X}}^2 = \frac{1}{n^2}\sigma^2 + \frac{1}{n^2}\sigma^2 + \cdots \frac{1}{n^2}\sigma^2 = \frac{\sigma^2}{n}. \tag{2.74}$$

We may summarize as follows: If X is distributed as $N(\mu, \sigma^2)$, then the mean $\bar{X}$ of a sample of size n is distributed as $N\left(\mu, \dfrac{\sigma^2}{n}\right)$. This result is the basis for confidence intervals for the normal parameter μ. Since a normal variable minus its population mean and divided by its standard deviation is distributed as $N(0, 1)$ we may write the probability statement:

$$\text{Prob}\left[z_{0.025} < \frac{\bar{X} - \mu}{\sigma/\sqrt{n}} < z_{0.975}\right] = 0.95.$$

Converting the two sides of the inequality into two inequalities on the unknown parameter μ and using $z_{0.025} = -z_{0.975}$ and $z_{0.975} = 1.96$ leads to:

$$\bar{X} - \frac{1.96\sigma}{\sqrt{n}} < \mu < \bar{X} + \frac{1.96\sigma}{\sqrt{n}}. \tag{2.75}$$

$\bar{X}$ denotes the observed value of the sample mean. Equation 2.75 represents a 95% confidence interval on the unknown parameter μ valid when the standard deviation is known. Using appropriate percentage points of the standard normal variable the confidence level of the statement may be changed as desired. The general expression may be written as:

$$\bar{X} \pm z_{1-\alpha/2} \frac{\sigma}{\sqrt{n}} \tag{2.76}$$

and is called a $100(1 - \alpha)\%$ two-sided confidence interval. As an example, if an 80% confidence interval is desired, $\alpha = 0.20$ and the appropriate z value is $z_{0.90} = 1.28$.

Sometimes only a one-sided confidence interval is needed. For example, one might wish only to have a lower bound for the average strength of a manufactured part. In that case the value of α is not divided between the two tails. So a lower 95% confidence limit would be written as:

$$\mu > \bar{X} - 1.645 \frac{\sigma}{\sqrt{n}}.$$

Example

A normally distributed random variable has a standard deviation of $\sigma = 2$. A sample of 25 values drawn randomly from this distribution had a sample mean of $\bar{X} = 20.0$. Compute a 90% confidence interval for the population mean μ. As already noted, there is a 90% probability that z will fall between -1.645 and $+1.645$. Analogous to the statement given above, we may calculate:

$$19.34 = 20 - 1.645 \frac{2}{\sqrt{25}} < \mu < 20 + 1.645 \frac{2}{\sqrt{25}} = 20.66.$$

An *upper* 90% confidence interval for μ is:

$$\mu < 20 + 1.28 \frac{2}{\sqrt{25}} = 20.51.$$

Example

The resistance X of a certain type of connector follows the normal distribution with a mean of 1000 ohms and a variance of 90 ohms^2. An inspection procedure requires that a sample of 10 resistors be taken and that the shipment be accepted if the average of these 10 values is between 995 and 1005. Find the probability that the shipment will be accepted:

$$Pa = \text{Prob}[995 < \bar{X}_{(10)} < 1005]$$

$$\text{where } \bar{X}_{(10)} \sim N(1000, 90/10)$$

$$Pa = \text{Prob}[\bar{X}_{(10)} < 1005] - \text{Prob}[\bar{X}_{(10)} < 995]$$

$$= \text{Prob}\left[Z < \frac{1005 - 1000}{3}\right] - \text{Prob}\left[Z < \frac{995 - 1000}{3}\right]$$

$$= \Phi(5/3) - \Phi(-5/3)$$

$$Pa = 0.9525 - 0.0475 = 0.9050.$$

Just as the sample mean $\bar{x}$ is an unbiased estimate of the population mean μ, so the sample variance, defined as,

$$s^2 = \sum_{i=1}^{n} [x_i - \bar{x}]^2 / (n-1). \tag{2.77}$$

is an unbiased estimate of σ^2. Loosely speaking, when an infinite number of samples of size n is considered and s^2 is computed for each sample, the average will be σ^2. In other words, for the sampling distribution of s^2, it is true that,

$$E(s^2) = \sigma^2.$$

It was seen above that confidence limits for μ can be found when σ is known by using the known distribution of the random variable $z = \dfrac{\bar{x} - \mu}{\sigma / \sqrt{n}}$. When σ is unknown it must be estimated by s, the sample standard deviation.

The random variable

$$t = \frac{\bar{x} - \mu}{s / \sqrt{n}} \tag{2.78}$$

is identical in form to z except that the sample standard deviation s replaces σ. It can be shown that this quantity varies from sample to sample in accordance with the one-parameter distribution having the following pdf:

$$f(t; v) = \frac{1}{\sqrt{\pi v}} \cdot \frac{\Gamma\left(\dfrac{v+1}{2}\right)}{\Gamma\left(\dfrac{v}{2}\right)} \cdot \left(1 + \frac{t^2}{v}\right)^{-\left(\frac{v+1}{2}\right)}. \tag{2.79}$$

The distribution of Equation 2.79 is called Student's t distribution. It is symmetrical about t = 0.0. Its parameter, v, is called the "degrees of freedom" of the distribution, and for t as defined in terms of the mean and standard deviation in Equation 2.78, v = n − 1. Percentage points of the t distribution may be found using Excel or any statistical software package.

Analogous to the case where σ is unknown, the construction of two-sided 100 $(1 - \alpha)\%$ confidence limits for μ, proceeds from the probability statement:

$$\text{Prob}\left[t_{\frac{\alpha}{2}} < \frac{\bar{x} - \mu}{\dfrac{s}{\sqrt{n}}} < t_{1-\frac{\alpha}{2}} \right] = 1 - \alpha.$$

By solving the inequalities for μ one finds

$$\bar{x} - \frac{st_{1-\alpha/2}}{\sqrt{n}} < \mu < \bar{x} - \frac{st_{1-\alpha/2}}{\sqrt{n}}. \tag{2.80}$$

Since $t_{1-\alpha} = -t_{\alpha/2}$, the upper and lower ends of the confidence interval may be expressed succinctly as $\bar{x} \pm st_{1-\alpha/2}/\sqrt{n}$ and wherein it is understood that t has $n-1$ degrees of freedom. Excel or a statistical software package may be used to find any desired percentage point of t with any prescribed degrees of freedom.

Example

Calculate two-sided 80% confidence limits for μ given that a random sample of size $n = 25$ led to a calculated mean $\bar{x} = 42.0$ and standard deviation of $s = 4.1$. From tables of the t distribution or using software it is found that $t_{0.90}$ with 24 degrees of freedom is 1.32.

The upper and lower 80% confidence limits are thus,

$$40.92 = 42 - \frac{1.32 \cdot 4.1}{5} < \mu < 42 + \frac{1.32 \cdot 4.1}{5} = 43.08.$$

Notice that had σ been known, the interval would have been shorter since $z_{0.90} = 1.28 < 1.32$. The greater width in this case is ascribed to the uncertainty in s due to sampling error. As the sample size increases the percentiles of t approach the corresponding percentiles of the standard normal distribution. When using simulation to determine the distribution of some random function, generally the sample size will be large (>1000) so one may safely set confidence limits on the population mean value of the simulated function using the computed mean and standard deviation and relevant percentage points of the z distribution. This is justified by the central limit theorem and the approach of the t distribution to the standard normal. In Chapter 5 the life of a parallel system is simulated 10,000 times and confidence limits are set on the system mean life using the sample mean and standard deviation. These limits are shown to include the true value obtained by numerical integration.

Confidence limits for σ^2 may be computed when the data are normally distributed and they follow from the fact that

$$Y = \frac{(n-1)s^2}{\sigma^2} \tag{2.81}$$

follows a chi-square distribution. The density function of the chi-square distribution is:

$$f(y; v) = \frac{1}{2^{v/2} \Gamma\left(\frac{v}{2}\right)} y^{\left(\frac{v}{2}-1\right)} \exp\left(-\frac{y}{2}\right); 0 < y < \infty, \tag{2.82}$$

where the parameter υ is known as the degrees of freedom. The random variable defined by Equation 2.81 follows the chi-square distribution with $\upsilon = n - 1$. Percentage points of the chi-square distribution may be found using Excel or a statistical software package.

Confidence limits for σ^2 derived from the chi-square distribution are heavily dependent of the underlying data being normally distributed, that is, they are not robust against departures from normality and so will not be of much use to us. We will encounter the chi-square distribution again when we discuss inference on the exponential distribution in Chapter 5. We note here that in the special case that $\upsilon = 2$ degrees of freedom the chi-square density function reduces to:

$$f(y) = \frac{1}{2}\exp\left(-\frac{y}{2}\right). \tag{2.83}$$

This is the density function of an exponential distribution having a mean of 2.

2.12.1 $P[X < Y]$ for Normal Variables

We take up this topic in Chapter 4 when both X and Y are Weibull distributed. If X represents the stress that a component will be subject to in service and Y is the strength of the population of components, then a component will survive if its X value is less than its Y value. If X is normally distributed with mean μ_x and variance σ_x^2 and likewise Y is normal with mean and variance μ_y and σ_y^2, then $P[X < Y]$ can be expressed as $P[X - Y < 0]$. Define D as $X - Y = X + (-1)Y$.

D is thus of the form:

$$D = c_1 X + c_2 Y.$$

with $c_1 = 1$ and $c_2 = -1$. As a linear combination of two normal variables, D will itself be normally distributed. Its mean will be the same linear combination of the means of X and Y and its variance will be the sum of the squares of the coefficients multiplied by the variances of X and Y as shown in Equation 2.73:

Thus,

$$D \sim N(\mu_x - \mu_y, \sigma_x^2 + \sigma_y^2). \tag{2.84}$$

Example

The stress X on aircraft rivets vary with location and geometrical factors as $N(3000, 900)$. Their strength varies because of manufacturing tolerances and variability in material properties as $N(3100, 1200)$. The probability that $X < Y$

may be thought of as the reliability since it represents probability that a random rivet will survive.

$$R = P[D < 0] = \Phi\left(\frac{0 - (3000 - 3100)}{\sqrt{900 + 1200}}\right) = \Phi(2.182) = 0.984.$$

2.13   THE LOGNORMAL DISTRIBUTION

The random variable X, whose natural logarithm is normally distributed, is called the lognormal distribution and is a popular life model. Since $\ln x \sim N(\mu, \sigma^2)$, the CDF can be written as:

$$F(x) = \text{Prob}[X < x] = \text{Prob}[\ln X < \ln x] = \Phi\left(\frac{\ln x - \mu}{\sigma}\right). \qquad (2.85)$$

In this equation, μ and σ are, respectively, the mean and variance of the logarithm of X. X is a positive random variable since $\ln x$ is undefined for $x < 0$, and it is unlimited on the right. As for any normally distributed variable, the p-th quantile of $\ln X$ is $\mu + z_p\sigma$. The p-th quantile of X is therefore:

$$x_p = \exp(\mu + z_p\sigma). \qquad (2.86)$$

The density function $f(x)$ can be found by differentiating $F(x)$:

$$f(x) = \frac{1}{\sigma x}\varphi\left(\frac{(\ln x - \mu)}{\sigma}\right). \qquad (2.87)$$

The expected value of X is:

$$E(X) = \exp\left(\mu + 0.5\sigma^2\right). \qquad (2.88)$$

The variance is:

$$var(X) = \exp\left(2\mu + \sigma^2\right) \cdot [\exp(\sigma^2) - 1]. \qquad (2.89)$$

The hazard function is readily found by dividing $f(x)$ by $R(x) = 1 - F(x)$. The hazard function is not monotonic. It increases from 0 at $x = 0$, reaches a maximum, and then decreases monotonically to 0 as x increases..

Example
Consider a lognormal random variable having $\mu = 6.908$ and $\sigma = 0.317$. The median value of x is:

$$x_{0.50} = \exp(6.908 + 0) = 1000.$$

The 90th percentile is:

$$x_{0.90} = \exp(6.908 + 1.28 \cdot 0.317) = 1500$$

As the reader no doubt guessed, the values of μ and σ were "reverse engineered" to give these nice round numbers for the median and the 90th percentile.

The expected value for this example is:

$$E(X) = \exp(6.908 + 0.5(0.371)^2) = 1071.5.$$

The variance is:

$$var(X) = \exp(2 \cdot 6.908 + (0.317)^2) \cdot [\exp(0.317^2) - 1] = 1.17E5.$$

The probability that X is less than 1300 may be computed from:

$$\text{Prob}[X < 1300] = F(1300) = \Phi\left(\frac{\ln(1300) - 6.908}{0.317}\right) = \Phi(0.827) = 0.796.$$

In practice, data drawn from a lognormal population can be difficult to distinguish from Weibull distributed data. Some formal tests for deciding which of the two distributions gave rise to a given sample are described in Chapter 6.

2.14 SIMPLE LINEAR REGRESSION

Consider a random variable Y that has a distribution whose mean varies linearly with the value of a nonrandom independent variable X. The mean is often written $\mu_{Y|X}$ to denote this dependence and may be written as:

$$\mu_{Y|X} = \beta_0 + \beta_1 X. \tag{2.90}$$

The quantity β_0 is the intercept and β_1 the slope of the linear relationship between the mean of Y and X. The variable X is sometimes called the independent or explanatory variable. For a fixed value of the nonrandom variable X, Y will vary randomly about $\mu_{Y|X}$. The random departure of an observed value of Y from its mean is denoted ε. A random value of Y may therefore be written as:

$$Y = \beta_0 + \beta_1 X + \varepsilon. \tag{2.91}$$

Given n pairs of values of X and the associated observed values of Y, a statistical problem is to use these data to estimate the values of the slope and intercept of the linear relationship between the value of X and the mean of the distribution of Y. The usual estimation method is the method of least squares. The estimated slope and intercept are denoted b_0 and b_1, respectively, and are the values that make the following sum a minimum:

$$S = \sum_{i=1}^{n} (Y_i - b_0 - b_1 X_i)^2. \tag{2.92}$$

The solutions are linear functions of the Y values:

$$b_0 = \sum_{i=1}^{n} a_i Y_i. \tag{2.93}$$

and

$$b_1 = \sum_{i=1}^{n} c_i Y_i. \tag{2.94}$$

The coefficients are given by:

$$a_i = \frac{1}{n} - \frac{(X_i - \bar{X})}{\sum_{i=1}^{n}(X_i - \bar{X})^2} \bar{X}. \tag{2.95}$$

and

$$c_i = \frac{(X_i - \bar{X})}{\sum_{i=1}^{n}(X_i - \bar{X})^2}. \tag{2.96}$$

If the errors ε_i at each X_i value have an expected value of zero and are uncorrelated, then the Gauss–Markov theorem asserts that these linear estimates are unbiased and have the smallest possible variance among the class of linear estimators. There is no explicit assumption of the nature of the distribution of Y about its mean (cf.Draper and Smith 1998). Excel and all statistical software can perform the calculations to compute the least squares estimates of the slope and intercept in terms of a set of pairs of values of Y and X. Linear regression is often used in fitting probability distributions using transformed scales in which the CDF plots as a straight line against the value of the random variable. The conditions for the Gauss–Markov theorem to hold are not met in this application of regression, but the method nevertheless provides a non-subjective way of fitting a straight line.

When Y varies about its mean in accordance with a normal distribution it is possible to compute confidence limits and test hypotheses about the true slope and intercept.

REFERENCES

Draper, N. and H. Smith. 1998. *Applied Regression Analysis*. 3rd ed. John Wiley and Sons, Inc., New York.

Marsaglia, M. and M. McLaren. 1965. Uniform random number generators. *Journal of the Association of Computing Machinery* 12: 83–89.

Mosteller, F., R. Rourke, and G. Thomas. 1961. *Probability and Statistics*. Addison-Wesley, Reading, MA.

EXERCISES

1. A test consists of 100 multiple-choice questions each with four possible answers. A passing grade is 70 or more correct. If a person guesses at each question, randomly choosing a response, compute the probability that:

   a. the person passes the test.

   b. the first correct answer occurred on question No. 4.

2. An oil drilling company drills at a large number of locations in search of oil. The probability of success at any location is 0.25 and the locations may be regarded as independent.

   a. What is the probability that the driller will experience 1 success if 10 locations are drilled?

   b. The driller feels that she will go bankrupt if she drills 10 times before experiencing her first success.(first success occurs on trial 10). What is the probability that she will go bankrupt?

3. A double sampling plan used in incoming acceptance sampling consists of numbers, n_1, n_2, c_1, and c_2 and functions as follows: A random sample of size n_1 is drawn from the lot. If the sample contains c_1 or fewer defective items, the lot is accepted. If the sample contains more than c_2 defectives, the lot is rejected. If the sample contains more than c_1 but less than c_2 defectives, an additional sample of size n_2 is taken. If the combined sample contains c_2 or fewer defectives, the lot is accepted. If it contains more than c_2 defectives, the lot is rejected. For the plan $n_1 = 50$, $n_2 = 100$, $c_1 = 2$, and $c_2 = 6$ compute the following assuming that the samples are drawn from a large lot that contains 4% defectives.

   a. The probability of acceptance on the first sample

   b. The probability of rejection on the first sample

   c. The total probability of acceptance.

4. How large a sample must survive without failure to demonstrate that the reliability of a component exceeds 0.997 with 80% confidence.

5. A distribution defined over the interval $(0, 3)$ has pdf:$f(x) = cx^2; 0 < x < 3$

   a. Find the value of c.

   b. Compute the mean and variance of x.

   c. Determine the CDF of x.

   d. Compute the median and the 75th percentile of x.

6. The Fit-Is-Us Spa guarantees that after a month of their special diet and exercise program a client will lose at least 10 pounds. Their experience shows that X, the actual number of pounds that their clients will lose, is a discrete random variable ranging from 10 to 14 pounds with the probability distribution tabled below.

X	10	11	12	13	14
$P(x)$	0.35	0.25	0.20	0.15	0.05

   a. Compute the mean and variance of a client's weight loss.

   b. What is the distribution of L, the percentage weight loss in a week for a 200-pound client. Compute the mean and variance of L directly from the distribution of L.

   c. Compute the mean and variance of L using the results from (a) above and the fact that L is a linear function of X.

7. The travel time experienced by a driver going from City A to City B along U.S. Highway 95 can be viewed as a normally distributed random variable with a mean of 3.50 hours and a standard deviation of 0.45 hours.

   a. Calculate the probability that a randomly selected driver experiences a travel time in excess of 4.0 hours.

   b. Consider an experiment involving the measurement of the travel time of 50 different randomly and independently selected drivers going from City A to City B.

   i Calculate the probability that the *average* travel time (taken over the 50 observations) is between 3.4 and 3.65 hours.

   ii What is the expected number of drivers in this sample who will have experienced a travel time in excess of 4.0 hours? What is the standard deviation of the number of drivers in the sample who experienced a travel time in excess of 4.0 hours?

   iii Some drivers, after arriving at City B, may decide to continue on to City C. We can assume that the travel time from City B to City C is normally distributed with a mean of 1.25 hours and a standard deviation of 0.75 hours. We can further assume that the travel time between City B and City C is independent of that between City A and City B. For a randomly selected driver going from A to B and then continuing on to C, calculate the probability that the total travel time (assuming no stops along the way) is in excess of 5 hours but less than 6 hours.

8. I have three items of business to transact in an adjacent building. Let X_i be the amount of time in minutes required for each transaction ($i = 1, 2, 3$) and let X_4 denote the total time to walk over and back including my travel time from one transaction to the other. Suppose the X_i's are independent and normally distributed with the following means and standard deviations: $\mu_1 = 15$, $\mu_2 = 5$, $\mu_3 = 8$, $\sigma_1 = 4$, $\sigma_2 = 1$, $\sigma_3 = 2$. I plan to leave my office precisely at 10:00 a.m. and wish to put a note on my door that says I will return by t a.m., What time t should I write down if I want the probability that I return after t to be less than 0.01?

9. A sample of 25 pieces of laminate used in the manufacture of circuit boards was selected and the amount of warpage in inches was determined for each piece, resulting in a mean of 0.0635 and a sample standard deviation of 0.0065. Assuming normality, calculate an 80% confidence interval for the mean of the population from which the sample was taken.

10. The quality department of a candy manufacturer uses a demerit system in the inspection of boxes of candy. Each defective piece of candy counts as five demerits and each blemish in the wrapping counts as three. Thus, for each inspected candy box the inspector determines a value $Q = 5X + 3Y$. Y follows a Poisson distribution with mean of 1.5 blemishes. Consider boxes containing 25 pieces of candy and assume the probability of a defective piece of candy is 0.1 independently for every piece in the box.

    a. Compute the theoretical mean and variance of Q.
    b. Using Excel (or other appropriate software) simulate 1000 boxes of candy by sampling from the distribution of X and Y and computing Q.
    c. Compute the mean and variance of the 1000 values and compare with the theoretical values.
    d. Sort the 1000 Q values and determine what proportion of them are less than 15.
    e. Now approximate the distribution of Q as a normal distribution with the same theoretical mean and variance and compute $P[Q < 15]$.

11. Consider a disease whose presence can be detected by means of a blood test. A potentially economical way to conduct such tests is to take a portion of a blood sample from each of n people and combine them. If a test on the combined sample proves negative, all n patients are free of the disease and no further testing is performed. If the test on the combined sample is positive, then the test is conducted on the n individual samples. (This group testing method was introduced by the army during World War II to test for syphilis in inductees.) For $n = 3$ and prevalence $p = 0.10$,

compute the expected number of tests to be performed on each group of three people.

12. The reliability of each of 10 identical components is 0.95. If these components are part of a system for which at least six components must function for the system to function, compute the system reliability. If this system could be replaced by a parallel combination of five identical components, what would the reliability of those components have to be to give the same system reliability as the 6 out of 10 system?

13. The life of a product follows a lognormal distribution. The median life is 1000 hours. The probability that the product will survive a life of 2000 hours is 10%. Compute the expected life.

14. The hazard function for a certain product is a linear function of the product's age x, that is, $\lambda(x) = kx$. Compute the reliability function $R(x)$.

CHAPTER 3

Properties of the Weibull Distribution

3.1 THE WEIBULL CUMULATIVE DISTRIBUTION FUNCTION (CDF), PERCENTILES, MOMENTS, AND HAZARD FUNCTION

There are two forms of the Weibull distribution distinguished by the presence of either two or three parameters. Unlike the normal distribution, the Weibull CDF is expressible in closed form. The CDF of the three parameter version of the Weibull distribution may be written as:

$$F(x) = 1 - \exp\left[-\left(\frac{x-\gamma}{\eta}\right)^{\beta}\right]; x > \gamma. \tag{3.1}$$

γ, the location parameter, is also known as the threshold parameter, or, in life testing applications, as the guarantee time, since failure cannot occur until x exceeds γ. When γ is zero the three-parameter Weibull distribution specializes to the much more widely employed two-parameter version. η, the scale parameter, is known as the characteristic value or, in life testing applications, as the characteristic life, for a reason explained further below. β is called the shape parameter and for $\beta = 1$ the two-parameter Weibull distribution specializes to the exponential distribution. The parameters η and β are positive. γ is generally taken to be positive although there is no mathematical reason that this is necessary. In this chapter we will deal, except where noted, with the two-parameter Weibull model. Other parameterizations of the Weibull are possible. For example, some writers replace $1/\eta^{\beta}$ by λ for the typographical benefit of writing the CDF on a single line. The version we have adopted has the useful advantage that the parameter η is expressed in the same units as the random variable itself.

Using the Weibull Distribution: Reliability, Modeling, and Inference, First Edition. John I. McCool.
© 2012 John Wiley & Sons, Inc. Published 2012 by John Wiley & Sons, Inc.

The reliability function $R(x)$, known more usually as the survivorship function in biomedical applications, expresses the probability that the life of a device or subject will exceed a given value. For the two-parameter Weibull distribution this function has the form:

$$R(x) = \text{Prob}[X > x] = exp\left[-\left(\frac{x}{\eta}\right)^{\beta}\right]; x > 0. \tag{3.2}$$

Taking logarithms twice gives

$$\ln\ln\left(\frac{1}{R(x)}\right) = \beta\ln x - \beta\ln\eta. \tag{3.3}$$

On graph paper with a vertical scale ruled proportionally to the values of $\ln\ln(1/R(x))$ and with a logarithmic horizontal scale, the Weibull CDF plots against x as a straight line. Graph paper so constructed is often called Weibull paper. Figure 3.1 shows the line corresponding to the Weibull distribution having $\beta = 1.5$ and $\eta = 100$. This plot can be used to determine $F(x)$ to within graphical accuracy for x values of interest. The principal use of Weibull paper, however, is in the graphical estimation of the Weibull parameters, a topic that we will take up in Chapter 5. With one choice of the scales used in the construction of Weibull paper, the slope of the straight line representation becomes

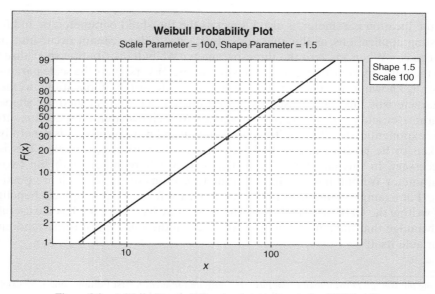

Figure 3.1   A Weibull probability plot for the population $W(100, 15)$.

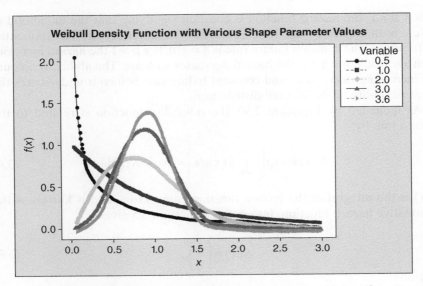

Figure 3.2   Weibull pdf for various values of the shape parameter β.

numerically equal to the Weibull shape parameter. In some engineering literature β is therefore known as the Weibull slope. In some other graph paper
designs in popular use, an auxiliary scale is provided to set the shape parameter appropriately. See for example Nelson (1967) and Nelson and Thompson
(1971).

The probability density function (pdf) of the two parameter Weibull distribution is:

$$f(x; \eta, \beta) = \frac{dF(x)}{dx} = \frac{\beta}{\eta} \left[\frac{x}{\eta} \right]^{\beta-1} \cdot exp \left[-\left(\frac{x}{\eta} \right)^{\beta} \right]. \qquad (3.4)$$

Figure 3.2 is a plot of $f(x)$ for $\eta = 1$ and several values of β. It illustrates the
powerful role of β in determining the shape of the Weibull density and suggests
visually why the Weibull distribution can be fit to widely diverse kinds of
random phenomena.

3.1.1   Hazard Function

From Equation 2.49 in Chapter 2 the hazard function for the two-parameter
Weibull distribution is:

$$\lambda(x) = \frac{f(x)}{R(x)} = \frac{\beta x^{\beta-1}}{\eta^{\beta}}. \qquad (3.5)$$

When $\beta = 1$, the hazard function is constant signifying that the likelihood of failure is unaffected by age. In this case the Weibull reduces to the exponential distribution. The constant failure rate is $\lambda = 1/\eta$. For $\beta > 1$ the hazard increases with age, while for $\beta < 1$, the hazard decreases with age. The ability to account for increasing, decreasing, and constant failure rate behavior underscores the great flexibility of the Weibull distribution.

As indicated by Equation 2.50, the reliability function is related to the hazard rate by:

$$R(x) = \exp\left[-\int_0^x \lambda(x)dx\right] = \exp[-\Lambda(x)]. \tag{3.6}$$

$\Lambda(x)$ is the integral of the hazard function from 0 to x, and is known as the cumulative hazard function. Integrating Equation 3.5 gives:

$$\Lambda(x) = \left(\frac{x}{\eta}\right)^\beta. \tag{3.7}$$

Using Equation 2.52, one may express the average failure rate over an interval from x_1 to x_2 as:

$$\bar{\lambda} = \frac{\int_{x_1}^{x_2} \lambda(x)dx}{x_2 - x_2} = \frac{\Lambda(x_2) - \Lambda(x_1)}{x_2 - x_1} = \frac{x_2^\beta - x_1^\beta}{\eta^\beta[x_2 - x_1]}. \tag{3.8}$$

Example

The life of a product follows a Weibull distribution with a shape parameter of 1.5 and a scale parameter of 1000 hours. Compute the instantaneous failure rate at 500 hours and the average failure rate over the time interval from 500 to 1500 hours:

$$\lambda(500) = \frac{1.5 \cdot (500)^{1.5-1}}{(1000)^{1.5}} = 1.061 \times 10^{-3} = 1.061/1000 \text{ hours.}$$

The average failure rate from 500 to 1500 hours is:

$$\bar{\lambda} = \frac{(1500)^{1.5} - (500)^{1.5}}{1500 - 1000} = 1.484/1000 \text{ hours.}$$

If a limited mission life is at issue one could consider the approximation of replacing a Weibull model by an exponential model having a constant failure rate equal to the Weibull average failure rate over the mission life. This approximation could facilitate further calculations and is reasonable if the failure rates at the beginning and end of the interval in question are not greatly different.

3.1.2 The Mode

The mode of a distribution is the value of x at which the probability density function is largest. For a β value of 1.0 or less, the mode of the Weibull distribution occurs at $x = 0$. For $\beta \geq 1.0$ the mode occurs at:

$$x_m = \eta \left(\frac{\beta - 1}{\beta} \right)^{1/\beta} \tag{3.9}$$

For large values of the shape parameter the mode approaches the scale parameter η.

3.1.3 Quantiles

The p-th quantile of the two-parameter Weibull distribution, found as the solution of $F(x_p) = p$, is:

$$x_p = \left\{ \ln \left(\frac{1}{1-p} \right) \right\}^{1/\beta} \cdot \eta. \tag{3.10}$$

It is convenient to define

$$k_p \equiv \ln \left(\frac{1}{(1-p)} \right). \tag{3.11}$$

so that the p-th quantile may be written,

$$x_p = \eta k_p^{1/\beta}. \tag{3.12}$$

To graphical accuracy one may read quantiles from a Weibull plot by setting $F(x) = 100p$ on the ordinate and reading the corresponding abscissa.

Using Equation 3.12 in Equation 3.2, one may express the CDF in terms of β and a general quantile, that is,

$$F(x) = 1 - exp \left[-k_p \left(\frac{x}{x_p} \right)^{\beta} \right]. \tag{3.13}$$

This parameterization is convenient when a product is rated by the value of a specific quantile. The bearing industry, for example, rates its products by the tenth quantile $x_{0.10}$ and manufacturers' catalogs contain factors whereby $x_{0.10}$ may be computed for a given loading. It is therefore convenient for bearing engineers to calculate reliabilities directly in terms of $x_{0.10}$ rather than having. to first convert x_{010} to η.

An equivalent alternative form is:

$$F(x) = 1 - (1-p)^{\left(\frac{x}{x_p} \right)^{\beta}}. \tag{3.14}$$

For $p = 1 - 1/e = 0.632$, $x_p = \eta$, regardless of the value of β. Since all other percentiles depend on both η and β, $x_{0.632}$ or η is called the characteristic value or, in life testing applications, the characteristic life.

The ratio of two quantiles, say x_p and x_q, may be found using Equation 3.10 as:

$$\frac{x_p}{x_q} = \left[\frac{\ln(1-p)}{\ln(1-q)}\right]^{1/\beta}. \tag{3.15}$$

3.1.4 Moments

The expected value of x raised to an integer power, k, is conveniently expressible as:

$$E\left(X^k\right) = \int_0^\infty x^k dF(x) = \eta^k \Gamma\left(\frac{k}{\beta}+1\right) = \eta^k B_k. \tag{3.16}$$

where,

$$B_k = \Gamma\left(\frac{k}{\beta}+1\right). \tag{3.17}$$

And $\Gamma(\cdot)$ is the gamma function of applied mathematics defined as:

$$\Gamma(z) = \int_0^\infty t^{z-1} e^{-t} dt. \tag{3.18}$$

The gamma function is widely tabulated and otherwise available in computing software. Excel contains a function that computes the logarithm of the gamma function. $E(X^k)$ is called the k-th raw moment. The terminology *raw moment* distinguishes it from the k-th *central moment* defined as $E(x - \mu)^k$.

The mean $\mu = E(x)$ may therefore be written,

$$\mu = \eta B_1. \tag{3.19}$$

In life testing applications the mean is often designated MTTF for mean time to failure. It is sometimes called MTBF for mean time between failures, although most writers reserve this term for repairable systems. Bearing application engineers, knowing the computed value of $x_{0.10}$ for their product may, on occasion, encounter requests from reliability analysts of a client company to provide the MTTF for their bearings. They may use Equation 3.19 after first computing η in terms of $x_{0.10}$ using Equation 3.10:

$$\eta = \frac{x_{0.10}}{(-\ln(0.90))^{1/\beta}}. \tag{3.20}$$

The second raw moment is,

$$E\left(X^2\right) = \eta^2 B_2. \tag{3.21}$$

Table 3.1   Values of $B_1 = \Gamma(1/\beta + 1) =$ and $B_2 - B_1^2 = \Gamma(2/\beta+1) - \Gamma^2(1/\beta+1)$ as a Function of the Shape Parameter β

β	B_1	$B_2 - B_1^2$
1.0	1.0000	1.0000
1.1	0.9649	0.7714
1.2	0.9407	0.6197
1.3	0.9336	0.5133
1.4	0.9114	0.4351
1.5	0.9027	0.3757
1.6	0.8966	0.3292
1.7	0.8922	0.2919
1.8	0.8893	0.2614
1.9	0.8874	0.2360
2.0	0.8862	0.2146
2.5	0.8873	0.1441
3.0	0.8930	0.1053
3.5	0.8997	0.0811
4.0	0.9064	0.0647
5.0	0.9182	0.0442

The variance σ^2 is thus expressible as:

$$\sigma^2 = E(X^2) - \mu^2 = \eta^2[B_2 - B_1^2]. \tag{3.22}$$

Values of $B_1 \equiv \Gamma\left(\dfrac{1}{\beta}+1\right)$ and $B_2 - B_1^2 = \Gamma(2/\beta+1) - \Gamma^2(1/\beta+1)$ taken from Abramowitz and Stegun (1964) are listed in Table 3.1. It is seen from Table 3.1 that for a fixed value of η, σ^2 decreases with increasing β. β itself is therefore often used directly to characterize the dispersion of the Weibull distribution. As $\beta \to \infty$ the variance approaches 0 and the mean approaches η. Thus, the Weibull distribution with a very large shape parameter value acts like a constant equal to its scale parameter. Referring to the Weibull plot in Figure 3.1 as β increases, the line pivots about the point $F(x) = 0.632$, $x = \eta$, as it approaches the vertical.

For $\beta = 1$ the Weibull distribution becomes the single-parameter exponential distribution and its mean is the scale parameter η while its variance is η^2. As already noted its failure rate becomes a constant equal to $1/\eta$.

The coefficient of variation is:

$$cv = \frac{\sigma}{\mu} = [B_2 - B_1^2]^{0.5} / B_1. \tag{3.23}$$

The coefficient of variation depends only upon the shape parameter β. Once β is specified the standard deviation is set as a fixed fraction of the mean. The coefficient of variation for the exponential distribution is 1.0.

The skewness, sk is defined as,

$$sk = \frac{\mu_3}{\sigma^3} \tag{3.24}$$

μ_3 is the third moment about the mean $E(x - \mu)^3$. Expressing all terms in the numerator and denominator in terms of B_i gives:

$$sk = \frac{B_3 - 3B_1B_2 + 2B_1^2}{[B_2 - B_1^2]^{3/2}}. \tag{3.25}$$

Skewness decreases with β, becoming zero at about $\beta = 3.6$, at which value the Weibull distribution is a reasonable approximation to the normal. This does not mean that the Weibull can approximate any normal distribution. Once β is specified to be 3.6 in order to achieve zero skewness, the standard deviation becomes a fixed fraction of the mean as shown by the coefficient of variation expression in Equation 3.23, that is, $\sigma = (0.308)\mu$. Only normal distributions for which the standard deviation and the mean are in this proportion will be well approximated by the Weibull. Dubey (1967) has studied the comparative behavior of Weibull and, normal distributions in some detail.

Examples

1. An item is randomly drawn from a two-parameter Weibull population having a shape parameter $\beta = 1.5$ and a scale parameter $\eta = 100.0$ hours. What is the probability that the item fails before achieving a life of $x = 25$ hours?

   From Equation 3.2,

$$\text{Prob}[life < 25.0] = 1 - e^{-\left(\frac{25}{100}\right)^{1.5}}$$
$$= 1 - e^{-0.125} = 0.118.$$

   This result may be confirmed to graphical accuracy from the Weibull plot in Figure 3.1. Figure 3.3 shows the pdf with the calculated probability shown shaded.

2. Compute the tenth percentile $x_{0.10}$ for this distribution. From Equation 3.10,

$$x_{0.10} = [0.10536]^{\left(\frac{1}{1.5}\right)} \cdot 100.0 = 22.31.$$

   Entering the ordinate of Figure 3.1 at $F(x) = 10\%$ will confirm this computation to within graphical accuracy.

3. Compute the ratio of the 10th and 50th percentiles for this distribution. Taking $p = 0.50$ and $q = 0.10$ in Equation 3.14, the ratio is:

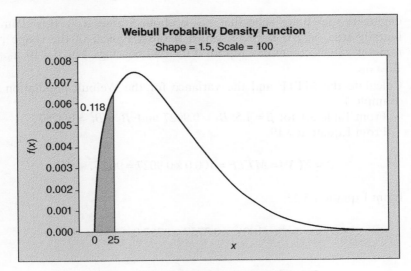

Figure 3.3 Weibull density showing $P[X < 25]$.

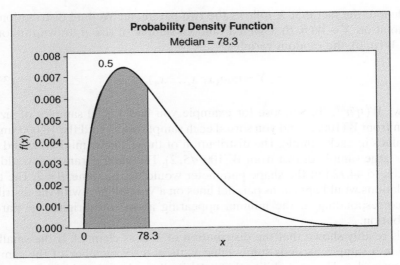

Figure 3.4 The Weibull median $x_{0.50}$.

$$\frac{x_{0.50}}{x_{0.10}} = \left(\frac{\ln(1-0.50)}{\ln(1-0.10)}\right)^{\frac{1}{1.5}} = 3.51.$$

The value of the median, or $x_{0.50}$ is therefore $3.51 \times 22.31 = 78.3$. The median is shown on the pdf in Figure 3.4.

Note that, like the coefficient of variation, the ratio of the 50th and the 10th percentiles depends only on the shape parameter β. Bearing engineers sometimes state that $x_{0.50}$ is about 5 times $x_{0.10}$. This claim is literally true only if $\beta = 1.17$, which is fairly typical of the shape parameter values commonly reported for ball bearings. For $\beta = 10$, $x_{0.50} = 1.2\ x_{0.10}$.

4. Calculate the MTTF and the variance for the Weibull population of example 1.

 From Table 3.1 for $\beta = 1.5$, $B_1 = 0.9027$ and $B_2 - B_1^2 = 0.3757$.
 From Equation 3.19,

$$\mu = E(X) = MTTF = 100.0 \times 0.9027 = 90.27.$$

From Equation 3.22,

$$\sigma^2 = (100)^2 \times 0.3757 = 3757.0.$$

3.2 THE MINIMA OF WEIBULL SAMPLES

We denote the fact that X follows the two-parameter Weibull distribution by the notation $X \sim W(\eta, \beta)$. Given a random sample of size n drawn randomly from $W(\eta, \beta)$, the random variable,

$$Y = \min(x_1, x_2, \ldots x_n). \tag{3.26}$$

follows $W(\eta/n^{1/\beta}, \beta)$. Suppose for example you had 10,000 samples of size 5 drawn from $W(100, 2)$ and you sorted each sample and found the 10,000 smallest values in each sample. The distribution of those 10,000 minima would act like a large sample drawn from $W(100/\sqrt{5}, 2)$. The scale parameter would be reduced to 44.72 but the shape parameter would be the same, $\beta = 2$. The two populations would appear as parallel lines on a Weibull plot with the distribution corresponding to the minima appearing above the original or parent distribution.

It is readily shown that the distribution of Y is as claimed. If the smallest value in a sample exceeds some value y, then *every* member of the sample must exceed y. The probability that a Weibull random variable exceeds y is $1 - F(y)$, where $F(y)$ is the Weibull CDF. Assuming independence of the observations in a sample, the probability that every sample observation exceeds y is then $[1 - F(y)]^n$.

Let Y denote the random variable representing the minimum of a sample of size n and having CDF $G(y)$, then,

$$P(Y > y) = 1 - G(y) = [1 - F(y)]^n = \left[exp\left(-\left(\frac{y}{\eta} \right)^\beta \right) \right]^n. \tag{3.27}$$

Thus, the CDF of the distribution of minima is:

$$G(y) = 1 - exp\left[-\left(\frac{y}{\eta / n^{1/\beta}}\right)^{\beta}\right].$$

(3.28)

That the Weibull is preserved under minimization reflects the role of the Weibull distribution as a limiting distribution of smallest extremes. This fact is often used to justify the Weibull model as appropriate to failure mechanisms governed by a "weakest link" behavior. This property will be discussed later in this chapter. The distribution of Weibull minima is the basis for a testing strategy called sudden death testing espoused by Johnson (1964) and discussed further in Section 7.6. It is also related to the life of series systems made of n identical Weibull elements and discussed in conjunction with system reliability in Chapter 7.

The mean, variance, and percentiles for the distribution of the minimum are computed in the same way as for any Weibull distribution after using $\eta / n^{1/\beta}$ in lieu of η.

3.3 TRANSFORMATIONS

3.3.1 The Power Transformation

If a random variable is Weibull distributed, that is, $X \sim W(\eta, \beta)$, we now show that the transformed variable $Y = aX^c$, where a and c are positive constants, is also Weibull distributed. Let $G(y)$ denote the CDF of Y.

$$G(y) = \text{Prob}[Y < y] = \text{Prob}\left[X < \left(\frac{y}{a}\right)^{\frac{1}{c}}\right] = F\left(\left(\frac{y}{a}\right)^{\frac{1}{c}}\right)$$

$$= 1 - \exp\left[-\left(\frac{\left(\frac{y}{a}\right)^{\frac{1}{c}}}{\eta}\right)^{\beta}\right].$$

(3.29)

Thus, $Y \sim W(a\eta^c, \beta/c)$.

This transformation for the Weibull is reminiscent of the linear transformation $Y = a + bX$ for the normal. In both cases the distribution of the transformed variable is in the same family as the untransformed variable, but with modified parameters. For the normal if $b = 1$, Y has the same variance as X but the mean is modified by an additive amount "a." In the Weibull case for $c = 1$, the Weibull distribution of Y has the same shape parameter as X but the scale parameter η is modified by the multiplicative factor a.

Choosing $a = 1$ and $c = \beta$ transforms any Weibull distribution to the exponential ($\beta = 1$) with mean value η. Suitably choosing a and c, one can transform one Weibull variable to any other, different, Weibull distribution having a more "convenient" shape parameter. Nelson (1994) made use of this fact to transform an exponential random variable (Weibull with $\beta = 1.0$) to a Weibull distribution having a shape parameter of $\beta = 3.6$ and for which the skewness is zero, using $a = 1$ and $c = 1/3.6 = 0.277$. As noted previously when the shape parameter is 3.6 the Weibull distribution is roughly symmetrical about its mean, and in a control chart of a Weibull variable thus transformed, one may reasonably employ auxiliary tests such as runs above and below the mean in assessing whether the process is in a state of control.

A purely multiplicative transformation, $Y = aX, (c = 1)$, does not affect the shape parameter and multiplies the scale parameter by the same factor. The practical benefit is that one may freely change units provided the change is purely multiplicative. If X in inches is $W(1, 3)$, then Y in millimeters will be $W(25.4, 3)$. If, however, temperature in centigrade follows the two-parameter Weibull, the temperature in Fahrenheit will not follow the two-parameter Weibull distribution since the transformation between the two scales is not purely multiplicative.

In the design and analysis of experiments with normally distributed response, it is customary to hypothesize that the effect of experimental factors is to change the mean by an additive amount while leaving the variance unchanged. With a Weibull response variable, the analogous assumption would be that external factors such as stress produce a multiplicative effect on the Weibull scale parameter while leaving the shape parameter unchanged. Such a model has been substantiated for bearing endurance life by Lieblein and Zelen (1956). In view of Equation 3.22, this implies heteroscedasticity; that is, the variance will increase with the mean response although β remains the same. The analysis of a one-way layout under the assumption of a constant shape parameter and multiplicative effects on the scale parameter was considered by McCool (1979) and extended to a two-way layout in McCool (1993). This topic is discussed further in Chapters 8 and 11.

3.3.2   The Logarithmic Transformation

The natural logarithm of a Weibull variable follows the type I distribution of smallest extremes, (cf. Gumbel 1958). There are many advantages of working with the logarithmically transformed Weibull variable as we shall discover. Under the logarithmic transformation

$$Z = \ln(X) \tag{3.30}$$

Z will be less than an arbitrary value z, as long as X is less than e^z. Therefore, the CDF $G(z)$ may be found as:

$$G(z) = \text{Prob}[Z < z] = F(e^z) = 1 - exp\left(-\left[\frac{e^z}{\eta}\right]^{\beta}\right). \tag{3.31}$$

Introducing the new parameters $\delta = \ln \eta$ and $\xi = 1/\beta$, $G(z)$ may be written in the following form:

$$G(z) = 1 - exp\left\{-exp\left[\frac{z-\delta}{\xi}\right]\right\}. \tag{3.32}$$

In this form the distribution has a location parameter δ and scale parameter ξ analogous to the two parameters of the normal distribution. It is clear from Equation 3.32 why this distribution is sometimes called the doubly exponential distribution. As with the normal, it is useful to define a standardized variable Y as:

$$Y = \frac{(Z - \delta)}{\xi}. \tag{3.33}$$

Y follows the extreme value distribution with location parameter 0 and scale parameter 1. The p-th quantile of Y is:

$$y_p = \ln \ln \left\{\frac{1}{(1-p)}\right\} = \ln(k_p). \tag{3.34}$$

The corresponding quantile for Z is

$$z_p = \delta + \xi y_p. \tag{3.35}$$

The same quantile for the Weibull variate is computed by retransforming:

$$x_p = \exp(z_p). \tag{3.36}$$

The mean of the standardized variate is the negative of Euler's constant $\gamma = 0.57721$, so in terms of the Weibull parameters,

$$E(Z) = \ln \eta - \frac{\gamma}{\beta}. \tag{3.37}$$

The variance of the standardized variable Y is $\pi^2/6$ so that

$$\sigma_y^2 = \frac{\sigma_z^2}{\xi^2} = \frac{\pi^2}{6} \tag{3.38}$$

and thus the variance of $Z = \ln X$ is

$$\sigma_z^2 = \frac{\pi^2}{6\beta^2}. \tag{3.39}$$

Equations 3.37 and 3.39 show that a multiplicative factor applied to the scale parameter, that is, replacing η by $a\eta$ has no effect on the variance of $Z = \ln X$

and adds an amount $\ln(a)$ to its expected value. Thus, a multiplicative Weibull model for the effects of one or more factors, combined with the assumption of a constant shape parameter, will result in data which when logarithmically transformed will follow an additive model with homogeneous variance. Except for normality, such data obey the principal assumptions of the analysis of variance. A reasonable approximate analysis of multifactor Weibull data applicable if censoring is absent would be to apply the analysis of variance to the logarithms of the observations. An exact approach for multifactor Weibull data is discussed in Chapter 11.

Menon (1963) recommends estimating the Weibull parameters by the method of moments applied to the logarithms of the data. Menon's approach and recent extensions to it are discussed in Section 5.3.1.

3.4 THE CONDITIONAL WEIBULL DISTRIBUTION

The distribution of a Weibull variate conditional on $X \geq x_0$ has been termed the conditional Weibull distribution by Aroian (1965). It is termed the truncated Weibull distribution by Harlow (1989). It is of particular use when the Weibull is used as a lifetime model although other applications are possible.

Let $A = \{x_0 < X < x\}$ and $B = \{X > x_0\}$. Then $P\,[A|B]$ is the CDF of X conditional on $X > x_0$:

$$F(x|x_0) = \frac{F(x) - F(x_0)}{1 - F(x_0)} = 1 - \exp\left[-\left(\frac{x}{\eta}\right)^\beta + \left(\frac{x_0}{\eta}\right)^\beta \right]; x > x_0. \qquad (3.40)$$

It is useful to define y as life lived beyond x_0, that is, $y = x - x_0$ is the residual life after having run for a period of x_0. This is the life as measured by a customer who has received a product that was "burned in" for a period of time x_0. To do so we replace x by $y + x_0$ in Equation 3.40.

$$F(y|x_0) = 1 - exp\left[-\left(\frac{y + x_0}{\eta}\right)^\beta + \left(\frac{x_0}{\eta}\right)^\beta \right]. \qquad (3.41)$$

This expression shows that when $\beta = 1.0$ the conditional exponential is the same as the unconditional exponential. The run-in or burn-in period has had no effect. This is a manifestation of the so-called memorylessness property of the exponential distribution. For $\beta \neq 1$ the conditional Weibull does not have the Weibull form.

A run-in time that is too short increases the risk of failure under warranty when the burned-in product is delivered to the customer. An overly long run-in time increases the cost due to the loss of product failing during run-in and the running costs. Determining the optimum run-in duration to minimize the total cost is considered in Section 4.9.

A simpler, equivalent way of expressing Equation 3.41 is in terms of the reliability function, namely:

$$R(y|x_0) = \frac{R(y+x_0)}{R(x_0)} = exp\left[-\left(\frac{y+x_0}{\eta}\right)^\beta + \left(\frac{x_0}{\eta}\right)^\beta\right].$$ (3.42)

For $x_0 = 0$ this reduces to the ordinary, unconditional, reliability function.

The p-th quantile of the conditional distribution y_p may be found by setting $F(y|x_0)$ equal to p and solving for y. It is expressible in terms of the p quantile of the original, unconditional, distribution as,

$$y_p = x_p\left[1+\left(\frac{x_0}{x_p}\right)^\beta\right]^{1/\beta} - x_0.$$ (3.43)

When equipment is "burned in," that is, subject to a period of operation of length x_0 to eliminate early failures (infant mortality), the customer's shipment is drawn from the population of survivors. Equation 3.43 gives the $100p$-th percentile as measured from when the customer puts the item into service.

When $\beta = 1.0$, $y_p = x_p$, signifying that no deleterious effect of aging has occurred. For $\beta > 1.0$, $y_p < x_p$; that is, aging occurs. For $\beta < 1$, $y_p > x_p$, indicating that the survivors of the run-in period are superior to the population as a whole.

The mean remaining life, also called the mean residual life, is obtained by integrating the conditional reliability function. It is a function of the run-in time x_0:

$$MRL(x_0) = \int_0^\infty R(y|x_0)dy.$$ (3.44)

Since the reliability function for $x_0 = 0$ is the unconditional reliability function, MRL(0) = MTTF.

MRL(x_0) may be computed in terms of the incomplete gamma function (cf. Leemis, 1995)

$$MRL(x_0) = \frac{\eta}{\beta}exp\left[\left(\frac{x_0}{\eta}\right)^\beta\right]\Gamma\left[\frac{1}{\beta}, \left(\frac{x_0}{\eta}\right)^\beta\right]$$ (3.45)

where $\Gamma(a, x)$ denotes the incomplete gamma function defined as:

$$\Gamma(a, x) = \int_x^\infty t^{a-1}e^{-t}dt.$$ (3.46)

The mean residual life increases with x_0 for $\beta < 1$ and decreases with x_0 for $\beta > 1$. Figure 3.5 is a plot of the mean residual life divided by MTTF versus x_0 for two cases: $\beta = 0.5$ and $\beta = 1.5$. In both cases $\eta = 1000$. The figure shows that

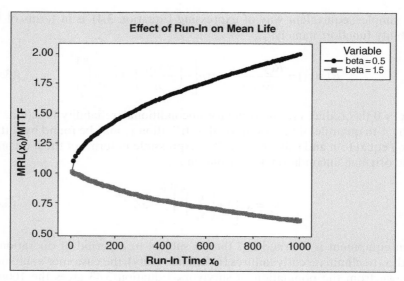

Figure 3.5 Mean residual life versus run-in time for two values of β.

when $\beta < 1$ the MRL increases with burn-in, while for $\beta > 1.0$ the MRL decreases with burn-in.

Examples
1. A product's life in hours follows the distribution $W(30E6, 1.5)$. Compute the first percentile. If the product is run for 2E6 hours and the survivors are sold, what is the first percentile of the surviving population from the point of view of the new owner?
From Equation 3.10 the 0.01 quantile, $x_{0.01}$ is:

$$x_{0.01} = 30E6 \cdot [-\ln(1-0.01)]^{1/1.5} = 6.51E4 \text{ hours.}$$

From Equation 3.43 the 0.01 quantile post run-in, $y_{0.01}$, is:

$$y_{0.10} = 6.51E4 \cdot \left[1 + \left(\frac{2E6}{6.51E4} \right)^{1.5} \right]^{1/1.5} - 2.0 = 7.857E3 \text{ hours.}$$

2. Consider an item that fails in accordance with a Weibull distribution with scale parameter $\eta = 240$ months and shape parameter $\beta = 0.6$. (Recall that the Weibull has a decreasing failure rate when $\beta < 1$.) The reliability at $t = 2$ months without burn-in is given by:

$$R(2) = \exp -\left(\frac{2}{240} \right)^{0.6} = 0.945.$$

Now calculate the reliability at 2 months in the population of survivors of a burn-in period of 3 months. Using Equation 3.42 results in:

$$R(2\,|\,3) = \exp-\left(\frac{2+3}{240}\right)^{0.6} \bigg/ \exp-\left(\frac{3}{240}\right)^{0.6} = \frac{0.9066}{0.9304} = 0.9745.$$

Thus the customer would see a sample from a population having reliability of 97.45% rather than 94.5% which would be the reliability without burn-in.

What fraction of the population would fail to survive the burn-in period?

$$\text{Prob}\,(life < 3\,months) = F(2) = 1 - \exp-\left(\frac{3}{240}\right)^{0.6} = 0.0696\ or\ 6.96\%.$$

The MTTF and MRL(3) are:

$$MTTF = 361.1$$
$$MRL(3) = 385.0.$$

3.5 QUANTILES FOR ORDER STATISTICS OF A WEIBULL SAMPLE

Order statistics are the individual values in a random sample, after they are sorted in ascending sequence; for example, the first order statistic is the smallest value in the sample, the second order statistic is the second smallest value, and so on. We have already seen that the first order statistic in Weibull samples varies from sample to sample in accordance with a Weibull distribution having a reduced scale parameter.

Order statistics are random variables varying from sample to sample in accordance with a probability distribution that depends on the underlying parent distribution, the sample size, and the order number under consideration. Consider a large number of samples of size n randomly drawn from a distribution whose CDF is $F(x)$. The r-th order statistic $x_{r,n}$ will vary from sample to sample. Now transform the values of $x_{r,n}$, the r-th ordered member of each sample by computing $F(x_{r,n})$ where $F(x)$ is the CDF of the distribution from which the samples were drawn. $F(x_{r,n})$ will now vary from sample to sample since $x_{r,n}$ varies from sample to sample. Amazingly the distribution of $F(x_{r,n})$ in repeated samples is the same regardless of the population from which the samples were taken, and it is known as the beta distribution. The beta distribution has two parameters customarily denoted a and b. $F(x_{r,n})$ follows the beta distribution with $a = r$ and $b = n - r + 1$. Let $X(p, a, b)$ denote the p-th quantile of the beta distribution having parameters a, b. In this notation a 90% probability interval for $F(x_{r,n})$ may be expressed as :

$$X(0.05, r, n-r+1) < F(x_{r,n}) < X(0.95, r, n-r+1). \tag{3.47}$$

Table 3.2   Values of $X(0.05, r, 6 - r)$ and $X(0.95, r, 6 - r)$ for $n = 5$

r	$X(0.05, r, 6 - r)$	$X(0.95, r, 6 - r)$	$X(0.50, r, 6 - r)$	$\dfrac{r - 0.3}{n + 0.4}$
1	0.01021	0.45072	0.12945	0.1296
2	0.07644	0.65741	0.31381	0.3148
3	0.18926	0.81074	0.50000	0.50
4	0.34259	0.92356	0.68619	0.6852
5	0.54928	0.98979	0.87055	0.8704

Values of $X(p, a, b)$ are tabled for various values of p, a, and b by Harter (1964). Commercial statistical software such as Minitab can provide percentage points of the beta distribution. Some values of these percentage points applicable for a sample size $n = 5$ and $r = 1(1)5$ are listed in Table 3.2. In the literature dealing with probability plotting, these values are known as the 5% and 95% ranks. The median value of $F(x_{r,n})$ is called the median rank and is denoted as $X(0.50, r, n - r + 1)$ in our notation. More extensive tables of the 5%, 50%, and 95% ranks are given in tables provided by Kapur and Lamberson (1977) and others. In plotting life data for the purpose of estimating the Weibull parameters, some software includes the plotting of the 5% and 95% ranks as the basis for approximate confidence limits. This is discussed further in Chapter 5.

The median ranks are often recommended as plotting positions for graphical estimation of the Weibull parameters. The following approximation to the median ranks was proposed by Benard and Bos-Levenbach (1953) and is in wide use:

$$X(0.50, r, n - r + 1) \approx \frac{r - 0.3}{n + 0.4}. \tag{3.48}$$

Values of this approximation for $n = 5$ are shown in column 5 of Table 3.2. It is clear that they are indistinguishable from the exact values to within graphical accuracy.

It is sometimes useful to be able to calculate an interval in which a given order statistic will fall with high probability. For example, if one wished to obtain five failed specimens for metallurgical investigation, and put 10 specimens on test, the waiting time to completion of the test is the fifth order statistic in the sample of size 10, designated $x_{5,10}$.

A 90% probability interval for the r-th order statistic in a sample of size n from any distribution may be calculated by taking the inverse of $F(x)$ in Equation 3.47 as follows,

$$F^{-1}[X(0.05, r, n - r + 1)] < x_{r,n} < F^{-1}[X(0.95, r, n - r + 1)]. \tag{3.49}$$

where $F^{-1}(\cdot)$ is the inverse of the parent distribution function.

Calculation of the above interval may be carried out graphically on a probability plot by entering the values of $X(p, a, b)$ on the probability ordinate and reading the interval values as the associated abscissa values (cf. McCool 1969).

It may also be performed analytically by using the mathematical expression for the inverse of the distribution function. For the two-parameter Weibull distribution this yields,

$$\eta\left[\ln\left(\frac{1}{1-X(0.05, r, n-r+1)}\right)\right]^{1/\beta} < x_{r,n} < \eta\left[\ln\left(\frac{1}{1-X(0.95, r, n-r+1)}\right)\right]^{1/\beta}. \tag{3.50}$$

Example

Calculate a 90% interval for the third-order statistic in a sample of size 5 drawn from the two-parameter Weibull population having $\beta = 1.5$ and $\eta = 100$.

Using Equation 3.50 and the 5th and 95th percentiles $X(0.05, 3, 3) = 0.18926$ and $X(0.95, 3, 3) = 0.81074$ gives the following interval:

$$35 < x_{3,5} < 140.$$

These calculations may be roughly corroborated by entering the ordinate of the Weibull plot of the population at these two percentile values (after multiplying by 100) and reading the corresponding abscissae. See Figure 3.6:

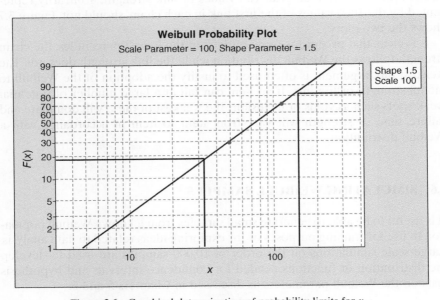

Figure 3.6 Graphical determination of probability limits for $x_{3,5}$.

Calculations of this type may be useful for rough estimation of the uncertainty in the time needed to complete a life test. To perform the calculation it is necessary to guess at what the life test results will reveal regarding the parameter values. The evident contradiction is that if the parameters were known there would be no need to conduct the life test.

3.5.1   The Weakest Link Phenomenon

Let X be an arbitrary positive random variable describing the strength of a population of chain links. Randomly assemble chains each consisting of n of these links. The strength of a chain will now be equal to the strength of its weakest link. If in the vicinity of $X = 0$, the CDF of the link strength distribution behaves like a function of the form $F(x) = cx^\beta$, then for a sufficiently large number of links, the chain strength will vary from chain to chain in accordance with a Weibull distribution with shape parameter β. This is shown in a somewhat general form by Epstein (1960) in a very readable survey of extreme value theory.

We will demonstrate the truth of this assertion by a simulation exercise. Let link strength X be uniformly distributed $U(0, 1000)$. We know from Equation 2.60 in Section 2.10 that the CDF is $F(x) = x/1000$, so that β, the exponent of x is 1.0. We will randomly sample 1000 sets of 50 observations of $U(0, 1000)$ to represent 1000 chains of 50 links per chain. We will then determine the minimum of each set of 50 to represent the strength of each of the 1000 chains. The 1000 values of chain strength should follow at least an approximate straight line on a Weibull grid. The values of link strength, arbitrarily represented by the 1000 values of the first link in each chain, should not. Figure 3.7 shows the two plots:

It is clear that except for some outliers at the lowest percentiles, the chain strength follows the Weibull distribution while the link strength does not. The weakest link argument is often used to justify the adoption of the Weibull to model strength or lifetime. If an item can fail at a large number of potential weak spots such as voids or inclusions the weakest of these will dominate and failure time or strength will vary from item to item in accordance with a Weibull distribution.

3.6   SIMULATING WEIBULL SAMPLES

It is useful to be able to generate small samples from a known Weibull population to use for training purposes or for exploring ideas related to data analysis. Large-scale simulations on the order of 10,000 samples are used to develop the distribution of functions needed for confidence intervals and hypothesis tests on the Weibull parameters as discussed in Chapters 5 and 6.

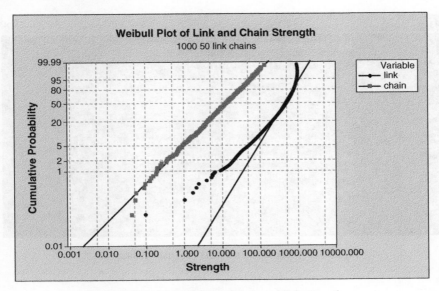

Figure 3.7 Weibull plots of chain and link strength.

The method described in Section 2.10 works quite well for the Weibull. The CDF $F(x)$ is equated to a value u randomly taken from the distribution which is uniformly distributed over the interval $(0, 1)$. The equation is then solved for x in terms of u and yields:

$$x = \eta(-\ln[u])^{1/\beta} + \gamma. \tag{3.51}$$

This expression uses the fact that the complement $1 - u$ of a uniform random variable is also uniform. Taking a series of n values of the uniform random variable u and applying Equation 3.51 to each produces a sample of size n from the Weibull population with a given set of parameters. As noted in Section 2.10, techniques for generating values that act like samples from a uniform distribution have been widely studied. The values generated are called pseudorandom since a deterministic set of calculations is used to compute them and hence they are not truly random. Excel may be used to generate a set of uniform random values and Equation 3.51 applied to yield Weibull observations. The disk operating system (DOS) program Weibsam may be used to generate small samples from a two-parameter Weibull ($\gamma = 0$). Figure 3.8 shows the input screen with the input needed to generate a single Weibull sample of size 10 from the distribution $W(100, 1.5)$.

The random values thus generated are shown in Figure 3.9.

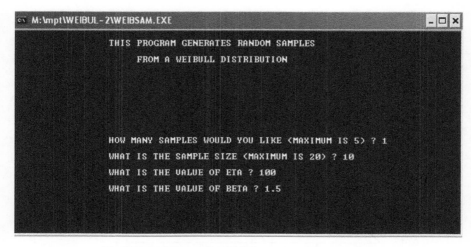

Figure 3.8   Input screen for DOS program Weibsam.exe.

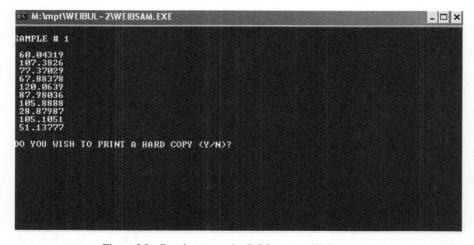

Figure 3.9   Results screen for DOS progam Weibsam.exe.

REFERENCES

Abramowitz, M. and I. Stegun, eds. 1964. *Handbook of Mathematical Functions*. Applied Mathematics. U.S. Department of Commerce, National Bureau of Standards, Washington, DC.

Aroian, L.A. 1965. Some Properties of the Conditional Weibull Distribution. ASQC Technical Conference Transactions, ASQC Milwaukee, WI.

Benard, A. and E.C. Bos-Levenbach. 1953. The plotting of observations on probability paper (in Dutch). *Statististica Neerlandica* 7: 163–173.

Dubey, S. 1967. Normal and Weibull distributions. *Naval Research Logistics Quarterly* 14: 69–79.

Epstein, B. 1960. Elements of the theory of extreme values. *Technometrics* 2(1): 27–41.

Gumbel, E.J. 1958. *Statistics of Extremes*. Columbia University Press, New York.

Harlow, D.G. 1989. The effect of proof-testing on the Weibull distribution. *Journal of Material Science* 24: 1467–1473.

Harter, H.L. 1964. *New Tables of the Incomplete Gamma-Function Ratio and of the Percentage Points of the Chi-Square and Beta Distributions*. U.S. Govt. Printing Office, Washington, D.C.

Johnson, L.G. 1964. *The Statistical Treatment of Fatigue Experiments*. Elsevier, New York, NY.

Kapur, K.C. and L.R. Lamberson. 1977. *Reliability in Engineering Design*. Wiley, New York, NY.

Leemis, L.M. 1995. *Reliability*. Prentice Hall, Englewood Cliffs, NJ. 07632.

Lieblein, J. and M. Zelen. 1956. Statistical investigation of the fatigue life of deep-groove ball bearings. *Journal of Research of the National Bureau of Standards* 57(5): 273–316.

McCool, J.I. 1969. Graphical determination and uses of order statistics quantiles. *Journal of Quality Technology* 1: 237–241.

McCool, J.I. 1979. Analysis of single classification experiments based on censored samples from the two-parameter Weibull distribution. *Journal of Statistical Planning and Inference* 3(1): 39–68.

McCool, J.I. 1993. The analysis of a two way layout with two parameter Weibull response. PhD Thesis, Department of Statistics, Temple University, Philadelphia, PA, p. 143.

Menon, M.V. 1963. Estimation of the shape and scale parameters of the Weibull distribution. *Technometrics* 5(2): 175–182.

Nelson, L.S. 1967. Weibull probability paper. *Industrial Quality Control* 23: 452–453.

Nelson, L.S. 1994. A control chart for parts-per-million nonconforming items. *Journal of Quality Technology* 26(3): 239–240.

Nelson, W.B. and V.C. Thompson. 1971. Weibull probability papers. *Journal of Quality Technology* 3(2): 45–50.

EXERCISES

1. Let $\eta = 100.0$, $\beta = 1.5$. Find, graphically and/or by computation,

    a. Prob$[X < 34]$

    b. The value of $x_{0.10}$

    c. The value of $x_{0.50}$

    d. Prob$[80 < X < 120]$

    e. Compute the average failure rate over the interval (80, 120).

2. For what β value is it true that $x_{0.50} = 4.5\ x_{0.10}$?

3. Compute the median of the distribution $W(100, 2)$. Use Weibsam or Excel to generate a sample of 20 values from this distribution. How many are below the population median? How many did you expect to be below the median?

4. The strength in pounds of a population of chain links is $W(200, 2)$. Chains assembled from 20 links randomly selected from this population will break under a load equal to the strength of its weakest link. Compute the mean and median of the chain strength distribution.

5. The radius r in inches of a population of tree trunks is distributed as $W(30, 4)$. Find the mean of the cross-sectional area $A = \pi r^2$ of this population.

6. Under a prescribed set of conditions the life of a product in hours follows $W(1000, 1.5)$. If a random sample of 5 such items are put on test under those conditions, compute a 90% probability interval for the life of the fourth failure in this sample.

7. Compute the median life of the product in problem 6. If the product is run-in for 500 hours what is the median of the population of survivors?

8. A sample of ball bearings is drawn from a Weibull population having $\beta = 1.3$ and a tenth percentile of $x_{0.10} = 10.0$ million revolutions The survivors after running for $x_0 = 5.0$ million revolutions will now have what value of the 10th percentile?

9. Using Equations 3.10 and 3.13 verify Equation 3.14.

CHAPTER 4

Weibull Probability Models

This chapter contains a number of models involving Weibull random variables with the assumption that the parameters are known or assumed for the purposes of *what-if* speculation. The parameters are not known of course and are never known with complete certainty. Estimating Weibull parameters from a data sample is considered in Chapter 5.

4.1 SYSTEM RELIABILITY

In Chapter 1 we considered component reliability to be a fixed constant value since the mission time was considered to be fixed. If the component reliabilities are regarded as functions of time, then the system reliability will likewise be a time function.

4.1.1 Series Systems

As we have seen, a system is called a series system when the system functions only if every one of its components functions. If the failure times of the components are statistically independent, then the system reliability is the product of the reliabilities of each of the components that comprise the system, that is,

$$R_{system} = R_1 \cdot R_2 \cdots \cdots R_n = \prod_{i=1}^{n} R_i. \qquad (4.1)$$

The system life is expressible as:

$$t_{system} = \min(t_1, t_2, \ldots t_n). \qquad (4.2)$$

Where t_i denotes the time of the i-th component failure.

Using the Weibull Distribution: Reliability, Modeling, and Inference, First Edition. John I. McCool.
© 2012 John Wiley & Sons, Inc. Published 2012 by John Wiley & Sons, Inc.

If the life of the i-th component follows a Weibull distribution with scale parameter η_i and the shape parameter β is the same for each component, the system reliability at time t becomes

$$R_{system}(t) = \prod_{i=1}^{n} \exp\left[-\left(\frac{t}{\eta_i}\right)^{\beta}\right] = \exp\left[-\left(\frac{t}{\eta_e}\right)^{\beta}\right], \tag{4.3}$$

in which the equivalent scale parameter value is:

$$\eta_e = \left\{\sum_{i=1}^{n} \frac{1}{\eta_i^{\beta}}\right\}^{-1/\beta}. \tag{4.4}$$

The life distribution of a series system of Weibull components is seen to have a Weibull distribution provided that the component shape parameters are identical. If the components have identical scale parameters as well, the equivalent scale parameter reduces to:

$$\eta_e = n^{-1/\beta} \cdot \eta. \tag{4.5}$$

It is easy to show that Equations 4.4 and 4.5 apply to any quantile and not just to $\eta = x_{0.632}$. The mean life of a series system for which all of the components lives are Weibull distributed with the same shape parameter is expressible using the expression for the mean of a Weibull distribution (cf. Equation 3.19):

$$E(t_{system}) = \eta_e B_1. \tag{4.6}$$

Equation 4.4 shows that the worst component will have the biggest effect on the system life. This is confirmed numerically in the following example.

Example 1
A system consists of a rotating shaft supported by three bearings. Failure of any one of the bearings is equivalent to failure of the system. The life of each bearing is assumed to follow the Weibull distribution with shape parameter $\beta = 2.0$ and scale parameter values of $\eta_1 = 1000$, $\eta_2 = 500$, and $\eta_3 = 100$.
   The system scale parameter is:

$$\eta_e = \left[\frac{1}{1000^2} + \frac{1}{500^2} + \frac{1}{100^2}\right]^{-1/2} = 97.6 \text{ hours.}$$

The mean life of this series system is:

$$E(t_{system}) = \eta_e B_1 = \eta_e \Gamma\left(1 + \frac{1}{\beta}\right) = 97.6 \cdot \Gamma\left(1 + \frac{1}{2}\right) = 86.5.$$

Suppose we are able to double the life of the longest lived bearing. The system scale parameter value becomes:

$$\eta_e = \left[\frac{1}{2000^2} + \frac{1}{500^2} + \frac{1}{100^2} \right]^{-1/2} = 97.9 \text{ hours.}$$

On the other hand doubling the life of the shortest lived bearing results in:

$$\eta_e = \left[\frac{1}{1000^2} + \frac{1}{500^2} + \frac{1}{200^2} \right]^{-1/2} = 182.6 \text{ hours.}$$

The message is clear. For the greatest improvement in system life, devote resources and effort to improving the life of the component with the smallest scale parameter value.

Example 2

Find the mean time to failure of a series system consisting of $n = 5$ components drawn from a population having a Weibull life distribution with a shape parameter of 2.0 and a scale parameter of 100 hours.

The scale parameter of the system is:

$$\eta_e = 5^{-1/2.0} \cdot 100 = 44.7 \text{ hours.}$$

The mean time to system failure is:

$$E(t_{system}) = 44.7 \Gamma \left(1 + \frac{1}{2.0} \right) = 39.6 \text{ hours.}$$

4.1.2 Parallel Systems

As discussed in Chapter 1, if a system contains n components but can function as long as only one of its components is functioning, the system is said to be a parallel system. The probability that the system fails is then the probability that all of its components fail. The failure probability is the complement of the reliability so, assuming the component failure times are statistically independent, the probability that the system fails may be expressed as:

$$1 - R_{system}(t) = \prod_{i=1}^{n} [1 - R_i(t)].$$

So that,

$$R_{system}(t) = 1 - \prod_{i=1}^{n} [1 - R_i(t)]. \tag{4.7}$$

The reliability of a parallel system exceeds the reliability of the most reliable component. For parallel combinations of Weibull distributed components the system reliability function will *not* be of the Weibull form except in the trivial case where $n = 1$.

When all the components follow the same distribution with reliability function $R(t)$, Equation 4.7 becomes:

$$R_{system}(t) = 1 - (1 - R(t))^n. \tag{4.8}$$

Using the binomial expansion Equation 4.8 may be expressed as:

$$R_{system}(t) = \sum_{k=1}^{n} \binom{n}{k} (-1)^{k+1} R^k(t). \tag{4.9}$$

Where $\binom{n}{k} = \dfrac{n!}{k!(n-k)!}$ is the binomial coefficient and represents the number of combinations of n things taken k at a time.

For the Weibull

$$R^k(t) = \exp\left(k\left[\frac{x}{\eta}\right]^\beta\right) = \exp([x/k^{-1/\beta}\eta]^\beta). \tag{4.10}$$

The k-th term of the expansion is of the Weibull form but with scale parameter $k^{-1/\beta}\eta$.

The mean time to failure of a parallel system with n identical Weibull components may be obtained by integrating $R_{system}(t)$ from 0 to infinity. Using Equations 4.9 and 4.10 and integrating term for term results in:

$$MTTF(n) = \eta B_1 \cdot \sum_{k=1}^{n} \binom{n}{k} (-1)^{k+1} k^{-1/\beta}. \tag{4.11}$$

Where ηB_1 is the MTTF for a single component.

Example

For $n = 2$ parallel components with time to failure distributed as $W(100, 1.5)$

$$MTTF(2) = MTTF(1)[2 \times 1^{-1/1.5} - 2^{-1/1.5}] = 1.37 \; MTTF(1).$$

Thus, the MTTF for two components in parallel is 37% higher than the MTTF for a single component.

With three components the result is:

$$MTTF(3) = MTTF(1) \, [3 \times 1^{-1/1.5} - 3 \times 2^{-1/1.5} + 3^{-1/1.5}] = 1.59 MTTF(1).$$

With four components,

$$MTTF(4) = MTTF(1)[4 - 6 \times 2^{-1/1.5} + 4 \times 3^{-1/1.5} - 4^{-1/1.5}] = 1.74 MTTF(1).$$

For $n = 5$ the factor is 1.87, for $n = 6$, 1.96, and for $n = 7$ it is 2.04. The marginal return on each additional component decreases as the number of components increases.

If the components in the previous series system example (Example 1) were instead used in a parallel system, the system reliability function would become:

$$R_{system}(t) = 1 - \left(1 - \exp\left[-\left(\frac{t}{100}\right)^2\right]\right) \cdot \left(1 - \exp\left[-\left(\frac{t}{500}\right)^2\right]\right) \cdot$$
$$\left(1 - \exp\left[-\left(\frac{t}{1000}\right)^2\right]\right). \tag{4.12}$$

This is *not* a Weibull reliability function and there is no closed form expression for the mean system life. One can, however, integrate the system reliability function numerically to find the system mean life, that is,

$$E\left(t_{system}\right) = \int_0^\infty R_{system}(t) dt. \tag{4.13}$$

Alternately one can use simulation via a spreadsheet to determine the system life distribution and its mean. For a parallel system, the system life t_{system} is the life of the longest lived component:

$$t_{system} = \max(t_1, t_2, \dots t_n). \tag{4.14}$$

To simulate system life, one may enter simulated lives for each component into separate columns of a spreadsheet and then find the maximum row by row to simulate the corresponding parallel system life. Figure 4.1 is a Weibull plot of the simulated life of the longest lived component and for comparison the life of the component which has a scale parameter of 1000. The system life is clearly not Weibull distributed. The mean system life calculated as the average of the 10,000 simulated values is 935.7. The standard deviation of the system life is 423. Using Equation 2.75 with $n = 10,000$ a 95% confidence interval for the mean is computed to be (927.4, 944.1). Using numerical integration of the reliability function of Equation 4.12, the exact value of the system mean life is computed to be 933.0 and falls within the confidence interval obtained from the simulation results.

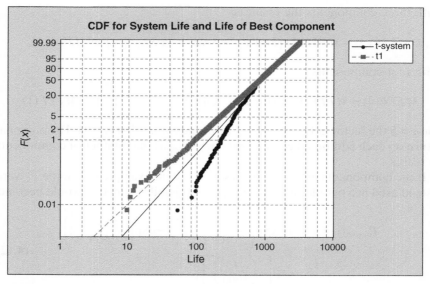

Figure 4.1 Simulated component and system life for a parallel arrangement of three components distributed as $W(100, 2)$, $W(500, 2)$, and $W(1000, 2)$.

4.1.3 Standby Parallel

Standby Parallel refers to the case where only a single component is employed at any time. Upon failure of that component another is switched into service. Thus, rather than having n components functioning simultaneously as in ordinary parallel redundancy, only one component is in service and the other $n - 1$ components are standing by ready to be employed. Automobile spare tires are an example, or were, before the "doughnut" became standard. In some applications such as when the components under discussion are engines, a distinction is made between hot, warm, and cold standby. In hot standby the units standing by are running and under load. In warm standby they are running but not fully loaded. In cold standby the standby units are not running. Some models include a finite probability that the switching from a failed unit to the replacement unit will cause system failure. The switching failure probability is generally assumed to be greatest for cold and least for hot standby.

Neglecting the possibility of switching failures, the time to system failure is the sum of the lives of the n components, that is,

$$t_{system} = \sum_{i=1}^{n} t_i. \tag{4.15}$$

The expected value of the system life is the sum of the expected lives of each component. For identical Weibull components the result is:

$$E\left(t_{system}\right) = n\eta\Gamma\left(1 + \frac{1}{\beta}\right). \tag{4.16}$$

The variance of the system life is the sum of the variances of the components or

$$var\left(t_{system}\right) = n\eta^2\left[B_2 - B_1^2\right]. \tag{4.17}$$

By the central limit theorem, the system life will be approximately normally distributed with these values of the mean and variance if n is sufficiently large. How large is *sufficiently large* will vary with the shape parameter β since the shape parameter governs how different the distribution is from the normal to begin with. For $\beta = 3.6$ one could expect the normal approximation to be valid at a smaller value of n than if $\beta = 1$. One could, of course, use simulation to verify the adequacy of the normal approximation.

Example

$N = 5$ components are in standby parallel, that is, one is running and four are standing by as replacements. The component life distribution is Weibull with a scale parameter of 1000 hours and a shape parameter of 2.0. Using the normal approximation, compute the system reliability at a life of 3500 hours.

The mean system life is $5 \times 1000 \times B_1 = 5000 \times 0.8862 = 4431$.

The variance of the system life is $5 \times 1000^2 [0.2146] = 1.073\text{E}6$.

The reliability is the probability that the system life exceeds 3500. Using the normal approximation and probability calculations summarized in Section 2.11 gives:

$$R(3500) = \text{Prob}\left[t_{system} > 3500\right] = 1 - \Phi\left[\frac{3500 - 4431}{1035.9}\right] = 1 - \Phi(-.08992) = 0.816.$$

In a simulation of 10,000 values of the sum of five observations from $W(1000, 2)$ 1849 sums were 3500 or less. The estimated value of $R(3500)$ is thus $1 - 1849/10,000 = 0.815$, in very good agreement with the normal approximation.

4.2 WEIBULL MIXTURES

When distributions are mixed in known proportions, the cumulative distributive function (CDF) of the mixture is readily obtained by applying the Law of Total Probability. So, for example, let us say a product is produced by three different vendors and the strength distribution is different for each vendor. Denote the CDFs of the three distributions by $F_1(x)$, $F_2(x)$, and $F_3(x)$. Let us say that the proportions purchased from each vendor are 20%, 30%, and 50%

respectively. To compute the probability that the strength of an item randomly selected from the mixture is less than x, we write:

$$\text{Prob}[X < x] = \text{Prob}[X < x|vendor\ 1] \cdot P(vendor\ 1) + \text{Prob}\{X < x\ |\ vendor\ 2]$$
$$\cdot P(vendor\ 2) + \text{Prob}[X < x|vendor\ 3] \cdot P(vendor\ 3).$$

More compactly,

$$F(x) = 0.20F_1(x) + 0.30F_2(x) + 0.50F_3(x).$$

The generalization to a greater number of vendors is apparent. The result applies for any distributions and they need not be of the same form. The mean of the mixture may also be computed in terms of the means for each population using the Law of Total Expectation:

$$E(X) = 0.20E(X|vendor\ 1) + 0.30E(X|vendor\ 2) + 0.50E(X|vendor\ 3).$$

This also generalizes in an obvious way to a greater number of mixture components.

If the distributions are Weibull whether with the same shape parameter or not, the overall distribution will not be Weibull. However, the distribution is readily simulated. In the present example, one could generate 2000 values from F_1, 3000 values from F_2, and 5000 values from F_3. Combining these would result in 10,000 random values of the mixture.

Let the three distributions be Weibull with the shape and scale parameters tabled below. Also shown in the table are the expected values and $F(100)$ for each distribution.

Distribution	Shape Parameter	Scale Parameter	Proportion	$F(100)$	$E(X|vendor)$
Vendor 1	0.7	100	0.20	0.632	126.6
Vendor 2	1.2	50	0.30	0.899	47.03
Vendor 3	2.5	200	0.50	0.162	177.45

The simulated distribution displayed on a Weibull grid is shown in Figure 4.2 and confirms that the mixture is *not* Weibull distributed.

Graphically it appears that $F(100)$ for the mixture is about 50%. Applying the Law of Total Probability to compute the exact value of $F(100)$ in terms of the values of $F(100)$ tabled for each distribution above gives:

$$F(100) = 0.20 \times 0.632 + 0.30 \times 0.899 + 0.50 \times 0.162 = 0.48.$$

The mean of the simulated values of the mixture was 126.99 with a 95% confidence interval of (124.65, 129.34). The true value of the mean of the mixture may be computed using the Law of Total Expectation:

$$E(X) = 0.20 \times 126.58 + 0.30 \times 47.03 + 0.50 \times 177.45 = 128.15.$$

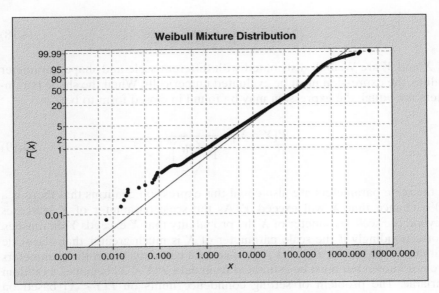

Figure 4.2 Simulated distribution of a mixture of Weibull distributions.

The actual expected value is within the confidence interval computed from the simulated mixture.

4.3 $P(Y < X)$

One probability problem that appears in a number of contexts is determining the probability that a value randomly drawn from one probability distribution and denoted X exceeds the value randomly and independently drawn from another distribution and denoted Y. If Y is the diameter of a shaft and X is the diameter of a housing into which the shaft must fit, $P(Y < X)$ is the probability that the shaft will fit without interference. If X is the strength of a component for example, a rivet, and Y is the random stress that the component will be subjected to in service, then $P(Y < X)$ is the reliability in the sense that it is the proportion of the population of rivets that will survive the application of the stress to which they are subjected. The idea of stress and strength generalize beyond their meanings in the realm of applied mechanics and would include physical quantities such as temperature and voltage for example.

The probability that Y is less than a specific value of x is $F_y(x)$. The subscript is used to denote that it is the CDF of the distribution of Y that is being evaluated at a value x. The total probability that Y is less than X comes from integrating this probability over the entire domain of x.

$$P\{Y < X\} = \int_0^\infty f_x(x) F_y(x) dx. \qquad (4.18)$$

If X and Y are Weibull distributed with different scale and shape parameter values, the expression must be evaluated numerically. When the shape parameters are equal, the expression may be found in closed form to be:

$$P[Y < X] = \frac{1}{1 + \left(\dfrac{\eta_y}{\eta_x}\right)^\beta}. \qquad (4.19)$$

If the scale parameters are also equal this expression confirms that there is a 50% chance that X will exceed Y. As the scale parameter of Y increases beyond the scale parameter of X the probability that X exceeds Y diminishes. Correspondingly if the scale parameter of X is much greater than the scale parameter of Y, then $P(Y < X)$ approaches a certainty. When the parameters are not known but must be estimated from data, $P[Y < X]$ becomes a random variable. The problem of setting confidence limits on $P[Y < X]$ based on Weibull parameters estimated from data samples is discussed in Chapter 8.

Example
The strength of a population of rivets follows a Weibull distribution with a scale parameter of 1000 psi and a shape parameter of 3.0. When deployed, each rivet will experience a different constant stress which varies randomly from rivet to rivet in accordance with a Weibull distribution with the same shape parameter but with a scale parameter of 750 psi. Compute the proportion of rivets that will endure the stress applied to them. The stress and strength density functions are shown plotted in Figure 4.3:
    Using Equation 4.19:

$$P[stress < strength] = \frac{1}{1 + \left(\dfrac{750}{1000}\right)^3} = 0.703.$$

The idea here is readily extended to additional populations. Let X, Y, and Z be Weibull distributed random variables with a common shape parameter β and scale parameters η_1, η_2 and η_3 respectively. We wish to compute the probability that Y is smaller than either X or Z. The question may be restated as:

$$P[Y < \min(X, Z)]. \qquad (4.20)$$

We know from Equation 4.4 of Section 4.1 that $\min(X, Z)$ is Weibull distributed with shape parameter β and scale parameter:

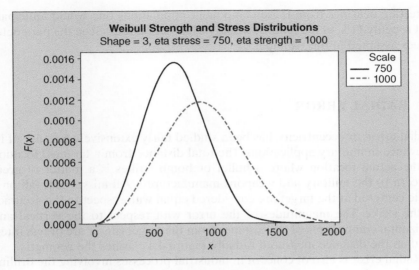

Figure 4.3 Weibull stress and strength distributions.

$$\eta_e = \left(\frac{1}{\eta_x^{\beta}} + \frac{1}{\eta_z^{\beta}} \right)^{-1/\beta}.$$

Thus, reapplying Equation 4.19 we have:

$$P[Y < \min(X, Z)] = \frac{1}{1 + \left(\dfrac{\eta_y}{\eta_e} \right)^{\beta}} = \frac{1}{1 + \left(\dfrac{\eta_y}{\eta_x} \right)^{\beta} + \left(\dfrac{\eta_y}{\eta_z} \right)^{\beta}}. \quad (4.21)$$

This result can be extended to any number of Weibull random variables if they have a common shape parameter.

Example

A ball bearing consists of an inner ring, an outer ring, and a set of rolling elements. The life of the rolling elements, taken as a set, is Weibull distributed with a shape parameter of 1.5 and a scale parameter of 1000 hours. The inner ring and outer ring are both Weibull distributed with the same shape parameter and with scale parameters of 300 and 600 hours, respectively. Compute the probability that a bearing fails by means of a ball set failure.

$$P[ball\ set\ life < ring\ lives] = \frac{1}{1 + \left(\dfrac{1000}{300} \right)^{1.5} + \left(\dfrac{1000}{600} \right)^{1.5}} = 0.1083.$$

In testing bearings from these component populations one would anticipate that roughly 11% of the failures would be ball set failures given the parameter values assumed in this example.

4.4   RADIAL ERROR

Radial error, or eccentricity, has been studied fairly extensively because of its relevance in military applications. The total distance from a targeted location to the actual location where a bullet or bomb strikes is a matter of great concern to the military and weapons manufacturers. All misses that fall on a circle centered at the target are considered equal with respect to the accuracy of the strike. The magnitudes of the error with respect to the vertical and horizontal components of the distance from the target center are of less interest than the distance measured radially. Figure 4.4 clarifies the geometry.

Radial error is also of concern in industrial processes involving the drilling of holes or the placement of components on a circuit board at a targeted site. Eccentricity of the outer diameter and the inner diameter of a bearing ring is known as radial runout and is carefully controlled. It is of obvious interest to a manufacturer that radial error be held within a specified tolerance.

In Figure 4.4 the targeted center is indicated by the origin of coordinates O, and because of errors in placement in the horizontal and vertical directions, the actual center is displaced by random distances X and Y, respectively, resulting in a radial error ε computed by the Pythagorean theorem as the square root of the sum of the squares of the two orthogonal components of the error.

In applications such as the hole drilling example cited above, the holes will generally be drilled by a robotic device which moves to the assigned x coordinate (say) and then moves in an orthogonal direction to the assigned y

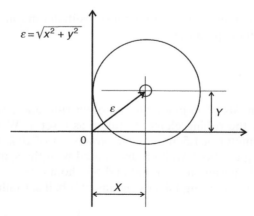

$$\varepsilon = \sqrt{x^2 + y^2}$$

Figure 4.4   Radial error. From *Quality Technology and Quantitative Management*, Vol. 3, No.3, Copyright 2006 by NCTU Publication Press. Reprinted by permission of NCTU Publication Press.

coordinate. It is reasonable to assume that the variability in locating the two coordinates is the same and that the deviations in the two directions are independent.

The case where the means are not offset and the variances are equal appears in Cramer (1946) for the general case of $p \geq 2$ dimensions. For $p = 2$ with $\sigma_x = \sigma_y = \sigma$, the radial error follows a Rayleigh distribution. The Rayleigh distribution is a special case of the Weibull distribution for which the shape parameter equals 2.0. The Weibull or Rayleigh distribution of radial error has a scale parameter of $\sigma\sqrt{2}$. The cumulative distribution of radial error is thus:

$$F(\varepsilon) = P[radial\ error < \varepsilon] = 1 - exp\left(-\frac{\varepsilon^2}{2\sigma^2}\right). \tag{4.22}$$

The expected value of the radial error is therefore:

$$E(\varepsilon) = \mu_\varepsilon = \eta\Gamma\left(1 + \frac{1}{\beta}\right). \tag{4.23}$$

Using $\beta = 2$ and $\eta = 1.414\sigma$, the mean radial error is expressible in terms of σ as:

$$\mu_\varepsilon = 1.253\sigma. \tag{4.24}$$

Thus the mean of the radial error distribution depends linearly upon σ.

The standard deviation of the radial error is:

$$\sigma_\varepsilon = \eta\sqrt{\Gamma\left(1 + \frac{2}{\beta}\right) - \Gamma^2\left(1 + \frac{1}{\beta}\right)}. \tag{4.25}$$

With $\beta = 2$ and $\eta = 1.414\sigma$ the standard deviation of the radial error ε becomes:

$$\sigma_\varepsilon = 0.655\sigma. \tag{4.26}$$

These expressions are useful if one wishes to keep control charts for radial error. For example, using the average eccentricity measured in subgroups of size n and assuming n is sufficiently large that the central limit theorem holds, the center line of a control chart for radial error can be set at 1.253σ. The upper and lower control limits would then be located at distances $\pm 3\sigma_\varepsilon / \sqrt{n} = 1.965\sigma / \sqrt{n}$ from the center line.

Example 1

In a production process, a robot arm moves in the horizontal and then vertical direction to the desired location and then drills a hole at that location. The error in locating the x coordinate of the desired center location is independent of the error in locating the y coordinate. The errors in the two directions are both normally distributed with a mean of zero and a standard deviation of 0.001 inch. Compute the probability that the radial error of a given hole location does not exceed 0.002 inch.

$$P[\varepsilon < 0.002] = 1 - \exp\left(-\left[\frac{0.002}{\sqrt{2}\cdot 0.001}\right]^2\right) = 0.865.$$

Example 2

For the same case, compute the center line and upper and lower three sigma limits for a control chart for eccentricity based on the means of samples of size $n = 5$. The center line is $0.655 \times 0.001 = 0.000655$ inch. The upper control limit is $0.000655 + 1.965 \times 0.001/\sqrt{5} = 0.0015338$. The lower control limit is zero since eccentricity cannot be less than zero.

Using control charts to distinguish a change in the standard deviation σ from a change in the coordinates of the center is discussed in McCool (2006a).

4.5   PRO RATA WARRANTY

Consider a purchased item that is sold with a warranty that specifies that the user will be reimbursed for the unused portion of a warranted amount W if failure should occur prior to the customer receiving W units of satisfactory use. What fraction of the purchase price can a customer expect to recover under such a warranty? A tire is an example and W might represent, say, 80,000 miles of usage. Assume the usage rate U is known; for example, in the tire example perhaps the purchaser drives $U = 15,000$ miles per year. Let X, the amount of usage until failure occurs, be a two-parameter Weibull distributed random variable. If failure occurs at X units of usage, the time since purchase will be X/U. The value at that point has diminished to $(1 - X/W)$ of the original amount or to $(1 - X/W)C_0$ if C_0 is the purchase price. A further question to a user considering the purchase of such a warranty is *what is the present value to me of the amount* $C_0 \times (1 - X/W)$ *received at a time* X/U *in the future since a lower amount invested now will become* $C_0 \times (1 - X/W)$ *at time* X/U *at some interest rate* α?

Using continual compounding for convenience, the present value of a cash amount C paid at t units in the future may be computed as $Ce^{-\alpha t}$. Thus, for item failure after X units of usage, the present value is:

$$C(X) = C_0\left(1 - \frac{X}{W}\right)e^{-\alpha X/U}. \tag{4.27}$$

Denoting by $R(X)$ the ratio of the present value to the original cost gives:

$$R(X) = \frac{C(X)}{C_0} = \left(1 - \frac{X}{W}\right)e^{-\alpha X/U}. \tag{4.28}$$

$R(X)$ is a decreasing function of X ranging from 1.0 at $X = 0$ to 0.0 at $X = W$.

The Weibull model should be able to account, at least broadly, for the failure life experience of most products. The exponential special case corresponds to a constant failure rate and thus accounts for a failure mode that does not vary with the amount of usage. Failure due to a random accident or shock would give rise to a constant failure rate mode. For $\beta > 1$ the model accounts for increased vulnerability with usage, that is, wearout-related failure.

The expected value of the present value fraction may be computed by integrating the product of $R(x)$ and $f(x)$, the Weibull density, over the range from 0 to W:

$$\bar{R} = E(R) = \int_0^W R(x)f(x)dx = \frac{\beta}{\eta^\beta}\int_0^W\left(1 - \frac{x}{W}\right)e^{-\frac{\alpha x}{U}}x^{\beta-1}e^{-\left(\frac{x}{\eta}\right)^\beta}dx. \tag{4.29}$$

A Mathcad module named rbarwarranty.xmcd will perform this integration. With $\beta = 2$, $\eta = 100,000$, $W = 80,000$, $U = 15,000$, and $\alpha = 0.06$, the average value of the R ratio is 0.153. Introduce the transformation $Y = X/\eta$. Y is a then a Weibull random variable with unit scale parameter and shape parameter β.

The distribution of R has finite mass at $r = 0$ since there is no payout when $X > W$. The probability that $R = 0$ may be computed as:

$$P[R = 0] = P[X > W] = e^{-\left(\frac{W}{\eta}\right)^\beta}. \tag{4.30}$$

Percentage points of the distribution of R are readily found. Since R is a monotonically decreasing function of X, then when X is less than some value x_0, R will be greater than $R(x_0)$.

Using the $100p$-th percentile of X one may write:

$$\text{Prob}[X < x_p] = p = \text{Prob}[R > r(x_p)]. \tag{4.31}$$

Therefore,

$$\text{Prob}[R < R(x_p)] = 1 - p. \tag{4.32}$$

Hence, the $100(1 - p)$th percentage point of R is $R(x_p)$. So, for example, to find the 70th percentile of R, one substitutes the 30th percentile of x into the function $R(x)$.

For the example discussed above the probability of no payoff is:

$$\text{Prob}[X > W] = \exp\left[-\left(\frac{W}{\eta}\right)^{\beta}\right] = \exp\left[-\left(\frac{80,000}{100,000}\right)^{2}\right] = 0.52.$$

The distribution of R has a discrete mass of 0.52 at $R = 0$. The median value of R is therefore 0. To find $R_{0.6}$ calculate $x_{0.40}$ using the formula:

$$x_p = \eta[-\ln(1-p)]^{1/\beta}. \tag{4.33}$$

The result is:

$$x_{0.40} = 100000 \times (-\ln(1-.4))^{1/2} = 71,472.$$

The 60th percentile of R is now calculated by substituting this value in the $R(x)$ function.

$$R_{0.60} = R(x_{0.40}) = \left(1 - \frac{x_{0.40}}{W}\right)e^{-\frac{\alpha x_{0.40}}{U}} = \left(1 - \frac{71,472}{80,000}\right) \times \exp\left[-\frac{0.06 \times 71,472}{15,000}\right].$$
$$= 0.080$$

There is a 60% chance that the purchaser will receive less than 8% of the purchase price as a return under the warranty. Further discussion and tables may be found in McCool (2006b).

4.6 OPTIMUM AGE REPLACEMENT

We have seen that when the Weibull shape parameter exceeds 1.0 the hazard function increases with time. In this circumstance, depending on the relative cost of a planned replacement and a failure, it may pay to replace the item preemptively prior to failure since new items are less prone to failure than used ones. It is easy to visualize situations wherein a failure can be extremely costly, resulting, for example, in lost production or damage to machinery or goods. Let c_1 be the cost of a failure and c_2 the cost of a planned replacement. This problem seems to have been first considered by Barlow and Hunter (1960). Assume an item is replaced at age t unless failure occurs prior to t. A cycle is considered to be the time between consecutive replacements due either to failure or to replacement. The expected cost of a cycle is $c_1 \times F(t) + c_2 \times (1 - F(t))$. The duration T of a cycle is a random variable. Its expected value is computable using the Law of Total Expectation:

$$E(T) = E(T \mid failure) \times F(t) + E(T \mid replaced) \times R(t). \tag{4.34}$$

Given that failure occurs at some time τ prior to replacement time t, the pdf of τ conditional on $\tau < t$ is $f(\tau)/F(t)$. Equation 4.34 thus becomes:

$$E(T) = \int_0^t \tau f(\tau) d\tau + t \cdot R(t). \tag{4.35}$$

This is equivalent to, as may be verified by integrating by parts,

$$E(T) = \int_0^t R(\tau) d\tau. \tag{4.36}$$

Thus, the expected cost per unit expected time, denoted $c(t)$ is given by:

$$c(t) = \frac{c_1 F(t) + c_2(1 - F(t))}{\int_0^t R(\tau) d\tau}. \tag{4.37}$$

One benchmark for assessing savings due to preventive maintenance is the cost per unit time of not doing any planned replacement. This is equivalent to letting the age at replacement approach infinity in the expression for $c(t)$. For large values of t, $F(t)$ approaches 1.0 and so the numerator approaches c_1 and the denominator approaches the mean $\mu = $ MTTF. Therefore:

$$c(\infty) = \frac{c_1}{\mu}.$$

The optimum scheduled replacement time, t^*, that is, the time that minimizes $c(t)$ is found as the value for which the derivative of $c(t)$ with respect to t is zero. Differentiating and simplifying gives:

$$\lambda(t^*) \int_0^{t^*} R(\tau) d\tau + R(t^*) = \frac{c_1}{c_1 - c_2}. \tag{4.38}$$

where $\lambda(t^*)$ is the hazard function evaluated at $t = t^*$. This expression is valid for any failure time distribution.

For the Weibull failure model the expression for $c(t)$ becomes:

$$c(t) = \frac{c_1 \left[1 - exp\left(-\left(\frac{t}{\eta}\right)^\beta \right) \right] + c_2 \, exp\left(-\left(\frac{t}{\eta}\right)^\beta \right)}{\int_0^t exp\left(-\left(\frac{\tau}{\eta}\right)^\beta \right) d\tau}. \tag{4.39}$$

This function is shown plotted in Figure 4.5 for $c_1 = 200$, $c_2 = 40$, $\beta = 2$, and $\eta = 100$.

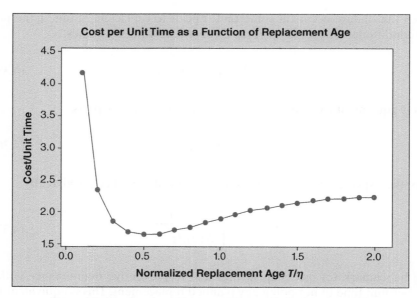

Figure 4.5 The function $c(t)$ for $c_1 = 200$, $c_2 = 40$, $\beta = 2$, and $\eta = 100$.

The solution for the optimum replacement time t^* may be found by solving:

$$\frac{\beta}{\eta}\left(\frac{t^*}{\eta}\right)^{\beta-1} \int_0^{t^*} e^{-\left(\frac{t}{\eta}\right)^{\beta}} dt + e^{-\left(\frac{t^*}{\eta}\right)^{\beta}} = \frac{c_1}{c_1 - c_2}. \tag{4.40}$$

It is convenient to introduce the change of variable $w = t/\eta$ into this expression and use W to represent the upper limit when $t = t^*$. The result is:

$$e^{-W^{\beta}} + \beta W^{\beta-1} \int_0^W e^{-w^{\beta}} dw = \frac{c_1}{c_1 - c_2}. \tag{4.41}$$

In this form it is clear that the optimum time t^* expressed as a fraction W of the scale parameter depends only on the shape parameter. For fixed β, but various values of η, it is only necessary to solve Equation 4.41 once. Moreover, the numerator and denominator on the right-hand side may be divided by c_2 (or c_1) which shows that the solution depends on the two costs only through their ratio. The units in which the costs are expressed (dollars, Euro) therefore do not matter, as long as both costs are expressed in the same units. Having solved Equation 4.41 for W, the cost per unit time may be computed by substituting $W = \left(\dfrac{t^*}{\eta}\right)$ and $w = \left(\dfrac{\tau}{\eta}\right)$ into Equation 4.39. The result may be expressed as,

$$\frac{\eta c(t^*)}{c_1} = \frac{[1 - \exp(-W^\beta) + c\,exp(W^\beta)]}{\int_0^W \exp\{-w^\beta\}dw} \qquad (4.42)$$

where $c \equiv c_2/c_1$.

Tadikamalla (1980) tabulates W and $\dfrac{\eta c(t^*)}{c_1}$ for various values of the Weibull shape parameter β and the cost ratio c.

For the example we are considering with $c = 40/200 = 0.2$, $\eta = 100$, and $\beta = 2$, the solution is $t^* = 51.07$. Substituting $t^* = 51.1$ in Equation 4.39 shows that the minimum cost per unit time is \$1.634. Lewis (1987) shows that when c_1 is much greater than c_2, necessitating very early replacement ($t^* \ll \eta$), the solution may be approximated by:

$$t^* = \eta \left[\frac{c_2}{(\beta - 1)c_1} \right]^{1/\beta}. \qquad (4.43)$$

For the example we are considering, the approximation gives $t^* = 44.721$. This corresponds to a cost per unit time of \$1.646. This is not substantially different from the optimum because the cost function has a broad minimum as seen in Figure 4.5.

For comparison, the cost per unit time with no planned replacement is:

$$c(\infty) = \frac{c_1}{\mu} = \frac{c_1}{\eta \Gamma\left(1 + \dfrac{1}{\beta}\right)} = \frac{200}{100 \cdot 0.886} = \$2.25.$$

The Mathcad worksheet below is titled agereplace.xmcd and may be downloaded from the author's website.

4.6.1 Age Replacement

c_1 is the cost of an unscheduled replacement and c_2 is the cost of a planned replacement. η is the Weibull scale parameter and β is the Weibull shape parameter.

Set values for example:

$$\beta = 2, \eta = 100, c_1 = 200, c_2 = 40$$

The expected cost of a replacement given scheduled replacement at time t is denoted $c(t)$:

$$c(t) = c1 \left[1 - e^{-\left(\frac{t}{\eta}\right)^\beta} \right] + c2 \left[e^{-\left(\frac{t}{\eta}\right)^\beta} \right].$$

The average length of time between replacements for any reason as a function of the planned replacement time is denoted $RP(t)$:

$$RP(t) = \int_0^t e^{-\left(\frac{\tau}{\eta}\right)^\beta} d\tau.$$

The cost per unit time is denoted $cc(t)$:

$$cc(t) = \frac{c(t)}{RP(t)}$$

One may use trial and error to find the t value for which $cc(t)$ is smallest. Alternatively one may solve the equation that results from setting the derivative of $cc(t)$ to zero. This is done below in dimensionless form. W is the optimum time $t*$ divided by η.

Define c:

$$c = \frac{c2}{c1} \quad \beta = 2$$

$$F(\beta, W) = \beta \times W^{\beta-1} \int_0^W e^{-\omega^\beta} d\omega + e^{-W^\beta}$$

$$W = 1 \quad \text{Initial guess at the solution}$$

$$W = \text{root}\left[\left(F(\beta, W) - \frac{1}{1-c}\right), W\right].$$

The optimum is:

$$W = 0.511$$

$$t* = W\eta = 0.511 \times 100 = 51.1$$

Compare to Lewis' Approximation of T denoted TT

$$TT = \left[\eta\left(\frac{c2}{\beta-1}\frac{1}{c1}\right)^{\frac{1}{\beta}}\right]$$

$$TT = 44.721$$

Calculate the optimum cost per unit time.

c_{opt} is equal to $\eta \times cc(t*)/c_1$:

$$c_{opt} = \frac{\left(1 - e^{-W^\beta}\right) + ce^{-W^\beta}}{\int_0^W e^{-\omega^\beta} d\omega}$$

$$c_{opt} = 0.817$$

$$\text{So } cc(t^*) = c1 \times \frac{c_{opt}}{\eta} = 1.634.$$

Substitute Lewis' Approximation into $cc(t)$:
$cc(44.721) = 1.646$ Only slightly higher than the minimum.

4.6.2 MTTF for a Maintained System

If under preventive maintenance a unit is replaced every T time units, thereby restoring it to its original condition, it is shown by Lewis that the MTTF becomes (Lewis, 1987):

$$MTTF = \frac{\int_0^T R(t)\,dt}{1 - R(T)}. \tag{4.44}$$

The verification of Equation 4.44 follows. The reliability of the maintained system $R_M(t)$ at some time $KT < t < (K+1)T$ for some integer value K is expressible as:

$$R_M(t) = R(T)^K \cdot R(t - KT). \tag{4.45}$$

This is the product of the probability of surviving K intervals of length T and the additional time interval of length $t - KT$.

The ratio of $R_M(t)$ to the unmaintained reliability $R(t)$ evaluated for simplicity at $t = KT$ is

$$\frac{R_M(KT)}{R(KT)} = \frac{\exp\left[-K\left(\frac{T}{\eta}\right)^\beta\right]}{\exp\left[-\left(\frac{KT}{\eta}\right)^\beta\right]} = \exp\left[K^\beta - K\right] \tag{4.46}$$

R_M will exceed R if K^β exceeds K or if

$$K^{\beta-1} > 1. \tag{4.47}$$

For $K > 1$ implying that maintenance has occurred R_M will exceed R provided $\beta > 1$. For $\beta = 1$, $R_M = R$, that is, maintenance has no effect. For $\beta < 1$ the maintained system is less reliable than the unmaintained system.

The MTTF is the integral of $R_M(t)$ as t ranges from 0 to ∞. This integral breaks into an infinite sum of integrals over the range $[KT, (K+1)T]$ as follows:

$$MTTF = \sum_{K=0}^{\infty} R(T)^K \int_{KT}^{(K+1)T} R(t - KT)\,dt. \tag{4.48}$$

Introducing the change of variable $\tau = t - KT$, we have:

$$MTTF = \int_0^T R(\tau)d\tau \cdot \sum_{K=0}^{\infty} R(T)^K. \tag{4.49}$$

Since $R(T) < 1$, the geometric series on the right may be summed to $1/(1 - R(T))$, thus giving the result shown earlier as Equation 4.44.

For the Weibull distribution $\int_0^T R(\tau)d\tau$ may be expressed in terms of the lower incomplete gamma function $\gamma(a, x)$ defined as:

$$\gamma(a, x) = \int_0^x t^{a-1}e^{-t}dt \tag{4.50}$$

The result is:

$$MTTF = \frac{\eta}{\beta}\gamma\left[\frac{1}{\beta},\left(\frac{T}{\eta}\right)^{\beta}\right] \Big/ \left[1 - \exp\left(-\left(\frac{T}{\eta}\right)^{\beta}\right)\right]. \tag{4.51}$$

When $\beta = 1$, that is, the Weibull reduces to the exponential, the MTTF of the maintained system becomes simply η, indicating that periodic replacement does not increase the mean time to failure when the time to failure is exponentially distributed.

Figure 4.6 is a plot of MTTF as a multiple of η versus the replacement interval also scaled by η for $\beta = 0.5, 1, 2$, and 3.

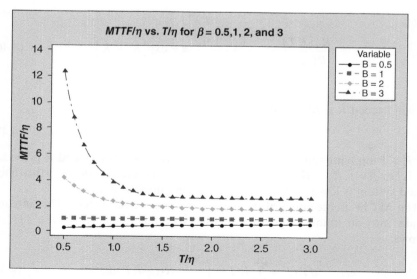

Figure 4.6 $MTTF/\eta$ versus the dimensionless replacement interval T/η for $\beta = 0.5, 1, 2$, and 3.

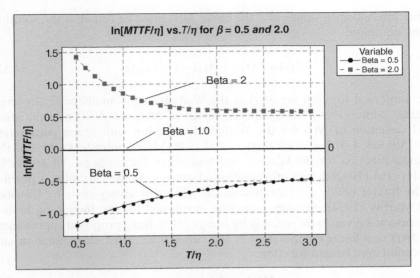

Figure 4.7 ln[$MTTF/\eta$] versus dimensionless replacement interval T/η for $\beta = 0.5$ and 2.0.

Figure 4.6 confirms that when $\beta = 1$ the MTTF remains equal to η regardless of how often replacements take place. The larger the value of β when it exceeds 1.0, the greater the MTTF. Also the shorter the replacement interval, the larger the MTTF. For $\beta < 1$ frequent replacements (small T) decrease the MTTF because fewer infant mortals are then removed. This may be thought of as insufficient burn-in. The scale in the figure does not reflect adequately the extent to which the MTTF is decreased for small T when $\beta < 1$.

Figure 4.7 shows the logarithm of $MTTF/\eta$ plotted against T/η for $\beta = 0.2$ and 3.0, which highlights the extent to which $\beta < 1$ impacts the MTTF.

4.7 RENEWAL THEORY

Consider a component whose life distribution under constant environmental conditions such as load, humidity, and stress is the two-parameter Weibull. Assume this component is observed until it fails, whereupon it is replaced in negligible time by a new item from the same population. This sequence of failure and replacement is considered to continue indefinitely. At time t the number of failures and hence the number of renewals, $N(t)$, is a discrete random variable. For large t the average number of renewals $M(t) = E(N(t))$ approaches t/μ. For very small $t(t<<\mu)$ the number of failures is either 0 or 1 so its approximate expected value for small t is:

$$M(t) = 1 \cdot F(t) + 0 \cdot [1 - F(t)] = F(t). \tag{4.52}$$

For intermediate values of t, the mean $M(t)$ must be found as the solution of the equation:

$$M(t) = F(t) + \int_0^t M(t-x)f(x)dx. \tag{4.53}$$

The numerical solution for $M(t)$ in the Weibull case is mathematically challenging. White (1964) has computed tables of $M(t) = E(N(t))$ and the standard deviation of $N(t)$ for the Weibull distribution with shape parameters $\beta = 0.5(0.5)3, 4, 5, 7$, and 10. Baxter et al. (1981) have published tables of the renewal function for the Weibull and four other distributions. An article by Baxter et al. (1982) details other published sources of renewal tables. A portion of the Baxter tables are reprinted in the text on warranty costs by Blischke and Murthy (1994). An approximation developed for the increasing failure rate case was given more recently by Jiang (2010). Jiang proposed approximating $M(t)$ as a linear combination of the cumulative distribution function and the cumulative hazard function:

$$M(t) \approx pF(t) + q\Lambda(t). \tag{4.54}$$

Where $p + q = 1$ and p depends on the Weibull shape parameter and may be computed from the equation:

$$p = 1 - \exp\left[-\left(\frac{\beta-1}{0.8731}\right)^{0.9269}\right]. \tag{4.55}$$

Using the Weibull CDF and cumulative hazard function from Equation 3.7, Jiang's approximation of the Weibull renewal function becomes:

$$M(t) \approx p\left[1 - \exp-\left[\left(\frac{t}{\eta}\right)^\beta\right]\right] + q\left(\frac{t}{\eta}\right)^\beta. \tag{4.56}$$

Jiang's approximation should apply to the Weibull when $\beta > 1$. In the special case where the shape parameter is 1.0 and the Weibull reduces to the exponential distribution, the number of replacements $N(t)$ at time t follows a Poisson Process:

$$\text{Prob}(N(t) = k) = \frac{e^{-\lambda t}(\lambda t)^k}{k!}; k = 0, 1.. \tag{4.57}$$

λ is the constant failure rate and is equal to the reciprocal of the scale parameter, that is, $\lambda = 1/\eta$.

The expected value is:

$$E(N(t)) = M(t) = \lambda t. \tag{4.58}$$

Renewal theory has obvious application to spare parts provisioning and in the evaluation of free replacement warranties. In Section 4.10 we illustrate the use of Equation 4.57 for computing the number of spare parts required for achieving a specified level of reliability when component failure lives are exponentially distributed. Another application of renewal theory discussed in the next section is in the analysis of a preventive maintenance policy known as block replacement. The simple exponential distribution result is of no value for preventive maintenance calculations since, as we have seen, with $\beta = 1.0$ there is no aging and hence no advantage to preventive maintenance.

4.7.1 Block Replacement

Block replacement is the name of a preventive replacement policy that is more costly but administratively simpler than age replacement. It is more costly with respect to the number of replacements per unit time but may ultimately be less expensive when the cost of recordkeeping is considered. Under the block replacement policy, items are replaced at some constant time interval t, regardless of whether a failure replacement had just recently occurred. Under block replacement there is no need to keep track of the age of each item in service; they are all replaced at the same time. Thus, the age of the item being replaced does not matter, although most replaced items will not have failed and therefore will be t units old when they are removed from service.

Again let c_1 be the cost of replacing a failed part and c_2 ($<c_1$) the cost of a planned replacement. If t is the time between replacements, $M(t)$ will be the expected number of failures replaced during that time and there will be one unfailed replacement at the end of that time. The average replacement cost is therefore $c_1 M(t) + c_2$ and will be incurred every t time units. The cost per unit time is therefore:

$$c(t) = \frac{c_1 M(t) + c_2}{t}. \tag{4.59}$$

Dividing both sides by c_1 and introducing $c = c_2/c_1$ gives:

$$\frac{c(t)}{c_1} = \frac{M(t) + c}{t}. \tag{4.60}$$

The next step is to use tabled values of $M(t)$ to compute $c(t)/c_1$ as a function of t to find the value of $t = T$ that makes it a minimum. To illustrate, we will use the same parameters as used in the age replacement example namely, $\beta = 2$, $\eta = 100$, and $c = c_2/c_1 = 0.2$. Table 4.1 gives several values of $M(t)$ for $\beta = 2$, taken from White's tables. Jiang's approximation is also shown for comparison. The last column shows the computed value of $c(t)/c_1$ for each value of t.

To the accuracy allowed by White's tables, in which t/η is incremented in steps of 0.05, the minimum occurs at $T = 50$. Multiplying by $c_1 = 200$, the cost per unit time is \$1.72. This is somewhat more than the \$1.634 minimum found

Table 4.1 Computation of the Optimum Block Replacement Interval and Comparison of Approximations to the Weibull Renewal Function

t/η	t	$M(t)$	Jiang's Approx.	$[M(t) + c]/t$
0.3	30	0.087379	0.087	9.579E-3
0.4	40	0.151903	0.152	8.798E-3
0.45	45	0.189707	0.189	8.660E-3
0.50	50	0.230794	0.230	8.616E-3
0.55	55	0.274843	0.274	8.634E-3
0.6	60	0.321526	0.321	8.692E-3
0.7	70	0.421508	0.420	8.879E-3
0.8	80	0.528267	0.527	9.103E-3
0.9	90	0.639605	0.637	9.329E-3

in the age replacement optimization. If $F(t)$ is used as an approximation to $M(t)$, the optimum t may be found by setting the derivative of $C(t)$ with respect to t equal to 0. The optimum $w = t/\eta$ may then be computed as the root of the following equation:

$$\beta w^\beta e^{-w^\beta} - \left(1 - e^{-w^\beta}\right) = c. \qquad (4.61)$$

Solving this equation for the example case gave $T = 60$ rather than 50. Using the value of $(M(60) + 0.2)/60$ from the above table shows that the cost would have been $1.74 per unit time rather than $1.72 had the maintenance interval been taken to be 60. Jiang's approximation is seen to be quite adequate in this example over the range examined.

4.7.2 Free Replacement Warranty

Under a free replacement warranty the manufacturer agrees to replace free of charge any item that fails with a new one until, ultimately, the customer has the use of a functioning item for the full length of a warranty period of length W. The expected number of times that replacement will be necessary is therefore $M(W)$. The cost to the manufacturer is the production cost c times $(1 + M(W))$. Suppose a product has a lifetime that follows the two-parameter Weibull distribution with $\beta = 2$ and $\eta = 2$ years, and a free replacement warranty of length $W = 1$ year is offered. The warranty period expressed as a fraction of η is $1/2 = 0.5$. From the previous tabulation for $\beta = 2$, $M(W) = 0.231$, so the manufacturer's cost is now $1.231c$. Thus, offering a free replacement warranty increases the manufacturer's costs by 23%.

4.7.3 A Renewing Free Replacement Warranty

Under a renewing warranty, a failure during the 11th month of a one-year warranty is replaced by a new item which itself then enjoys a full year of warranty protection. The manufacturer is not free of obligation until one of his products outlives the warranty period. Let X equal the total number of items installed until one lives beyond W. Let $F(x)$ denote the CDF of the product. The probability that $X = 1$ is the probability that the original item failed beyond the warranty life W, that is,

$$P[X = 1] = 1 - F(W) = R(W). \qquad (4.62)$$

The number of items will equal 2 if one item failed before W and the second survived beyond W:

$$P[X = 2] = F(W)R(W). \qquad (4.63)$$

It follows that the probability that $X = k$ is:

$$P\{X = k\} = F^{k-1}(W)R(W). \qquad (4.64)$$

This is a geometric distribution in which $R(W)$ plays the role of the success probability p and $F(W)$ is $1 - p$ (cf. Section 2.7 of Chapter 2). Its expected value is:

$$E(X) = \frac{1}{R(W)}. \qquad (4.65)$$

If, as in the previous example, product life was distributed as $W(2, 2)$ the expected total number of items a customer would use, including the original, is:

$$E(X) = \frac{1}{exp\left[-\left(\frac{1}{2}\right)^2\right]} = exp\left[\left(\frac{1}{2}\right)^2\right] = 1.284.$$

Thus, extending a renewing free replacement warranty increases the manufacturer's cost by 28.4%.

The text by Blischke and Murthy (1994) is recommended for readers with a particular interest in warranties and their analysis.

4.8 OPTIMUM BIDDING

This is an old problem in the operations research literature adapted here to utilize a Weibull distribution.

In a series of sealed bid competitions assume the winning bid B, expressed as a markup, follows the Weibull distribution with known parameters. That is, $Y = (B - K)/K$ follows $W(\eta, \beta)$, where K is the bidder's cost. The problem is to find the optimum amount to bid to maximize the bidder's expected profit.

Let A be a bid. A will be a winning bid if $A < B$. Assuming equal cost K for all participants, the probability that bid A wins is:

$$\text{Prob}[win \mid A] = \text{Prob}\left[Y = \frac{(B-K)}{K} > \frac{(A-K)}{K}\right] = \exp\left[-\left(\frac{A-K}{K\eta}\right)^{\beta}\right]. \quad (4.66)$$

If the bid in the amount A wins the competition, the bidder's profit will be:

$$E[Profit \mid A] = (A-K)P[win \mid A] + 0 \cdot P[lose \mid A]$$

or

$$E[Profit \mid A] = (A-K) \cdot \exp\left[-\left(\frac{A-K}{K\eta}\right)^{\beta}\right] \quad (4.67)$$

When A is large, the profit $(A - K)$ is large but the probability of winning is small. When A is small, the probability of winning is large but the amount won is small. The optimum bid amount A^* is found by setting the derivative of the expected profit with respect to A equal to zero and solving. The result is that the maximum expected profit occurs when A is such that the markup is:

$$\frac{(A^* - K)}{K} = \eta\left(\frac{1}{\beta}\right)^{1/\beta}. \quad (4.68)$$

Example

By studying the pattern of winning bids in a certain market it was found that the winning markup followed a two-parameter Weibull distribution with $\eta = 0.73$ and $\beta = 1.5$. What markup optimizes expected profit assuming this pattern of bidding persists among the competition?

Optimum markup = $0.73(1/1.5)^{1/1.5} = 0.557$.

Implicit in this model is the assumption that all the bidders have equal cost and that the number of bidders remains fairly constant over time.

4.9   OPTIMUM BURN-IN

We have seen in Section 3.4 that if the Weibull shape parameter is less than 1.0, the hazard function decreases with time and the survivors of a period x_0 of running will be more reliable than the original population. However, the improved reliability comes at a cost. There is the cost of running, which includes a setup cost and a cost that increases with the length of the burn-in period x_0. There is also the cost of the product itself if it fails during the run-in process. These costs are balanced against the decreased cost of failing prior to a warranty period W.

Define the following costs:

c_0 = setup cost for performing burn-in
c_1 = cost per unit time spent running
c_2 = cost per failure during the run-in period
c_3 = penalty cost if a burn-in survivor fails prior to a warranty period W.

The expected cost per item as a function of burn-in period x_0 is:

$$f(x_0) = c_0 + c_1 x_0 + c_2 \left(1 - \exp\left[-\left(\frac{x_0}{\eta}\right)^\beta\right]\right) + c_3 \left[1 - \frac{\exp\left[-\left(\frac{x_0 + w}{\eta}\right)^\beta\right]}{\exp\left[-\left(\frac{x_0}{\eta}\right)^\beta\right]}\right]. \quad (4.69)$$

Example

A product life is distributed as $W(5000, 0.8)$, where the scale parameter is specified in hours. The product has to survive a warranty period of $W = 700$ hours. The costs are $c_0 = \$0.50$, $c_1 = \$0.07/\text{hour}$, and $c_2 = \$3/\text{unit}$ failed during burn-in, and $c_3 = \$500/\text{unit}$ failed under warranty. The optimum value of x_0 is found to be 46.9 hours using a Mathcad module. The associated expected cost is \$92.3/unit. To see if it is more economical not to conduct burn-in, it is necessary to let the setup cost $c_0 = 0$ and evaluate the cost function using $x_0 = 0$. In the present problem this results in a cost of \$93.7/unit so burn-in is worthwhile. Figure 4.8 shows the cost plotted against x_0.

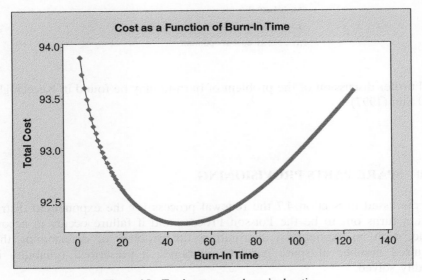

Figure 4.8 Total cost versus burn-in duration.

The Mathcad module for burn-in optimization (burninoptimizer.xmcd) is listed below:

Burn-In Optimizer

Compute the optimum duration t of burn in so that the combined cost of setup, testing time, failures during burn-in and failures under warranty of the burned-in product.

c_0=set up cost (value does not affect optimum burn-in time
c_1= testing cost per unit time
c_2 = cost of failing a specimen during burn-in
c_3= cost of a failure of the burned-in specimen under warranty after it is delivered to the customer.
β =Weibull shape parameter
η =Weibull scale parameter

Example

$c0 := 0.500 \quad c1 := 0.07 \quad c2 := 3 \quad\quad c3 := 500 \quad$ **Cost Coefficients**

$\eta := 5000 \quad\quad \beta := 0.8 \quad$ **Weibull Parameters**

$w := 700 \quad$ **Warranty Period**

$$f(t) := c0 + t \cdot c1 + c2 \cdot \left[1 - e^{-\left(\frac{t}{\eta}\right)^{\beta}}\right] + c3 \cdot \left[1 - \frac{e^{-\left(\frac{t+w}{\eta}\right)^{\beta}}}{e^{-\left(\frac{t}{\eta}\right)^{\beta}}}\right]$$

$t := 40 \quad$ **Initial guess at optimum**

Given

$t > 0$

$t = \text{Minimize}(f, t)$

$t = 46.948 \quad$ **optimum duration**

$f(t) = 92.283 \;$ **cost at optimum**

Further discussion of the problem of burn-in may be found in Kececioglu and Sun (1997).

4.10 SPARE PARTS PROVISIONING

As discussed in Section 4.7 the renewal process for the exponential distribution turns out to be the Poisson Process, and if failure occurs in accordance with an exponential distribution, the problem of determining the required number of spare parts to guarantee a prescribed reliability is readily solved.

If one component is running and n components are on hand, the reliability at time t, $R(t)$ is the probability that the cumulative number of failures $N(t)$ is less than n. Because we are assuming the exponential distribution, it does not matter how long the component in service has been running since there is no wearout. Using the Poisson distribution of $N(t)$, the reliability function is:

$$R(t) = \sum_{i=0}^{n} \frac{e^{-\lambda t}(\lambda t)^{k}}{k!}. \tag{4.70}$$

Example

Suppose a part has an exponential life distribution with $\eta = 100$ hours and hence the failure rate is $\lambda = 0.01$/hour. It is desired that the probability of surviving a mission of 135 hours be no less than 0.95. Using $\lambda t = 0.01 \times 135 = 1.35$ as the Poisson parameter, one sums the terms until the total is greater than 0.95. The corresponding value of k is the required number of spare parts. The individual and cumulative terms as computed using the Poisson function in Excel is given below:

k	$p(k)$	$\Sigma p(k)$
0	0.25924	0.25924
1	0.349974	0.609214
2	0.236233	0.845447
3	0.106305	0.951752
4	0.035878	0.98763

It is seen that $k = 3$ satisfies the reliability requirement so $n = 3$ spares are needed.

REFERENCES

Barlow, R. and L. Hunter. 1960. Optimum preventive maintenance policies. *Operations Research* 8(1): 90–100.

Baxter, L., E. Scheuer, et al. 1981. *Renewal Tables: Tables of Functions Arising in Renewal Theory*. University of Southern California, ADA108264.

Baxter, L., E. Scheuer, et al. 1982. On the tabulation of the renewal function. *Technometrics* 24(2): 151–156.

Blischke, W.R. and D.N.P. Murthy. 1994. *Warranty Cost Analysis*. Dekker, New York.

Cramer, H. 1946. *Mathematical Methods of Statistics*. Princeton Univ Press, Princeton, NJ.

Jiang, R. 2010. A simple approximation for the renewal function with and increasing failure rate. *Reliability Engineering and System Safety* 95: 963–969.

Kececioglu, D. and F. Sun. 1997. *Burn-In Testing: Its Quantification and Optimization*. Prentice Hall, Upper Saddle River, NJ.

Lewis, E. 1987. *Introduction to Reliability Engineering*. Wiley, New York, NY.

McCool, J. 2006a. Control charts for radial error. *Quality Technology and Quantitative Management* 3: 283–292.

McCool, J. 2006b. The distribution of the present value of a pro rata warranty policy. *International Journal of Industrial Engineering* 13(4): 395.

Tadikamalla, P.R. 1980. Age replacement policies for Weibull failure times. *IEEE Transactions on Reliability* R-29(1): 88–90.

White, J. 1964. Weibull renewal analysis. General Motors Corporation Research Report GMR-597.

EXERCISES

1. Find the smallest number of components that must be in standby parallel if each component life is distributed as $W(100, 2)$ and it is desired that the reliability at 1000 hours exceed 0.9. Assume that n is large enough that the time to failure is approximately normally distributed.

2. A population is a mixture of $W(100, 1)$, $W(200, 1.5)$, and $W(150, 1.8)$ in the proportions 20%, 30%, and 50%, respectively. The scale parameters are expressed in hours. Compute the population reliability at a life of 120 hours.

3. The stress distribution Y is $W(1000, \beta)$. The strength distribution X is $W(1200, \beta)$. Find the shape parameter value for which $P(Y < X)$ is 0.9.

4. If Y is distributed as $W(100, 2)$, X is distributed as $W(120, 2)$, and Z is distributed as $W(80, 2)$, compute the probability that Y is less than both X and Z.

5. A bomb is aimed at a target with coordinates $X = 1000$, $Y = 500$. The X location actually struck is normally distributed $N(1000, 2^2)$ expressed in miles. The Y location struck is $N(500, 2^2)$, also in miles. Assuming the X and Y locations are statistically independent, find the radial distance from the desired target location that will contain 90% of the bomb strikes.

6. A tire is sold with an optional pro rata warranty of 60,000 miles and the purchase price is $120. Tire life in miles is Weibull distributed with scale parameter of 100,000 miles and shape parameter is 1.2. The buyer's utilization rate is 15,000 miles/year. Assuming a discount rate of 5%, find the 70-th percentile of the distribution of the warranty buyer's value.

7. A drill bit has a life in revolutions that follows a Weibull distribution $W(10^6, 1.8)$. It costs \$25 to replace the drill bit on a planned replacement basis. However, if it fails in service, the ancillary damage is estimated to cost a total of \$250. Find the the optimum age (in revolutions) for planned replacement of the drill bit. What is the cost in dollars per revolution for this optimum policy? What is the cost if there is no planned replacement?

CHAPTER 5

Estimation in Single Samples

5.1 POINT AND INTERVAL ESTIMATION

All of the calculations described in the preceding two Chapters are applicable when the Weibull parameters are known or assumed. The present chapter is concerned with the problem of estimating one or more parameters or some functions of the parameters from a single data sample obtained from the results of an experiment, a life test or from field data.

A distinction is made between point and interval estimation. A point estimate is a single numerical value computed from the observations in a data sample in such a way as to be a good guess at what the true but unknown parameter value is. Point estimates, being functions of the values in a random sample, are themselves random variables and will vary in successive random samples drawn from the same distribution in accordance with a probability distribution known as the sampling distribution of the estimate.

An interval estimate consists of two numbers calculated so that one is highly confident in stating that the true parameter value is contained between them, provided that the distribution sampled has the form assumed. The degree of confidence is expressed as the proportion of times that the statement would be true in similar calculations performed on an indefinitely large number of independent random samples drawn from the same population.

5.2 CENSORING

A complete or uncensored sample is one for which the exact value of the random variable is observed for each member of the sample. In a censored sample the exact value of the random variable is only bounded for some of the members of the sample. An observation is said to be right censored if it is

Using the Weibull Distribution: Reliability, Modeling, and Inference, First Edition. John I. McCool.
© 2012 John Wiley & Sons, Inc. Published 2012 by John Wiley & Sons, Inc.

known that the random variable's exact value is greater than a bounding value. This is a common occurrence in life testing when a test is terminated prior to failure for some of the specimens. For those specimens it is known only that the exact time to failure exceeds the time at which testing was suspended. In strength testing, samples will ordinarily be uncensored.

Interval censoring occurs in life testing when a life test is monitored periodically, for example, once a day, and the exact lives of any failures that are found are known only to have taken place at some time since the last inspection. Specimens found to be failed at the time of the first inspection are said to be left censored. In principle left censoring could occur in strength testing if a specimen breaks at a load that is below the smallest load that the test machine is capable of registering. The discussion that follows will be in the context of life testing.

Censoring may be by design or random. Random censoring occurs when an arbitrary mechanism other than the failure mechanism under study acts to remove an item from test. A competing failure mode is an example. In testing rolling bearings most failures occur on the more highly stressed bearing inner ring rather than on the outer ring or on one of the rolling elements. When it is necessary to test a new, expensive, or difficult to machine material, a common practice is to make the inner ring from the new material while using standard outer rings and ball or roller sets. When the test is conducted failures of the outer ring or of one or more rolling elements are regarded as censored observations with respect to the primary test element, the inner ring. Censoring occurs in biomedical studies when a subject drops out of the study or succumbs to an accident or a fatal disease other than the one under study. Random censoring also occurs when subjects enter the study at random times but the study is of fixed duration. At the end of the study surviving members will have experienced various lengths of observation or treatment.

Planned censoring may be at a predetermined life (type I) or at a predetermined number of failures (type II). Planned censoring may also be single or multiple (cf. Nelson, 1982) depending, in the case of type I censoring, upon whether all items are removed from test when a prespecified life is achieved or whether a subset of the items are removed at each of several prespecified lives. Multiple type I censoring occurs in medical studies when subjects enter the study at various times but the study concludes on a fixed date. Similarly, in multiple type II censoring, prescribed numbers of items are randomly removed from test at the occurrence of prespecified numbers of failures. Multiple type II censoring is sometimes also called progressive censoring. Sudden death testing, a term coined by Johnson (1964), is a special case of progressive censoring. In sudden death testing, a subgroup of test elements are tested together on a machine that can simultaneously accommodate a number of test elements. When the first failure occurs on each of the machines, testing is suspended on the other elements under test on that machine. The set of first failures on each machine and the subgroup size permit estimation of the parameters of the original population. The design and analysis of the results of a sudden death test are discussed in Chapter 7. In progressive testing, upon the occurrence of a failure, a preselected number of items are removed from testing. These suspensions have the same accumulated life as the associated failures but are not necessarily tested on the same test unit. Rausand and Høyland (2004) define type III censoring in which testing terminates when either a prespecified time has elapsed or a prespecified number of failures have occurred, whichever comes first. Additionally, although no one has yet named the practice, the author has witnessed occasions when, after a fixed time has elapsed, it is judged that an insufficient number of failures have occurred and so testing resumes until an acceptable number of failures do occur.

5.3 ESTIMATION METHODS

Many methods are available for estimating the parameters of the Weibull distribution (cf. Lawless, 2002; Meeker and Escobar, 1998; Nelson, 1982) each having its advantages and disadvantages.

In the selection of a method of estimation it is relevant to consider the following characteristics:

1. Applicability to censored samples
2. Precision. This refers to the scatter in the sampling distribution of the estimate; the less scatter, the greater the precision.
3. Applicability to interval estimation
4. Degree of bias. An estimator is a biased estimator if the average value of the estimate over repeated samples is not equal to the true value of the quantity being estimated.
5. Simplicity of calculation.

To give concrete meaning to these five desiderata we next discuss two estimators that are lacking in some of them. Neither is recommended but both may prove useful on occasion.

5.3.1 Menon's Method

Menon's method (Menon, 1963) is an example of an estimation method that is easy to calculate but not applicable to censored sampling. Until very recently (Phan and McCool, 2009) it was not applicable to interval estimation. It will be shown to be biased but the bias is now correctible. Interval estimation and bias correction for Menon's method is deferred until Chapter 6, where it will be shown that while generally lacking the precision of the method of the method of maximum likelihood (ML), it is surprisingly precise for some purposes.

From Section 3.3.1 the natural logarithm Z of a Weibull distributed random variable has an expected value given by:

$$E(Z) = \ln \eta - \frac{\gamma}{\beta} \tag{5.1}$$

The population variance of Z is:

$$\sigma_z^2 = \frac{\pi^2}{6\beta^2}. \tag{5.2}$$

Menon suggested equating the mean and variance of the logarithms of the observations in a complete random sample from the two-parameter Weibull distribution to the corresponding population values. Equating the sample variance of $\ln x$, that is, $s_{\ln x}^2$, to the population value and solving for the shape parameter gives Menon's estimate of the Weibull shape parameter:

$$\tilde{\beta} = \frac{\pi}{\sqrt{6} \cdot s_{\ln x}} = \frac{1.283}{s_{\ln x}} \tag{5.3}$$

Equating the population mean $E(Z)$ to the mean of the logarithms of the sample, $\overline{\ln x}$, and substituting $\tilde{\beta}$ for β leads to Menon's estimate of the scale parameter:

$$\tilde{\eta} = \exp\left[\overline{\ln x} + \frac{\gamma s_{\ln x}}{\pi} \sqrt{6} \right] = \exp[\overline{\ln x} + 0.450 s_{\ln x}]. \tag{5.4}$$

Having estimated the scale and shape parameter, they may be substituted into Equation 3.12 to produce the following estimate of the $100p$-th percentile.

$$\tilde{x}_p = (k_p)^{1/\tilde{\beta}} \cdot \tilde{\eta} \tag{5.5}$$

Example
The following uncensored sample of size $n = 10$ was randomly drawn from a Weibull population for which the true shape parameter was 1.3 and the scale parameter was 56.46. This choice made the population tenth percentile equal to 10.0:

14.01, 15.38, 20.94, 29.44, 31.15, 36.72, 40.32, 48.61, 56.42, 56.97

The mean of the natural logarithm of the data is computed to be 3.450. The standard deviation of the logarithms is 0.5061. Menon's estimate of the shape parameter is calculated to be 2.539. The estimated value of the scale parameter is 39.56. The 10th percentile estimate is 16.04 and the 50th percentile (median) is estimated to be 34.24.

To check on the precision and possible bias of Menon's method of estimation, 10,000 uncensored samples of size 10 were generated from this same Weibull population and Menon's estimates of the shape and scale parameters as well as $x_{0.10}$ were computed for each sample. The mean of the 10,000 shape parameter estimates was 1.509 and the standard deviation was 0.4930. Since by the central limit theorem the mean of a sample of size 10,000 is sure to be normally distributed, a 95% confidence interval on the true mean may be calculated as:

$$1.509 - \frac{1.96 \times 0.4930}{\sqrt{10,000}} < E(\tilde{\beta}) < 1.509 + \frac{1.96 \times 0.4930}{\sqrt{10,000}}.$$

So that with high confidence the mean of Menon's shape parameter estimate falls in the interval 1.509 ± 0.0097. Since this interval does not include the true value of 1.3, we may conclude that Menon's estimate is biased high. In a similar way, the interval for the scale parameter estimate was computed to be 57.405 ± 0.290. Again, the interval does not include the true value, 56.46, so Menon's estimate of the scale parameter is also biased high. The corresponding interval for the estimator of the 10th percentile is 12.701 ± 0.140. Since the true value of the 10th percentile was 10.0, we see that the $x_{0.10}$ estimator is likewise biased high.

5.3.2 An Order Statistic Estimate of $x_{0.10}$

An intuitive estimator of the 10th percentile in a sample of size 10 is just the smallest observation, since 90% of the sample exceeds it. Using the same 10,000 simulated samples described above, the smallest value was determined for each sample of size 10 and the 10,000 values of the first failure were regarded as the sampling distribution of this estimate of $x_{0.10}$. The average of 10,000 estimates and the width of the associated 95% confidence interval on the mean was computed to be 8.932 ± 0.135. Despite its intuitive appeal, this estimate is accordingly seen to be biased low since the interval does not contain the true value 10.0. The size of the uncertainty interval is comparable to that of Menon's estimator of $x_{0.10}$, although it is calculated from only one member (the smallest) of the sample and Menon's estimate uses all 10 values in the sample.

Histograms showing the comparative distributions of Menon's estimate of $x_{0.10}$ and the first-order statistic estimate are shown in Figure 5.1.

The histograms clearly show the bias in both estimators and that the first-order statistic estimator is much more likely to result in an underestimate of the true value. The bias in the first-order statistic estimate can be assessed analytically. The distribution of the first-order statistic in Weibull samples is

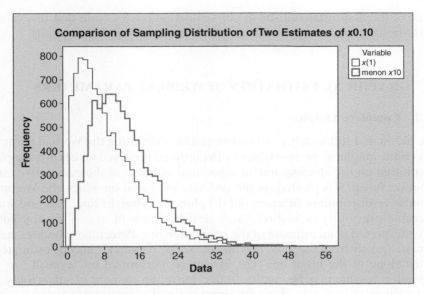

Figure 5.1 Sampling distribution of $x_{1,10}$ and Menon's estimate of $x_{0.10}$.

itself Weibull distributed as will be recalled from Section 3.2. The scale parameter is equal to the population scale parameter divided by $n = 10$ raised to the $1/\beta$ power, and the shape parameter is equal to the shape parameter of the Weibull population from which the sample was drawn. Thus, the mean of the sample first failure is:

$$E(X_1) = \frac{\eta}{10^{1/\beta}} \cdot \Gamma\left(1 + \frac{1}{\beta}\right). \tag{5.6}$$

Expressing the scale parameter in terms of $x_{0.10}$ and with $k_{0.10} = \ln(1/0.9) = 0.105361$, the mean of the first-order statistic is:

$$E(X_1) = x_{0.10} \cdot \frac{\Gamma\left(1 + \frac{1}{\beta}\right)}{(1.0536)^{1/\beta}} \tag{5.7}$$

For x_1 to be an unbiased estimate of $x_{0.10}$ it would be necessary for the quantity $\Gamma\left(1 + \frac{1}{\beta}\right)\Big/(0.10536)^{1/\beta}$ to be unity. This factor is approximately 1.0 when the true shape parameter is 0.88. For $\beta = 1.0$ it has the value 0.947. It then decreases slowly with β to a value of 0.862 at about 2.0 and rises very slowly thereafter, asymptotically approaching unity. At $\beta = 3.0$ its value is 0.878. At $\beta = 100$ it has risen to 0.994. At $\beta = 1.3$ it is 0.885, and so the expected value of the first-order statistic in our simulation is 8.85. Note that this value is within the confidence interval computed from the simulation results.

5.4 GRAPHICAL ESTIMATION OF WEIBULL PARAMETERS

5.4.1 Complete Samples

The oldest and still widely practiced means for estimating the Weibull parameters is the graphical method whereby the ordered observed values in a sample are plotted on the abscissa and an associated estimate of the cumulative distribution function is plotted on the ordinate on a grid on which the Weibull cumulative distribution function (CDF) plots as a straight line. This grid was discussed previously in Section 3.1. A straight line is fit to this set of points and interpreted as an estimate of the population line. Percentile estimates may be read directly from the fitted line and the shape parameter may be estimated as the slope of the fitted line. The fit may be performed by "eyeball," or a regression routine may be employed to remove the element of subjectivity.

As noted, the essence of the graphical method of parameter estimation is to use the sample data to construct an estimate of $F(x)$ and hence its parameters, using either Weibull graph paper as described in Chapter 3 or by transforming the data and the $F(x)$ estimates and plotting the transformed values on an ordinary linear grid.

Let $x_{(1)} < x_{(2)} < \cdots x_{(n)}$ denote the ordered observations in a complete sample of size n. Then, even though $F(x)$ is not known, we may write, since $F(x)$ is a nondecreasing function of x, $F(x_{(1)}) < F(x_{(2)}) \cdots < F(x_{(n)})$. For the i-th ordered observation in the sample, $x_{(i)}$, it is necessary to compute some estimate of the CDF evaluated at that order number. The set of values i/n $(i = 1..n)$ is known as the empirical distribution function. Its disadvantage as a plotting position is that its value is 1 or 100% for $i = n$, and this value cannot be plotted on a Weibull grid. However, just before the i-th observation the empirical distribution function is $(i - 1)/n$, and just after it, it is i/n, so some writers have proposed using the average, $(i - 1/2)/n$ as plotting position, that is, as the estimate of $F(x_{(i)})$.

Now as noted before, the quantities $F(x_{(i)})$ are an ordered set of random variables. If one took many samples, $F(x_{(i)})$ would vary from sample to sample in accordance with a probability distribution. It happens that this probability distribution, known as the beta distribution, depends as on i and n but not $F(x)$ as explained in Section 3.5. In particular the expected value of this distribution is simply $i/(n + 1)$. Some writers have therefore proposed using $i/(n + 1)$ as the plotting position on the $F(x)$ axis for $x_{(i)}$ on the strength of the argument that $F(x_{(i)})$ is equal to $i/(n + 1)$ "on the average." Kapur and Lamberson (1977) argue against this choice on the grounds that the beta distributions are skewed such that the mean will be greater than the median at early failure times and less than the median for later failure times. The net effect of using the mean

Table 5.1 Ordered Observations, Median, 5%, 95% Ranks, and Benard's Approximation

Failure Order No. (i)	Observation $X_{(i)}$	$\hat{F}(x_{(i)})$ Median Rank	$\dfrac{i-0.3}{n+0.4}$	5% Rank	95% Rank
1	14.01	0.06697	0.06731	0.00512	0.25887
2	15.38	0.16226	0.16348	0.03677	0.39416
3	20.94	0.25857	0.25962	0.08726	0.50690
4	29.44	0.35510	0.35577	0.15003	0.60662
5	31.15	0.45169	0.45192	0.22244	0.69646
6	36.72	0.54831	0.54808	0.30354	0.77756
7	40.32	0.64490	0.64423	0.39338	0.84997
8	48.61	0.74142	0.74038	0.49310	0.91274
9	56.42	0.83774	0.83654	0.60584	0.96323
10	56.97	0.93303	0.93269	0.74113	0.99488

compared with using the median would be a clockwise rotation of the fitted line. The medians of $F(x_{(i)})$ have been dubbed median ranks and are a popular choice of plotting position for Weibull data. As noted in Section 3.5, published tables of median ranks are available but they may readily be computed using statistical software or Excel to find the median of the beta distribution with appropriate parameters. An excellent numerical approximation to the i-th median rank is $(i - 0.3)/(n + 0.4)$. This approximation is attributed to Benard and Bos-Levenbach (1953). The various proposed plotting positions discussed above are all of the form $(i - a)/(n - 2a + 1)$ for $a = 0$, 0.5 and 0.3 Another plotting position that shares this property is $(i - 3/8)/(n + 1/4)$ due to Blom (1958).

Table 5.1 contains, in ascending order of magnitude, the same simulated uncensored random sample of $n = 10$ observations used in our example of Menon's method of estimation. Also shown in Table 5.1 are the true median ranks correct to five decimal places, the value of the approximation $(i - 0.3)/(n + 0.4)$, and the 5% and 95% percentiles of the rank distribution for each order number.

To within graphical plotting accuracy, the true median ranks and the approximation are indistinguishable.

To remove the subjectivity in fitting the straight line, one may formally use the method of least squares, although ordered life data are not consistent with the assumptions of the Gauss–Markov theorem. The ordered lives are correlated and their variance is not constant. Nevertheless, ordinary least squares is a reasonable, nonsubjective, way of fitting a straight line. Generalized least squares accounts for the different variances of the order statistics and the covariances among them and is the basis for the best linear unbiased estimates (BLUE) first developed for the Weibull distribution by Lieblein and Zelen

(1956). BLUE estimates require tables of coefficients which are difficult to compute and available for limited sample sizes. Their use for Weibull parameter estimation has largely been supplanted by the ML method discussed later in this chapter.

As noted in Section 3.1, the Weibull CDF transforms to linear form as:

$$\ln \ln \left(\frac{1}{1-F(x)} \right) = \beta \ln x - \beta \ln \eta. \tag{5.8}$$

For each ordered value $x_{(i)}$ in a sample of data we have the associated value of $\ln \ln \left(\dfrac{1}{1-\hat{F}(x_{(i)})} \right)$. These values should plot against $\ln(x_{(i)})$ as an approximate straight line with a slope equal to the shape parameter and an intercept equal to minus the product of the shape parameter and the log of the scale parameter. Defining:

$$Y_i = \ln \ln \left(\frac{1}{1-\hat{F}(x_{(i)})} \right). \tag{5.9}$$

and T_i as $\ln x_{(i)}$ one may use simple linear regression of Y against T to obtain a nonsubjective straight line fit. The slope of the fitted line will then be an estimate of the shape parameter and the intercept will be an estimate of $-\beta \ln(\eta)$. This will minimize the sum of the squared deviations of the points Y_i from the fitted straight line. Some argue that the squared deviations should be minimized in the other direction since the values of Y_i, being functions of the order number in the sample and the sample size, are not random. In that case T_i and Y_i are linked by the straight line equation:

$$T_i = \ln(\eta) + \frac{1}{\beta} Y_i. \tag{5.10}$$

When this choice is made, the slope of the fitted line is an estimate of the reciprocal of the shape parameter and the intercept is an estimate of $\ln(\eta)$. For the data in Table 5.1 the two fitted regression lines are

$$Y = -7.975 + 2.160T.$$

and

$$T = 3.682 + 0.4451Y.$$

The first leads to the estimates $\beta = 2.160$ and $\eta = 40.13$ and the second to $\beta = 2.246$ and $\eta = 39.13$. The associated estimates of $x_{0.10}$ are 14.158 and 14.587. Figure 5.2 shows the fitted line plot for the second of these regressions.

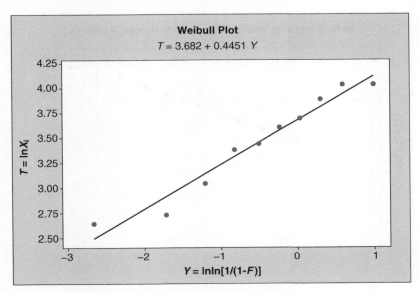

Figure 5.2 Weibull plot of sample data and fitted regression line.

In this figure the ordinate and abscissa are reversed from their usual representation on a Weibull probability plot since regression software customarily plots the dependent or response variable on the vertical axis.

In addition to the median ranks, one may also plot the transformed values of the 5% and 95% ranks to convey a sense of the uncertainty in the fit. Plotting these around the observed life values will not result in a smooth band, so the custom is to adjust the observed values so that they fall on the fitted straight line. Figure 5.3 shows the adjusted data points and the transformed values of the median, 5% and 95% ranks with the observed values shown on the abscissa consistent with custom.

Curves of this type are often used to compute a kind of confidence limit for percentiles of interest as the abscissa values where a horizontal line intersects the 5% and 95% curves. For example, recalling that the CDF evaluated at the scale parameter η, that is, $F(\eta) = 1 - 1/e$, it is clear that $\ln\ln[1/(1 - F(\eta))] = 0$. Drawing a horizontal line at the ordinate value of 0, the intersection with the median rank plot gives an abscissa value that may be considered a point estimate of the logarithm of the scale parameter η. The intersection with the other two 5% and 95% curves defines an uncertainty interval for $\ln \eta$. To within graphical accuracy these seem to be roughly 3.325 and 3.95 on the log scale, which translates to an interval of 27.8 to 51.9 in original units. This procedure is tantamount to assuming that the estimated straight line represents the true population CDF and not just an estimate of it.

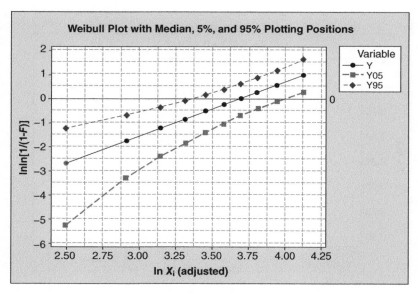

Figure 5.3 Weibull plot with approximate 90% limits.

If the location parameter γ is not zero, that is, the sample is drawn from the three parameter Weibull distribution, it is possible to estimate the location parameter graphically by iteratively assuming a value of γ, subtracting it from each data point, and constructing a probability plot, then revising γ and replotting until there is no obvious systematic curvature in the plot. When the trial value of γ is too small, the Weibull plot appears to be a concave function. The concavity of the plotted data when sampling from a three-parameter Weibull distribution is the basis for an analytic approach to estimating the location parameter that will be discussed in Chapter 10.

5.4.2 Graphical Estimation in Censored Samples

When the censoring is simple type I or type II censoring, probability plotting can proceed as above, except that the lives associated with the unfailed elements are unknown and hence cannot be plotted. One can plot the failures that have occurred using their order numbers in the complete sample and perform the fitting operation using just these points. However, when samples have progressive or random censoring some suspensions will occur between other failures, causing ambiguity in the order number applicable to the failures. Some methods are in existence for such cases notably those of Kaplan and Meier (1958), Johnson (1964), and Nelson (1969). They each involve plotting only failures at plotting positions determined by an algorithm specific for each author's method. The method due to Nelson is simplest to use and will be illustrated here. When applied to the uncensored case, it results in larger values of the plotting positions than median ranks and thus Weibull quantiles estimated by fitting in this way will be systematically smaller than those estimated using median rank plotting positions.

It is recalled from Equation 2.50 of Section 2.8.1 that for any distribution, the cumulative hazard function is related to the CDF as follows:

$$\Lambda(x) = -\ln(1 - F(x)) \tag{5.11}$$

The relationship between the cumulative probability $F(x)$ and the cumulative hazard $\Lambda(x)$ for a distribution can thus be rewritten as:

$$F(x) = 1 - \exp[-\Lambda(x)]. \tag{5.12}$$

where $F(x)$ represents the probability that a randomly selected item will fail prior to life x.

In Nelson's method, estimates of $\Lambda(x)$ constructed from a data sample are plotted against x on a special grid devised to display the theoretical relationship as a straight line. A straight line is then fitted to the test data and used as an estimate of $\Lambda(x)$. The CDF $F(x)$ may then be estimated using Equation 5.12.

For the Weibull distribution, the cumulative hazard function is related to the parameters as follows:

$$\Lambda(x) = \left(\frac{x}{\eta}\right)^{\beta}. \tag{5.13}$$

Taking logarithms in Equation 5.13 and rearranging gives the following relationship between the logarithm of life and the logarithm of the cumulative hazard function:

$$\ln(x) = \frac{1}{\beta}\ln(\Lambda(x)) + \ln(\eta). \tag{5.14}$$

$\ln(x)$ is a thus a linear function of $\ln(\Lambda(x))$. Weibull hazard paper in which estimated values of $\Lambda(x)$ plot linearly against x is seen to be simply log-log graph paper.

Given a sample of size n and denoting the ordered times to failure or test suspension $x_{(1)} < x_{(2)} < \ldots x_{(k)} < \ldots x_{(n)}$, one may form a stepwise continuous estimate of the hazard function. If $x_{(k)}$ is a failure time, then, just prior to time $x_{(k)}$ there are $n - k + 1$ items that have not yet failed or been suspended. Let Δx_k denote the time until the next failure. Further assume that any suspensions that take place during this interval occur at the end of the interval. With this assumption we may state that the number at risk over the interval is $n - k + 1$. Of these, only one fails. The failure rate measured as failures per unit time per number of items at risk can be expressed over the time between the failure at $x_{(k)}$ and just prior to the next failure as:

$$\hat{\lambda}(x) = \frac{1}{(n-k+1)\Delta x_{(k)}}; \; x_{(k)} < x < x_{(k)} + \Delta x_{(k)}. \qquad (5.15)$$

The integrated hazard $\Lambda(x)$ over that same interval may be approximated by summing the areas of the rectangles that form the stepwise approximation, over the intervals corresponding to ordered failure times $x_{(j)}$ for which $j \leq k$.

$$\hat{\Lambda}(x) = \sum_{j=1}^{k} \hat{\lambda}(x_j)\Delta x_j = \sum_{j}^{k} \frac{1}{n-j+1}; \; x_{(k)} < x < x_{(k)} + \Delta x_{(k)}. \qquad (5.16)$$

The sum takes place over the values of j corresponding to the order numbers of just the failures in the ordered list of failures and suspensions. Only failures contribute to the sum, but suspensions affect the amount of their contribution through their effect on the value of j that each failure assumes.

The CDF may be estimated as:

$$\hat{F}(x) = 1 - \exp[\hat{\Lambda}(x)]. \qquad (5.17)$$

The computations for Nelson's method is best illustrated by an example. Column 1 of Table 5.2 shows the sorted lives obtained in a life test of 10 specimens in which an extraneous failure mode claimed six of the test specimens before they could fail by the mode of interest. The lives of these specimens at the point of test suspension is indicated by the letter (S). The lives of the test elements that failed due to the primary mode are indicated by an (F).

The next column in Table 5.2, headed "Reverse Rank," contains the order number of each failure or suspension in reverse order, for example, the small-

Table 5.2   Hazard and $K - M$ Computations

Life	Reverse Rank	Hazard – λ	Cumulative Hazard Λ	$F(x) = 1 - e - \Lambda$	$K - M$
0.569 S	10	–	–	–	–
8.91 F	9	0.1111	0.1111	0.1052	0.111
21.41 S	8	–	–	–	–
21.96 F	7	0.1429	0.2540	0.2243	0.238
32.62 S	6	–	–	–	–
39.29 F	5	0.2000	0.4540	0.3649	0.390
42.99 S	4	–	–	–	–
50.40 F	3	0.3333	0.7873	0.5449	0.593
53.27 S	2	–	–	–	–
102.6 S	1	–	–	–	–

est member of the sample is assigned the value 10. For the j-th life, be it a failure or a suspension, the reverse rank's value is $n - j + 1$.

The entries in the third column, headed "Hazard," are the reciprocals of the corresponding reverse rank entries and are only computed for the lives labeled (F), that is, for the primary failure mode. Each entry in the fourth column, labeled "Cumulative Hazard," and designated by the symbol Λ, is the sum of the associated entry in the third column and all other entries above it in the third column. The fifth and final column is the plotting position F obtained from the associated value of Λ by the transformation $1 - e^{-\Lambda}$. These values of F may then be plotted against the primary mode failure lives on an ordinary Weibull grid. The fitted straight line represents the graphical estimate of the two-parameter Weibull distribution of primary mode life.

Another approach to nonparametric estimation of the CDF in the presence of multiple censoring is due to Kaplan and Meier (1958). Their method is easily explained with reference to Table 5.2. Prior to the first failure at a life of 8.91 there were nine items at risk since one had been withdrawn at life 0.569. Of these nine, one failed and eight survived beyond life 8.91. Therefore, 8/9 is the estimated reliability at life = 8.91. The corresponding estimate of the probability of failing prior to 8.91 is $F(8.9) = 1 - 8/9 = 0.1111$. This remains the estimated value of F until another failure occurs. At the second failure at life 21.96 there were seven items at risk of which six survived beyond life 21.96. The conditional reliability of surviving beyond a life of 21.98 given survival up to that point is therefore estimated as 6/7. The overall estimated reliability of surviving beyond a life of 21.96 is the product of surviving beyond 8.91 times the probability of surviving beyond 21.96 conditional on having survived beyond 8.91. This is the product $6/7 \times 8/9 = 0.7619$. The estimated CDF at 21.96 is therefore $1 - 0.7619 = 0.2381$. The rest of the table follows in a similar way. Typically the $K - M$ estimates are not much different from Nelson's. Like Nelson's, the $K - M$ estimates are easier to illustrate than to write in formal notation. The usual formal notation is to define a binary variable for each item in the ordered list of failures and suspensions as $\delta_j = 0$ for a suspension and 1 for a failure. The CDF at the k-th failure is then expressible as:

$$F(x_k) = 1 - \prod_{j=1}^{k} \left(\frac{n-j}{n-j+1} \right)^{\delta_j}. \tag{5.18}$$

For the censored observations the contribution to the product is 1.0 so only the failed items affect the computation.

The $K - M$ estimates may be used in the usual way in graphical estimation of the CDF. For the cumulative hazard estimates we need to fit a straight line to the linear relation between $\ln \Lambda(x)$ and $\ln x$.

As with ordinary graphical estimation one may use least squares as a fitting procedure to remove the subjectivity of fitting a line to the data points. Minimizing the scatter of the life values about the fitted line, we regress the natural

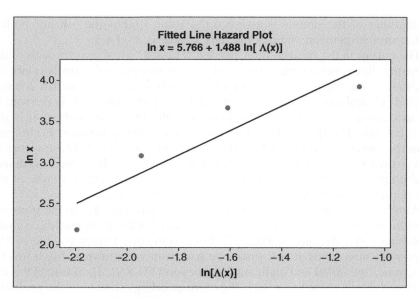

Figure 5.4   Regression fit to Weibull plot.

logarithm of life upon the logarithm of the estimated cumulative hazard function.

The intercept of the fitted straight line will be an estimate of $\ln(\eta)$ and the slope of the fitted straight line will be an estimate of $1/\beta$. The fitted straight line and the data points are shown in Figure 5.4.

The shape parameter is therefore estimated to be $1/1.488 = 0.672$ and the scale parameter estimate is $\exp(5.766) = 319.3$.

Romanowski (1998) has studied the behavior of the estimates obtained by hazard plotting for the case where primary and secondary Weibull failure modes act as competing risks to produce censored observations. For example, choosing a sample size of $n = 10$, a total of 10 observations are generated by simulation from the primary failure mode Weibull distribution having shape parameter β_1 and scale parameter η_1. Ten additional observations are then generated from the secondary Weibull distribution having shape and scale parameters of β_2 and η_2. The failure lives of the two modes are randomly paired as shown in Table 5.3. If the primary mode life is smaller, it is taken as the failure time. If the secondary mode distribution yields the smaller life, it is taken as the censoring life. The hazard plotting procedure is then conducted as usual. It was necessary to restrict the study to cases wherein at least three primary mode failures occurred because the estimates became unstable with only two.

For this example, the line fit by regressing $\ln(x)$ against $\ln\Lambda(x)$ had a slope of 0.9853 and an intercept of 0.2416. The computation of the scale and shape

Table 5.3 Hazard Calculations

x_1	x_2	Time to Failure = Minute(x_1, x_2)	Mode	Reverse Rank	Hazard Values $\lambda(x)$	Cumulative Hazard Value Mode 1 $\Lambda(x)$
0.35434	0.02969	0.02969	2	10	0.100000	
1.09545	0.04827	0.04827	2	9	0.111111	
0.19065	0.29876	0.19065	1	8	0.125000	0.125000
0.31964	1.35455	0.31964	1	7	0.142857	0.267857
0.38159	0.98788	0.38159	1	6	0.166667	0.434524
1.86767	1.00301	1.00301	2	5	0.200000	
1.20585	1.94545	1.20585	1	4	0.250000	0.684524
1.42434	1.25575	1.25575	2	3	0.333333	
1.83543	1.35151	1.35151	2	2	0.500000	
2.02333	1.86398	1.86398	2	1	1.000000	

parameter estimates in terms of the estimated slope and intercept of the fitted line are as follows:

$$\hat{\beta} = \frac{1}{b} = \frac{1}{0.9853} = 1.015; \quad \hat{\eta} = \exp(0.2416) = 1.273.$$

The, 10th percentile is computed as:

$$x_{0.10} = 1.273 \cdot \left[\ln\left(\frac{1}{0.90} \right) \right]^{1/1.015} = 0.1387.$$

Some findings of the study were as follows:

1. When the target failure mode is rare, simulations exhibit wide scatter at small sample sizes due to a few occasional extremely large values.
2. For sample sizes ≥30 and primary mode fraction ≥50% the shape parameter is underestimated by about 3% and the scale parameter is biased low by 1.5–2%.
3. For $n = 50$ and equal scale parameters there is a negligible effect on the shape and scale parameters when the primary failure mode distribution has unit shape parameter as the competing mode shape parameter value varies from 0.5 to 3. Over that range the proportion of mode 1 failures increases from 45% to 57%.

5.5 MAXIMUM LIKELIHOOD ESTIMATION

The method of ML is an estimation method due to Fisher (1934) that has the properties of unbiasedness and minimum variance in large samples. Loosely speaking the method of maximum likelihood yields estimates of the distribution parameters that make the probability of occurrence of the observed sample the largest.

Given a complete (uncensored) random sample $x_1, x_2, \ldots x_n$, the likelihood function is the product of the density function evaluated at each value of the observed data:

$$L = f(x_1) \cdot f(x_2) \cdots f(x_n). \tag{5.19}$$

With the sample in hand the values of the x's are known so L depends only on the unknown parameters of the distribution function. For the two-parameter Weibull distribution the values of the parameters that make the likelihood a maximum need to be found numerically. On the other hand, for the normal distribution, the ML estimates of the mean and variance are just the usual sample mean and variance. In general, under certain regularity conditions, the parameter values that maximize the likelihood function may be found by solving the system of equations that result from differentiating the likelihood function with respect to each of the parameters and setting the derivatives equal to zero. Given k parameters, $\alpha_1, \alpha_2 \ldots \alpha_k$, their ML estimates are the solution of the k equations that result when the derivatives of L with respect to each parameter are equated to zero. That is,

$$\frac{\partial L}{\partial \alpha_1} = \frac{\partial L}{\partial \alpha_2} = \cdots \frac{\partial L}{\partial \alpha_k} = 0. \tag{5.20}$$

When the sample is right censored and the number of failures is designated r and the number of censored observation is $n - r$, it is convenient to renumber the observations so that the first r, $x_1, x_2, \ldots, x_r$, represent the failure times and the remaining $x_{r+1}, x_{r+2}, \ldots x_n$ are the times at censoring. With this notational convention, the likelihood function may be written as:

$$L = C \prod_{i=1}^{r} f(x_i) \cdot \prod_{i=r+1}^{n} [1 - F(x_i)]. \tag{5.21}$$

The constant C varies with the type of censoring but is independent of the parameters of the distribution and not relevant to finding the maximizing values of the parameters. A failure contributes a term $f(x)$ evaluated at the failure life. A censored observation contributes a term $[1 - F(x)]$ evaluated at the censoring time. It is generally convenient to deal with the logarithm of L rather than L directly. The parameter values that maximize $\ln L$ also maximize L. Designating the natural logarithm of L by l, it may be expressed as:

$$\ln L = l = \ln C + \sum_{i=1}^{r} \ln[f(x_i)] + \sum_{i=r+1}^{n} \ln[1 - F(x_i)]. \tag{5.22}$$

5.5.1 The Exponential Distribution

In the special case that $\beta = 1$ the Weibull distribution reduces to the exponential distribution. The exponential distribution is very prominent in reliability engineering and merits discussion in its own right and not just as a special case of the Weibull distribution. Accordingly in this section we will alter notation and use t in lieu of x and denote the scale parameter by θ instead of η.

The pdf and CDF are therefore written:

$$f(t) = \frac{1}{\theta} e^{-\frac{t}{\theta}}. \tag{5.23}$$

$$F(t) = 1 - e^{-\frac{t}{\theta}}. \tag{5.24}$$

Using these expressions appropriately evaluated at observed lifetimes or at observed censoring times, the log likelihood becomes after simplifying:

$$l = \ln C - r \ln \theta - \frac{1}{\theta} \sum_{i=1}^{r} t_i - \frac{1}{\theta} \sum_{i=r+1}^{n} t_i. \tag{5.25}$$

Note that the failure lives and the lives at censoring may be combined into a single summation. With that simplification, taking the derivative with respect to θ results in:

$$\frac{d \ln L}{d\theta} = \frac{-r}{\theta} + \frac{1}{\theta^2} \sum_{i=1}^{n} t_i = 0.$$

The solution for θ is denoted by a caret overstrike to indicate that it is the ML estimate of θ:

$$\hat{\theta} = \frac{\sum_{i=1}^{n} t_i}{r}. \tag{5.26}$$

This result applies no matter what the censoring mode. The numerator of this expression is termed the total time on test and is often represented symbolically as TTT.

5.5.2 Confidence Intervals for the Exponential Distribution— Type II Censoring

When testing is continued until the r-th failure occurs, Epstein and Sobel (1953) found that the quantity $Y = \dfrac{2r\hat{\theta}}{\theta} = \dfrac{2 \times TTT}{\theta}$ will vary from sample to sample in accordance with a chi-square distribution having $2r$ degrees of freedom. The random variable Y is said to be a pivotal quantity; it contains the actual parameter value as well as the ML estimate of θ, but its distribution depends only on the number of failures and not on the parameter itself. Confidence intervals on θ may be computed using appropriate percentage points of the chi-square distribution with $2r$ degrees of freedom. For example, with $r = 10$, tables of percentage points for the chi-square distribution tell us that if Y follows a chi-square distribution with $2r = 20$ degrees of freedom, the following probability statement is true:

$$\text{Prob}[9.591 < Y < 34.170] = 0.95.$$

9.591 is the 2.5th percentile of the chi-square distribution with 20 degrees of freedom and 34.170 is the 97.5th percentile. There is thus a combined 5% chance of Y being above the upper or below the lower limit. Thus, the probability that a chi-square variable with 20 degrees of freedom will be observed to fall between these two values is 95%. Substituting for Y in terms of θ and its ML estimate,

$$\text{Prob}\left[9.591 < \frac{2r\hat{\theta}}{\theta} < 34.170\right] = 0.95.$$

The two inequalities in this expression can be solved for θ and leads to the following two-sided 95% confidence interval statement:

$$\frac{2r\hat{\theta}}{34.170} < \theta < \frac{2r\hat{\theta}}{9.591}.$$

Using the uncensored data in Table 5.1 and treating it as if it came from an exponential distribution, the ML estimate of θ is:

$$\hat{\theta} = \frac{\sum_{i=1}^{10} t_i}{10} = \frac{349.96}{10} = 35.0.$$

The 95% confidence interval for θ is therefore:

$$20.48 = \frac{2 \times 10}{34.170} \times 35 < \theta < \frac{2 \times 10}{9.591} \times 35 = 72.98.$$

In a given sample this statement will be either true or false; that is, the true value of θ will either fall between these two limits or it will not. However, in an indefinitely large number of statements made from a large set of samples of the same size, 5% of the statements will be wrong; the true θ will *not* be within the calculated limits. The error rate is thus 0.05. When data are simulated from a known population, the truth of a given confidence interval statement will be known. This is not so when the data come from testing rather than from simulation. It should therefore be borne in mind when declaring that an unknown parameter lies within calculated bounds that there is a risk, controllable by the experimenter, that the statement is wrong.

In general the expected fraction of incorrect intervals is designated by the Greek letter α. The two values of chi-square are chosen so that there is a chance of $\alpha/2$ of exceeding the upper and $\alpha/2$ of falling below the lower. The resultant confidence interval is said to be a $100(1 - \alpha)\%$ interval. A general $100(1 - \alpha)\%$ interval for the exponential parameter θ is expressed as follows:

$$\left[\frac{2r}{\chi^2_{1-\alpha/2,2r}}\right]\hat{\theta} < \theta < \left[\frac{2r}{\chi^2_{\alpha/2,2r}}\right]\hat{\theta}. \tag{5.27}$$

$\chi^2_{1-\alpha/2,2r}$ denotes the upper $100(1 - \alpha/2)$ percentile and $\chi^2_{\alpha/2,2r}$ the lower $100(\alpha/2)$ percentile of the chi-square distribution with $2r$ degrees of freedom.

The experimenter controls the error rate α and can set it to a smaller value, say 0.01, to lower the risk of making an incorrect statement. The result will be a wider set of limits having a greater chance of enclosing the true value of the population parameter. Regrettably, if the experimenter is unwilling to take any risk at all and so sets $\alpha = 0$, the limits become infinitely wide. Thus, to make any kind of meaningful statement, some risk must be tolerated. It should be borne in mind that the error rates associated with confidence intervals do not account for the error that stems from misidentifying the distribution. In the example above we used data from a known Weibull distribution to illustrate computations applicable to data from an exponential distribution. In practice one assumes a distribution form based on history or policy and perhaps verifies that the data do not contradict the choice using some kind of goodness-of-fit test. There is rarely any certainty that the distribution chosen is the correct one. Goodness-of-fit tests will be considered in Chapter 6.

Sometimes only one-sided statements are necessary. If, for example, a lower 95% confidence limit was wanted, one could write:

$$\left[\frac{2r}{\chi^2_{0.95,2r}}\right]\hat{\theta} < \theta. \tag{5.28}$$

In our numerical example, this leads to:

$22.3 = \left(\dfrac{20}{31.41}\right) \cdot 35 < \theta$. This is often the case in reliability where a customer may want to know with a high level of confidence that a certain low percentile of a life distribution exceeds a minimum acceptable value.

Since the $100p$-th percentile for the exponential distribution is equal to $k_p\theta$, a confidence interval for θ can be multiplied by k_p to compute a confidence limit for x_p. Thus, in the numerical example above, a two-sided 95% interval for $x_{0.10}$ is:

$$2.16 = 0.105361 \times 20.48 < x_{0.10} < 0.105361 \times 72.98 = 7.69.$$

The estimated reliability at a given life x may be computed as:

$$\hat{R}(x) = \exp[-x/\hat{\theta}].$$ (5.29)

For the example above, the estimated probability of surviving a life of $x = 50$ is:

$$\hat{R}(50) = \exp\left[-\frac{50}{35}\right] = 0.240,$$

Substituting a lower confidence limit for θ will give a corresponding lower limit for $R(x)$. In our example, a lower 95% confidence interval for $R(50)$ is:

$$R_{0.95}(50) = \exp\left[-\frac{50}{22.3}\right] = 0.106.$$

The failure rate for the exponential is the reciprocal of the scale parameter and so the ML estimate of the failure rate is the reciprocal of the ML estimate of the scale parameter. Thus, in the example above, $\hat{\lambda} = \frac{1}{35} = 0.0286$.

Because a large scale parameter means a small failure rate, the confidence intervals for failure rate are the reciprocals of the corresponding limits for θ but with the upper and lower ends interchanged. Thus, for the example:

$$0.0137 = \frac{1}{72.98} < \lambda < \frac{1}{20.48} = 0.0488.$$

In type II censoring the ML estimate of the mean is unbiased. This follows from the fact that the ML estimate can be expressed as:

$$\hat{\theta} = \frac{Y\theta}{2r}.$$

where Y follows the chi-square distribution with $2r$ degrees of freedom. Taking expected values and using the fact that the expected value of a chi-square variable is just its degrees of freedom shows that $E(\hat{\theta}) = \theta$.

5.5.3 Estimation for the Exponential Distribution—Interval Censoring

Data that results from interval censoring is sometimes called grouped data. It arises when an ongoing life test is monitored periodically and the number of failures since the last inspection is recorded. The intervals do not have to be the same length but ordinarily will be. The likelihood function for interval data is easy to write. As an illustration consider what would have resulted if the life test in Table 5.1 had been monitored every 15 hours. The table below shows the interval data that would have resulted.

Time Interval	No. of Failures
0–15	1
15–30	3
30–45	3
45–60	3

The likelihood function is the product of the probability of having these numbers of failures occur in each interval as a function of the scale parameter θ.

$$L(\theta) = F(15) \times [F(30) - F(15)]^3 \times [F(45) - F(30)]^3 \times [F(60) - F(45)]^3.$$

Had there been an open ended interval on the right, say >60, a further term $R(60) = [1 - F(60)]$ would have been included for every failure occurring beyond 60. The function F is the CDF:

$$F(t) = 1 - \exp\left[-\frac{t}{\theta}\right].$$

So the likelihood is a function of θ and the maximizing value may be found graphically or by the use of optimization software. It is generally easier to solve if one uses the logarithm of the likelihood:

$$l(\theta) = \ln F(15) + 3\ln[F(30) - F(15)] + 3\ln[F(45) - F(30)] + 3\ln[F(60) - F(45)].$$

The maximum value of $l(\theta)$ was found to be -18.249 and occurred at $\hat{\theta} = 33.949$. A plot of the logarithm of the likelihood function versus θ is shown in Figure 5.5.

Approximating the data by the midpoints of the intervals results in: 7.5, 22.5, 22.5, 22.5, 37.5, 37.5, 37.5, 52.5, 52.5, and 52.5. Using these values as if they were exact data gives the estimate:

$$\hat{\theta} = \frac{1}{10} \sum_{i=1}^{10} x_i = 34.5.$$

The value computed in this manner is at a good starting value to begin the iterative search for the ML estimate and may suffice in itself. Inference based on grouped data has not been widely studied. If the intervals are not too wide one might base confidence intervals and tests on the use of the midpoints as if they were exact data.

5.5.4 Estimation for the Exponential Distribution—Type I Censoring

It will be recalled that in type I censoring n items are tested for a fixed time t_0. One then observes the number of failures r and their associated lives. The

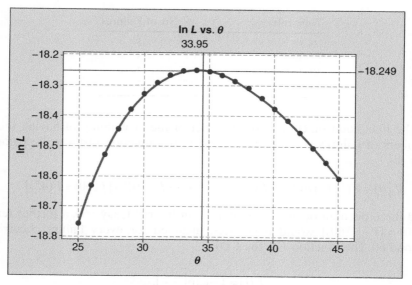

Figure 5.5   Log likelihood versus θ for an interval censored sample.

expression for the ML estimate of θ remains the same in this case, but both the number of failures r and the total time on test are random. Bartholomew (1963) has found the distribution of the ML estimate in this case to be a weighted sum of chi-square integrals and too complex for routine use. He offers two not particularly simple approximate solutions. An exact solution for the type I censored case is possible if the test is conducted in a specific way: Put a single item on test and replace it when it fails until the fixed time t_0 elapses. The total time on test will be t_0 and r will be the observed number of failures. Alternatively, if there are n test stands available, put an item on each and replace all failures as they occur. In this case the total time on test will be nt_0 and the number of failures r will be the total number of failures observed. In this case the number of failures will follow the Poisson Process with parameter nt_0/θ and the two sided $100(1 - \alpha)\%$ confidence limits for θ are:

$$\frac{2r\hat{\theta}}{\chi^2_{1-\frac{\alpha}{2},2r+2}} < \theta < \frac{2r\hat{\theta}}{\chi^2_{\frac{\alpha}{2},2r}}. \qquad (5.30)$$

This confidence interval is similar to the interval for type II censoring, except that the degrees of freedom is $2r + 2$ for the lower confidence limit. Many writers suggest this as an approximation for the type I censored case even when the Poisson protocol is not followed. Bartholomew offers two examples of data from type I censored experiments and applies his two, more complex, approximate solutions to each. In one example a considerable proportion (15/20) of the specimens failed prior to the time t_0. In the second example only 10 of 40 failed prior to t_0.

Example 1

Twenty items are observed for a time period $t_0 = 150$ hours and 15 fail at the following lives: 3, 19, 23, 26, 27, 37, 38, 41, 45, 58, 84, 90, 99, 109, and 138. The ML estimate of θ is 105.7.

Ninety-five percent limits calculated using the approximation of Equation 5.30 are:

$$64.1 = \frac{2 \times 15 \times 105.7}{49.48} < \theta < \frac{2 \times 15 \times 105.7}{16.791} = 188.9$$

Bartholomew's limits using his two approximations are (69, 184) and (68, 187). The Poisson approximation appears to be conservative (limits are wider) compared with Bartholomew's approximations and not substantially different.

Example 2

In Bartholomew's second example 40 items were tested for $t_0 = 300$ hours and 10 failed at the following lives: 6, 23, 31, 91, 94, 102, 119, 160, 170, and 241. The ML estimate of θ is 1003.7. A 95% confidence interval computed from Equation 5.30 is:

$$545.8 = \frac{2 \times 10 \times 1003.7}{36.781} < \theta < \frac{2 \times 10 \times 1003.7}{9.591} = 2093.$$

Bartholomew's two approximations gave (570, 1862) and (558, 1906). Again, the Poisson Process approximation is seen to be conservative.

5.5.5 Estimation for the Exponential Distribution— The Zero Failures Case

When a type I censored test results in zero failures the lower confidence limit in Equation 5.30 may still be calculated since the chi-square percentage point has 2 degrees of freedom when $r = 0$. Although the ML estimate of θ cannot be calculated, the product of r and $\hat{\theta} = TTT = nt_0$ is known. The result is further simplified by the fact that $\chi^2_{2,1-\alpha} = -2\ln(\alpha)$. Hence, a lower $100(1 - \alpha)\%$ limit for θ may be written as:

$$\frac{nt_0}{-\ln \alpha} < \theta. \tag{5.31}$$

This is readily converted to a lower bound on a percentile or the reliability at a prescribed life. As an example consider Bartholomew's second example above but suppose that $t_0 = 5$ hours instead of 300 hours. In that case there would have been no failures and the total time on test would be $40 \times 5 = 200$ hours. The lower 95% limit for θ would be:

$$66.8 = \frac{200}{-\ln(0.05)} < \theta.$$

Although the ML estimate of θ is undefined when $r = 0$, some use the lower 50% confidence interval as a way of producing some kind of estimate. In the present case that would be:

$$288.5 = \frac{200}{-\ln(0.50)} < \theta.$$

Equation 5.31 is sometimes used as the basis for a zero failures reliability demonstration test. As an example consider the problem of demonstrating with 95% confidence that the mean of an exponential distribution exceeds 1000 hours by means of a test of duration t_0 in which no failures occur. From Equation 5.31 this implies that

$$1000 = \frac{nt_0}{-\ln(0.05)} < \theta$$

So that

$$nt_0 = 2995.7$$

Any combination of n and t_0 having a product of 2995.7 will suffice with the proviso that n must be an integer. If, say, it were desired to run the test for 300 hours a total of 10 specimens would need to endure without failure to validate the claim.

5.6 ML ESTIMATION FOR THE WEIBULL DISTRIBUTION

5.6.1 Shape Parameter Known

It is recalled that the transformation $t = x^\beta$ applied to a Weibull random variable produces an exponential random variable whose scale parameter θ is equal to η^β. Estimation and hypothesis testing methods applicable to the exponential distribution may be applied to the transformed Weibull data when the shape parameter is believed to be known.

The ML method of estimation for censored exponential data leads to the following estimate of the exponential scale parameter:

$$\hat{\theta} = \frac{\sum_{i=1}^{n} t_i}{r} = \frac{\sum_{i=1}^{n} x_i^{\beta}}{r}. \tag{5.32}$$

Where $\sum_{i=1}^{n} t_i$ is the total transformed time on test summed over the life of both failures and suspensions and r is the number of failures. As before, this result applies no matter what censoring type gave rise to the data. Note that there is no specific dependence on sample size, only on the number of failures and the total time on test. Thus it does not matter if 10 items are tested until all fail or if 20 items are tested until the first 10 fail. If the total time on test happens to be the same so will the estimate of θ. The ML estimate of the Weibull scale parameter follows from the fact that ML estimates of functions are those functions of the ML estimates. Thus, since the scale parameter of the transformed Weibull is η^{β}, the ML estimate of the Weibull scale parameter is:

$$\hat{\eta} = \left(\frac{\sum_{i=1}^{n} t_i}{r} \right)^{1/\beta} = \left(\frac{\sum_{i=1}^{n} x_i^{\beta}}{r} \right)^{1/\beta}. \tag{5.33}$$

The estimate of the 100 p-th percentile is computable as:

$$\hat{x}_p = (k_p)^{1/\beta} \cdot \hat{\eta}. \tag{5.34}$$

For the data in Table 5.1, recalling that it was generated from a Weibull population having $\beta = 1.3$, we may estimate the scale parameter as:

$$\hat{\eta} = \left(\frac{1053.69}{10} \right)^{1/1.3} = 35.90.$$

The estimated value of $x_{0.10}$ is:

$$\hat{x}_{0.10} = 35.90 \cdot (0.105361)^{1/1.3} = 6.368.$$

5.6.2   Confidence Interval for the Weibull Scale Parameter—Shape Parameter Known, Type II Censoring

Although the ML estimate is the same regardless of the censoring mode, the computation of confidence limits is dependent on the type of censoring. Exact confidence limits can be computed for complete or type II censored samples.

For type II censoring at the r-th failure, Epstein and Sobel have shown that $\dfrac{2r\hat{\theta}}{\theta}$ follows a chi-square distribution with $2r$ degrees of freedom and thus so does $2r \cdot \left(\dfrac{\hat{\eta}}{\eta} \right)^{\beta}$ and, for that matter, $2r \left(\dfrac{\hat{x}_p}{x_p} \right)^{\beta}$. Harter and Moore (1965) use this fact to show that the ML estimate of the scale parameter or a percentile is biased but that the bias may be corrected by multiplying the raw estimate by the factor $C(r, \beta)$ defined below.

$$C(r, \beta) = \frac{r^{\frac{1}{\beta}}\Gamma(r)}{\Gamma\left(r + \frac{1}{\beta}\right)}. \tag{5.35}$$

For $\beta = 1.0$, this factor is 1.0 for any value of r, indicating, as we have shown, that the ML estimate of the exponential scale parameter is unbiased. For $\beta = 1.3$ and $r = 10$, the bias correction factor is 1.009. So the bias is negligible in this case.

With $2r = 20$ a 95% probability interval for $20\left(\dfrac{\hat{x}_p}{x_p}\right)^{\beta}$ is:

$$\text{Prob}\left[9.591 < 20\left(\frac{\hat{x}_p}{x_p}\right)^{\beta} < 34.170\right] = 0.95.$$

This leads, after some simplification, to the 95% confidence interval for x_p.

$$\left(\frac{20}{34.17}\right)^{1/\beta}\hat{x}_p < x_p < \left(\frac{20}{9.591}\right)^{1/\beta}\hat{x}_p.$$

For the example, a 95% confidence interval for $x_{0.10}$ is:

$$4.217 = 0.6623 \cdot 6.368 < x_{0.10} < 1.76 \cdot 6.368 = 11.20.$$

As noted previously, on the basis of a specific sample, a confidence interval will be either true or false; that is, the true value of x_p will either fall between the two limits or will not. However, in an indefinitely large number of statements made from a large set of samples of the same size, 5% of the statements will be wrong; the true x_p will *not* be within the calculated limits. The error rate is thus 0.05. As it happens the sample used here was generated from a population for which the true value of $x_{0.10}$ was 10.0 and so the computed confidence interval in this case is a correct statement.

A general $100(1 - \alpha)\%$ interval is expressed as follows:

$$\left[\frac{2r}{\chi^2_{1-\alpha/2}}\right]^{1/\beta}\hat{x}_p < x_p < \left[\frac{2r}{\chi^2_{\alpha/2}}\right]^{1/\beta}\hat{x}_p. \tag{5.36}$$

$\chi^2_{1-\alpha/2}$ denotes the upper $100(1 - \alpha/2)$ percentile and $\chi^2_{\alpha/2}$ the lower $100(\alpha/2)$ percentile of the chi-square distribution with $2r$ degrees of freedom.

A one-sided lower 95% confidence interval for x_p may be computed as:

$$\left[\frac{2r}{\chi^2_{0.95,2r}}\right]^{1/\beta} \hat{x}_p < x_p. \tag{5.37}$$

Similarly an upper 50% confidence interval could be written:

$$\hat{x}_p < \left[\frac{2r}{\chi^2_{0.50,2r}}\right]^{1/\beta} \hat{x}_p. \tag{5.38}$$

This expression suggests that the right-hand side is a median unbiased estimate of x_p. That is, there is a 50% chance that the modified estimate:

$$\hat{x}'_p = \left[\frac{2r}{\chi^2_{0.50,2r}}\right]^{1/\beta} \hat{x}_p. \tag{5.39}$$

will exceed the true value.

For the same 10,000 simulated samples as used previously in the evaluation of Menon's method and the first-order statistic estimate of $x_{0.10}$, ML estimates of $x_{0.10}$ were computed from Equation 5.34, although the exact sampling distribution of the ML estimate could have been established analytically in terms of the chi-square distribution as outlined above. For this case the bias correction factor is negligibly different from unity and so was not applied. The means, standard deviations, minima, maxima, and medians are tabled below:

Estimator	Mean	Std. Dev.	Minimum	Median	Maximum
Menon	12.701	7.140	0.397	11.434	21.0572
$X_{1:10}$	8.9474	6.9507	0.0127	7.1939	55.1643
ML	9.9447	2.4286	3.4347	9.7805	21.0572

The ML estimate has a substantially lower standard deviation and the mean and median are both closer to the true value of 10.0 than either of the other two estimators. The histogram below in Figure 5.6 substantiates the superiority of the ML estimate. Its distribution is very nearly symmetrical and has comparable variation on both sides of the true mean. A large part of the superiority of the ML estimate in this comparison is due to the use of a known value of β in computing the ML estimate but not the other two estimates.

5.6.3 ML Estimation for the Weibull Distribution— Shape Parameter Unknown

The ML method is applicable to arbitrary censoring and studies show it to be as precise as any other method in existence (cf. Gibbons and Vance, 1981 and Genschel and Meeker, 2010). In the important case of type II censoring, methods are available for constructing interval estimates for the Weibull

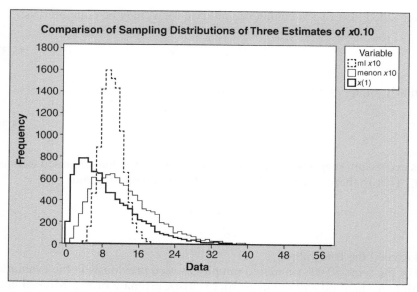

Figure 5.6   Sampling distribution of three estimators of $x_{0.10}$.

parameters. Implementation of the ML method involves the solution of a nonlinear transcendental equation that would generally be regarded as too difficult for hand calculations when more than two failures have occurred. It is, however, readily programmed for implementation on a computer.

Graphical methods are, however, simple to employ and unlike any other methods provide valuable insight into data discrepancies (outliers) and possible non-Weibull behavior that no other method does.

A reasonable approach to the analysis of Weibull data is to prepare a graphical estimate for a subjective appraisal of the validity of the data and the applicability of the Weibull model and then use the method of maximum likelihood for calculation of point and interval estimates. If there is doubt about the goodness of fit to a Weibull distribution, formal tests can be conducted. Such tests are discussed in Chapter 6.

Estimating both parameters of the two-parameter Weibull distribution by the ML method involves the iterative solution of two simultaneous nonlinear equations. These are the equations that result when the derivatives of the log likelihood with respect to β and η are set equal to zero. This includes those situations where the three-parameter model applies but the location parameter γ is known, for then one need only transform the data by subtracting the known location parameter from each observation prior to performing the analysis.

A statistical test of the hypothesis that the location parameter is nonzero is given by McCool (1998) and is described and illustrated in Chapter 10. This test must be performed using an iterative numerical procedure.

In the two-parameter Weibull case, the ML estimate of the shape parameter is given as the solution of the following equation (Cohen, 1965),

$$\left(\sum_{i=1}^{n} x_i^{\hat{\beta}}\right) \cdot \left(\frac{1}{\hat{\beta}} + \frac{1}{r}\sum_{i=1}^{r} \ln x_i\right) - \sum_{i=1}^{n} x_i^{\hat{\beta}} \ln x_i = 0. \qquad (5.40)$$

where $x_1 \dots x_r$ are the lives of the r failed test elements and $x_{r+1} \dots x_n$ are the lives at test suspension of the censored elements in a random sample of size n having r failures. Equation 5.40 applies irrespective of the type of censoring. There is no implication in Equation 5.40 that $x_r < x_n$, for example.

Having solved Equation 5.40 for $\hat{\beta}$, the ML estimate of the scale parameter is computed as:

$$\hat{\eta} = \left[\frac{1}{r}\sum_{i=1}^{n} x_i^{\hat{\beta}}\right]^{1/\hat{\beta}}. \qquad (5.41)$$

The p-th quantile x_p is calculated as,

$$\hat{x}_p = (k_p)^{1/\hat{\beta}}\,\hat{\eta} = \left\{\frac{k_p\sum_{i=1}^{n} x_i^{\hat{\beta}}}{r}\right\}^{1/\hat{\beta}}. \qquad (5.42)$$

The Mathcad module below, named weib1.xmcd, illustrates the computation of the ML estimates for the sample in Table 5.1.

Mathcad Module for ML Estimation of the Shape, Scale, and Tenth Percentile of the Two-Parameter Weibull Distribution

The vector **x** contains the data, ordered so that the failure times are listed first. n is the sample size and r is the number of failures:

$$\text{ORIGIN} \equiv 1$$

Set the sample size n, and number of failures r:

$$n := 10 \quad r := 10$$

$$\mathbf{x} := \begin{pmatrix} 14.01 \\ 15.38 \\ 20.94 \\ 29.44 \\ 31.15 \\ 36.72 \\ 40.32 \\ 48.61 \\ 56.42 \\ 56.97 \end{pmatrix}$$

Enter the r failure times and $(n - r)$ censoring times in the vector **x**. The first r components are the failure times and the next $n - r$ are the censoring times.

The ML estimate of the shape parameter is the root of the following equation:

$$f(b) := \left(\frac{1}{b}\right) + \left(\frac{1}{r}\right) \times \left(\sum_{i=1}^{n} \ln(x_i)\right) - \frac{\sum_{i=1}^{n}\left[(x_i)^b \times (\ln(x)_i)\right]}{\sum_{i=1}^{n}(x_i)^b}.$$

b = Initial guess of Weibull shape parameter $\qquad b := 1$

$$\beta := \mathrm{root}\,(f(b), b)$$

$$\beta = 2.582.$$

The scale parameter and 10th percentile estimates are computed as follows:

$$\eta := \left[\sum_{i=1}^{n} \frac{(x_i)^\beta}{r}\right]^{\frac{1}{\beta}} = 39.557 \quad x_{0.10} := \eta \times \ln\left(\frac{1}{0.9}\right)^{\frac{1}{\beta}} = 16.548.$$

A DOS program named Weib.exe will also compute the ML estimates. Figure 5.7 shows the input screen with the input for the data of Table 5.1 partially visible.

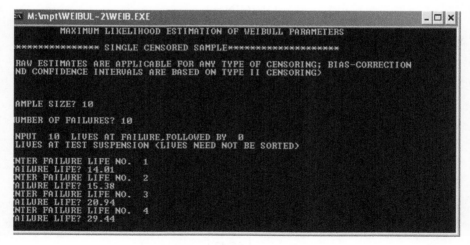

Figure 5.7 Input screen for Weib.exe.

The results screen for the complete sample shown as Figure 5.8 displays the raw ML estimates of the shape parameter $\hat{\beta}$ and the scale parameter $\hat{\eta}$ along with the estimated values of the 1st, 5th, 10th, 50th, 90th, 95th, and 99th percentiles. The lower half of the screen gives the 90% confidence limits and median unbiased estimates of the shape parameter and 10th percentile. These are computed using numerical approximations to percentage points of certain pivotal quantities discussed subsequently. The reader is advised to ignore these approximate values and to recompute the confidence limits using exact values of the relevant percentage points of the pivotal functions obtained using the simulation software program Pivotal.exe described in Chapter 7.

Figure 5.8 Results screen for Weib.exe.

If the statistical package Minitab is available, the ML estimates are given in the legend of the probability plot module. Figure 5.9 shows the plot for the test data we have used in the previous examples and confirms the estimates obtained with the Mathcad module and with the program Weib.exe.

The AD value refers to the Anderson Darling statistic used as a measure of goodness of fit. Goodness-of-fit testing for the Weibull is discussed in Chapter 6, where software for computing critical values of the AD statistic will be described. As will be explained in Chapter 6, the p-value shown in the legend of Figure 5.9 indicates that there is no basis to reject that the data were drawn from a two-parameter Weibull distribution.

A useful property of ML estimates is that the ML estimate of any function of a set of parameters is formed directly as that function of the ML estimates of those parameters. This property is termed the invariance property of ML

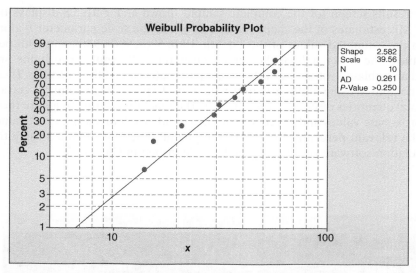

Figure 5.9   Weibull plot of sample data.

estimates. Thus, the ML estimate of the reliability at life x for the two-parameter Weibull case may be computed in terms of the ML estimates of the parameters:

$$\hat{R}(x) = exp - \left(\frac{x}{\hat{\eta}}\right)^{\hat{\beta}}. \tag{5.43}$$

Thus, using the estimates just obtained, we could estimate the reliability at the life of $x = 15$ to be:

$$\hat{R}(15) = \exp - \left(\frac{15}{39.56}\right)^{2.58} = 0.92.$$

5.6.4   Confidence Intervals for Weibull Parameters— Complete and Type II Censored Samples

With simple type II censoring the lives of the failed elements correspond to the first r order statistics of the sample and the other $n - r$ items are suspended at the failure time x_r of the r-th item.

In this case, exact confidence intervals on β and x_p can be computed because of the existence of a pair of random variables that are functions of the parameters and the ML estimates, and which follow distributions that depend only on sample size n, number of failures r, and for one of the functions, the quantile of interest p (cf. Thoman et al., 1969; McCool, 1970a,b). These functions, known as pivotal quantities, are:

$$v(r, n) = \frac{\hat{\beta}}{\beta}. \qquad (5.44)$$

and

$$u(r, n, p) = \hat{\beta} \ln\left(\frac{\hat{x}_p}{x_p}\right). \qquad (5.45)$$

The distribution of these random variables can be determined by Monte Carlo sampling and values of their quantiles tabulated for various values of r, n, and p. With these quantiles available one can, for example, set a one-sided upper 90% confidence interval on β as:

$$\beta < \frac{\hat{\beta}}{v_{0.10}(r, n)}. \qquad (5.46)$$

or a two-sided 90% confidence interval as,

$$\frac{\hat{\beta}}{v_{0.95}(r, n)} < \beta < \frac{\hat{\beta}}{v_{0.05}(r, n)}. \qquad (5.47)$$

where $v_{0.90}(r, n)$, $v_{0.05}(r, n)$, and $v_{0.95}(r, n)$ denote the 90th, 5th, and 95th percentiles of the distribution of $v(r, n)$.

Similarly a one-sided 90% confidence interval for x_p is computed as,

$$x_p < \left[\exp - \left(\frac{u_{0.10}(r, n, p)}{\hat{\beta}}\right)\right] \cdot \hat{x}_p. \qquad (5.48)$$

A two-sided 90% confidence interval is found as,

$$\left[\exp - \left(\frac{u_{0.95}(r, n, p)}{\hat{\beta}}\right)\right] \cdot \hat{x}_p < x_p < \left[\exp - \left(\frac{u_{0.05}(r, n, p)}{\hat{\beta}}\right)\right] \cdot \hat{x}_p. \qquad (5.49)$$

In general two-sided $100(1 - \alpha)\%$ confidence intervals for β and x_p may be found as follows,

$$\frac{\hat{\beta}}{v_{1-\alpha/2}(r, n)} < \beta < \frac{\hat{\beta}}{v_{\alpha/2}(r, n)}. \qquad (5.50)$$

$$\left[\exp-\left(\frac{u_{1-\alpha/2}(r,n,p)}{\hat{\beta}}\right)\right]\cdot\hat{x}_p < x_p < \left[\exp-\left(\frac{u_{\alpha/2}(r,n,p)}{\hat{\beta}}\right)\right]\cdot\hat{x}_p. \quad (5.51)$$

Fifty percent one-sided confidence intervals are a special case of the above and represent values that the true but unknown values of β and x_p are as likely to exceed as to fall short of. Defining,

$$\widehat{\beta}' = \frac{\hat{\beta}}{v_{0.50}(r,n)} \quad (5.52)$$

$$\hat{x}'_p = \left[\exp-\left(\frac{u_{0.50}(r,n,p)}{\hat{\beta}}\right)\right]\cdot\hat{x}_p. \quad (5.53)$$

It follows that $\hat{\beta}'$ and $\hat{x}'_p$ are median unbiased estimates of the Weibull parameters β and x_p. That is, calculation of modified point estimates of β and x_p according to Equations 5.52 and 5.53 results in estimates that 50% of the time will be smaller than the true (but unknown) values and 50% of the time larger than the true values. These estimates are said to be median unbiased. The factor by which the raw ML estimates are multiplied in calculating them may be regarded as bias correction factors.

Some percentage points for $v(r,n)$ and $u(r,n,p=0.632)$ for the uncensored case $(r=n)$ covering the range $5 < n < 120$ are given by Thoman et al. (1969).

McCool (1974) gives values of 21 percentage points of $u(r,n,p)$ and $v(r,n)$ ranging from $\alpha=0.01$ to 0.99 for $p=0.01,0.10,0.50,$ and 0.632 and the following values of n and r,

$$n = 5; r = 3, 5$$

$$n = 10; r = 3, 5, 10$$

$$n = 15; r = 5, 10, 15$$

$$n = 20; r = 5, 10, 15, 20$$

$$n = 30; r = 5, 10, 15, 20, 30.$$

At one time computation of the distribution of pivotal functions required a substantial amount of time on a mainframe computer and is the principal reason that published tables are so limited in coverage. These same simulations can now be computed in seconds on a standard personal computer. An easy to use program for computing these distributions called Pivotal.exe, and written in Visual Basic, will be described in Chapter 7. Table 5.4 lists the 5th, 95th, and 50th percentage points of u and v for several combinations of n and r for the purpose of illustrating the computations. The last two columns in this table will be discussed in Chapter 6.

Table 5.4 Table of Percentage Points of $v(r, n)$ and $u(r, n, p)$ for Some n and r

		$v(r, n)$			$u(r, n, 0.10)$				
r	n	0.05	0.5	0.95	0.05	0.5	0.95	R	$R_{0.50}^{\beta}$
3	5	0.6351	1.6510	6.7596	−1.2672	0.8483	9.9607	10.60	898
5	5	0.6795	1.2346	2.8146	−1.1422	0.4465	4.4453	4.14	92.4
3	10	0.6208	1.7223	7.6478	−1.4304	0.4313	7.0208	12.30	135
5	10	0.6482	1.3117	3.2791	−0.9571	0.3737	3.7698	5.06	36.7
10	10	0.7361	1.1031	1.8363	−0.8794	0.2125	2.1304	2.49	15.3
5	15	0.6430	1.3321	3.3937	−0.9223	0.2435	2.9446	5.28	18.2
10	15	0.7130	1.1269	1.9428	−0.6184	0.1933	1.9477	2.72	9.75
15	15	0.7715	1.0679	1.5634	−0.7648	0.1393	0.5091	2.03	8.41
5	20	0.6432	1.3353	3.5078	−0.9601	0.1482	2.5445	5.45	13.8
10	20	0.7047	1.1328	1.9913	−0.7274	0.1604	1.7473	2.83	8.89
15	20	0.7459	1.0754	1.6327	−0.7055	0.1215	0.4431	2.19	7.37
20	20	0.7949	1.0476	1.4454	−0.6740	0.0958	1.2262	1.82	6.13
5	30	0.6430	1.3475	3.4437	−1.1306	0.0176	1.6920	5.36	8.12
10	30	0.6996	1.1309	2.0236	−0.6348	0.0931	1.3891	2.89	5.99
15	30	0.7410	1.0819	1.6771	−0.6038	0.0928	1.2125	2.26	5.39
20	30	0.7662	1.0569	1.5182	−0.5955	0.0790	1.1130	1.98	5.04
30	30	0.8259	1.0290	1.3353	−0.5672	0.0536	0.9147	1.62	4.22

From Harris, *Rolling Bearing Analysis*, 3rd edition, copyright 1991 by John Wiley and Sons, Inc. Reprinted by permission of John Wiley and Sons, Inc.

Example

Using the ML method, calculate the 90% confidence intervals and median unbiased estimates for the shape parameter, scale parameter, and the tenth quantile $x_{0.10}$ for the uncensored data sample of size $n = 10$ given in Table 5.1.

The following values are needed for this calculation:

$v_{0.05}(10, 10) = 0.736$	$u_{0.05}(10, 10, 0.10) = -0.879$	$u_{0.05}(10, 10, 0.632) = -0.665$
$v_{0.50}(10, 10) = 1.103$	$u_{0.50}(10, 10, 0.10) = 0.213$	$u_{0.50}(10, 10, 0.632) = -0.230$
$v_{0.95}(10, 10) = 1.836$	$u_{0.95}(10, 10, 0.10) = 2.130$	$u_{0.95}(10, 10, 0.632) = 0.638$

Referring to the results of the the Mathcad module shown above, the raw ML estimates of $x_{0.10}$, η, and β were calculated to be:

$$\hat{x}_{0.10} = 16.55$$

$$\hat{\eta} = 39.56$$

$$\hat{\beta} = 2.58.$$

From Equation 5.47 one has,

$$1.41 = \frac{2.581}{1.836} < \beta < \frac{2.581}{0.736} = 3.51.$$

And from Equation 5.49:

$$7.25 = \left[\exp-\left(\frac{2.130}{2.58}\right)\right] \cdot 16.55 < x_{0.10} < \left[\exp-\left(\frac{-0.879}{2.58}\right)\right] 16.55 = 23.2$$

Recalling that the Weibull scale parameter is the same as the 63.2^{th} percentile we compute a 90% confidence interval for the scale parameter η:

$$30.9 = \left[\exp-\left(\frac{0.638}{2.58}\right)\right] \times 39.55 < \eta < \left[\exp-\left(\frac{-0.665}{2.58}\right)\right] \times 39.55 = 51.17.$$

These confidence intervals illustrate an important point. It will be recalled that the data sample was drawn from a two-parameter Weibull population for which the true values of β, $x_{0.10}$, and η were 1.3, 10.0, and 56.46 respectively. Thus, the statements $1.41 < \beta < 3.51$ and $30.9 < \eta < 51.17$ are incorrect, while the statement $7.25 < x_{0.10} < 23.3$ is correct. In practice one never knows whether the confidence interval statement is or is not correct since he or she does not know the true parameter values. Our only assurance is that we know that the procedure followed in making the interval calculation will lead to correct statements 90% of the time; that is, in 1000 such calculations we can expect an average of 900 correct and 100 incorrect statements. In the example above we have seen a false statement emerge from the calculations for β and η. It should be borne in mind that a false statement will occur in 10% of all independent calculations when one uses 90% confidence interval statements. It should also be borne in mind that a set of confidence intervals computed for β and a number of percentiles from the same sample as we have done here are not statistically independent.

From Equations 5.52 and 5.53, the median unbiased estimates of β, $x_{0.10}$, and η are:

$$\hat{\beta}' = \frac{2.58}{1.103} = 2.15$$

$$\hat{x}'_{0.10} = \left[\exp-\left(\frac{0.213}{2.58}\right)\right] \times 16.55 = 15.4$$

$$\hat{\eta}' = \left[\exp-\left(\frac{0.213}{2.58}\right)\right] \cdot 39.56 = 39.22$$

It is noted that these median unbiased values are in quite good agreement with the respective values, 2.16, 14.2, and 40.1 estimated using regression on the Weibull plot in Figure 5.1.

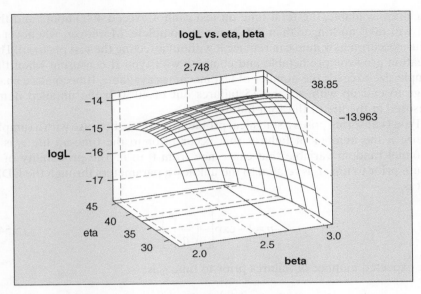

Figure 5.10 The Weibull log likelihood function for interval censoring.

5.6.5   Interval Censoring with the Weibull

The Weibull Likelihood function for interval censored data is constructed in the same manner as for the Exponential except that the CDF has two parameters. For the same data as used with the exponential example the Weibull likelihood function is maximized at $\eta = 38.849$ and $\beta = 2.748$. The maximized value of the log likelihood is -13.963. A surface plot of logL versus η and β is shown in Figure 5.10:

The ML estimates based on using the midpoints of the intervals as data are not far off: $\hat{\beta} = 2.556$ and $\hat{\eta} = 38.812$.

There is no theory yet available for setting confidence based on estimates computed from interval censored data. If the estimated parameters do not differ greatly from the values that result from treating the midpoints as exact observations, then one could reasonably use the methodology for exact data in setting confidence limits.

5.6.6   Confidence Limits for Weibull Parameters—Type I Censoring

Type I censoring is favored by experimenters and their management because the maximum amount of testing and hence the time to complete the test may be calculated in advance and so the testing costs may be determined and not just estimated. For a test of $n = 40$ items for $t_0 = 100$ hours and with 20 test machines available, the total time on test cannot exceed 4000 hours and the test will take no longer than 200 hours to complete. Moreover, whenever a failure occurs a new item can replace it without affecting the test protocol. The situation grows unpredictable and complex with type II censoring when the sample size exceeds the number of test machines available. It becomes a challenge to end up with, say, $r = 15$ failures out of 40 with the unfailed items censored at the life of the 15th failure.

Type I censoring poses a statistical problem, however, because with a sample of size n the number of failures that occur prior to the time-up life t_0 is a binomial random variable that can range from 0 to n. The probability of a failure prior to time t_0 is related to the unknown parameters through the CDF, that is,

$$p = 1 - \exp\left[-\left(\frac{t_0}{\eta}\right)^{\beta} \right]. \tag{5.54}$$

The expected number of failures prior to time t_0 is:

$$E(r) = np. \tag{5.55}$$

If in a given test r is less than $E(r)$, it means that more items than expected lived beyond t_0 with the result that such tests will tend to have greater than average life. When r is greater than $E(r)$, it means that more early failures occurred than expected and so such tests will tend to have lower than average life. If the time-up life t_0 is made small, p will be small, and the variability in r will diminish since the variance of a binomial random variable is $np(1 - p)$. Unfortunately with few failures one cannot estimate the distribution parameters with great precision.

The problem remains very much an open one and the best solution may well be to use testing strategies in which type II censored tests are performed in smaller subgroups and then combined statistically. Another useful strategy depending on the aims of the test program is to use the sudden death approach discussed in Chapter 7 to reduce test time with type II censoring.

Today the consensus approach for type I censored tests is to use asymptotic (large sample) ML theory for the construction of approximate confidence intervals for Weibull parameters and percentiles. There are a number of possibilities within this approach including the use of the distribution of the likelihood ratio and transformations of both the random variable and the parameter estimates to speed the approach to asymptotic normality. These techniques along with the use of Bootstrap methods are explained and compared in a comprehensive paper by Jeng and Meeker (2000) in which the comparative performance of no less than 10 approximate methods for computing confidence intervals on the Weibull parameters and percentiles with type I censoring is assessed. We will describe and illustrate the widely used asymptotic approach and compare its results in a few numerical examples to the expedient of treating the data as if the r value that resulted was the planned r value in a type II censored test.

The two-parameter Weibull distribution and the extreme value distribution followed by the logarithm of a Weibull random variable both satisfy the regularity conditions that assure that the ML estimates are the solution of the set of equations that result when the derivatives of the likelihood function with respect to each parameter are equated to zero. These regularity conditions guarantee that as the sample size increases without limit the ML estimates become unbiased and normally distributed. Because of the invariance property of ML estimates, the same numerical estimates result whether or not the likelihood function is expressed in terms of the Weibull distribution using the original data or the extreme value distribution using the logarithms of the observed lives and censoring times. In the latter case the ML estimates of the extreme value distribution are first computed and then transformed to the corresponding Weibull parameters. Lawless, Meeker, and Escobar, and others recommend working with the extreme value distribution on the grounds that the parameter estimates approach normality faster, that is, at smaller sample sizes in this metric. The confidence limits on the Weibull distribution parameters may then be obtained by transforming the confidence limits for the extreme value distribution parameters.

Recall that $Z = \ln X$ has the CDF:

$$F(z) = 1 - \exp\left[-\exp\left(\frac{z-\delta}{\xi}\right)\right]. \tag{5.56}$$

where the parameters are related to the corresponding Weibull parameters as $\delta = \ln(\eta)$ and $\xi = 1/\beta$. The density function is:

$$f(z) = \frac{dF(z)}{dz} = \frac{1}{\xi}\exp\left[\frac{z-\delta}{\xi}\right] \cdot \exp\left[-\exp\left(\frac{z-\delta}{\xi}\right)\right]. \tag{5.57}$$

The reliability function is:

$$R(z) = \exp\left[-\exp\left(\frac{z-\delta}{\xi}\right)\right]. \tag{5.58}$$

Let $z_1, z_2 \ldots z_r$ represent the logarithms of the time to failure of the r items that failed and $z_{r+1}, z_{r+2}, \ldots z_n$, denote the logarithms of the attained lives at which testing was suspended on the $(n-r)$ unfailed items. No ordering is implied other than that the failures are listed first and the suspensions second.

Except for a constant multiplier which we omit as inconsequential to the estimation process, the likelihood function is:

$$L = \prod_{i=1}^{r} f(z_i) \cdot \prod_{i=r+1}^{n} R(z_i). \tag{5.59}$$

Denote the natural log of the likelihood as $l(\delta, \xi)$:

$$l(\delta, \xi) = \sum_{i=1}^{r} \ln f(z_i) + \sum_{i=r+1}^{n} \ln R(z_i). \qquad (5.60)$$

The ML estimates are the values of δ and ξ for which this expression is maximum and are found by solving the two equations:

$$\frac{\partial l(\delta, \xi)}{\partial \delta} = \frac{\partial l(\delta, \xi)}{\partial \xi} = 0. \qquad (5.61)$$

The information matrix is defined as:

$$I(\delta, \xi) = \begin{vmatrix} -E\left[\dfrac{\partial^2 l}{\partial \delta^2}\right] & -E\left[\dfrac{\partial^2 l}{\partial \delta \partial \xi}\right] \\ -E\left[\dfrac{\partial^2 l}{\partial \delta \partial \xi}\right] & -E\left[\dfrac{\partial^2 l}{\partial \xi^2}\right] \end{vmatrix}. \qquad (5.62)$$

The expected values of the quantities in this matrix depend on how the censoring is carried out and on the true parameter values. The inverse of the information matrix is the 2×2 symmetric covariance matrix **V**:

$$\mathbf{V} = \begin{vmatrix} \mathbf{v}_{11} & \mathbf{v}_{12} \\ \mathbf{v}_{21} & \mathbf{v}_{22} \end{vmatrix} = I^{-1}(\delta, \xi). \qquad (5.63)$$

The component v_{11} is the variance of the ML estimate of δ, v_{22} is the variance of the ML estimate of ξ and $v_{12} = v_{21}$ is the covariance of the ML estimates of δ and ξ. Harter and Moore (1967) have computed the covariance matrices for various specified proportions of left and right censoring, For the uncensored case their results imply that the information matrix is:

$$I(\delta, \xi) = \frac{n}{\xi^2} \begin{vmatrix} 1 & 0.423 \\ 0.423 & 1.824 \end{vmatrix}$$

and depends only on ξ and not δ. When ML estimates are substituted for the actual parameters in the information matrix it becomes the estimated, or local, information matrix. The inverse of the local information matrix gives approximate values of the variance of the estimates. The covariance matrix thus obtained in the present case is:

$$V(\hat{\delta}, \hat{\xi}) = \frac{\hat{\xi}^2}{n} \begin{vmatrix} 1.109 & -0.257 \\ -0.257 & 0.608 \end{vmatrix}.$$

The elements of this matrix are actually what Harter and Moore published. The information matrix given above was reverse engineered.

When the expected values of the terms in the information matrix are not available as is the case for complex censoring patterns, the values of the second partial derivatives calculated from the data are substituted for the expected values. This matrix is sometimes called the *observed* information matrix and its inverse may be used as a further, less exact, approximation of the covariance matrix. (Some writers use observed and local information matrix interchangeably.) The observed information matrix for the extreme value distribution is shown by Lawless to be expressible as:

$$I\left(\hat{\delta}, \hat{\xi}\right) = \frac{1}{\hat{\xi}^2} \begin{vmatrix} r & \sum_{i=1}^{n} \hat{y}_i e^{\hat{y}_i} \\ \sum_{i=1}^{n} \hat{y}_i e^{\hat{y}_i} & r + \sum_{i=1}^{n} \hat{y}_i^2 e^{\hat{y}_i} \end{vmatrix}. \tag{5.64}$$

The terms y_i are the standardized values of the logarithm of the i-th failure or censoring time and $\hat{y}_i$ is the standardized value after replacing the parameters by their ML estimates:

$$\hat{y}_i = \frac{z_i - \hat{\delta}}{\hat{\xi}} = \frac{\ln x_i - \hat{\delta}}{\hat{\xi}}. \tag{5.65}$$

The estimates of the variance and covariance are the customary elements of the inverse of this matrix. For the uncensored sample of size 10 given in Table 5.1, the ML estimates of the extreme value distribution are $\hat{\delta} = \ln \hat{\eta} = 3.678$ and $\hat{\xi} = \frac{1}{\hat{\beta}} = 0.387$. The computed value of $I\left(\hat{\delta}, \hat{\xi}\right)$ is:

$$I\left(\hat{\delta}, \hat{\xi}\right) = \frac{10}{\hat{\xi}^2} \begin{vmatrix} 1 & 0.4106 \\ 0.4106 & 1.6904 \end{vmatrix}.$$

In this form, with the sample size factored out and shown as a scalar multiplier, the components of the matrix may be compared to the exact expected values in the matrix $I(\delta, \xi)$ shown above for an uncensored sample following Equation 5.63. We will compare the confidence intervals based on the local and observed information matrices, further below.

Inverting the observed information matrix after multiplying through by the scalar factor gives the following approximate covariance matrix:

$$V = \begin{vmatrix} 0.017 & -4.046E - 3 \\ -4.046E - 3 & 9.855E - 3 \end{vmatrix}$$

Approximate 90% confidence limits for δ based on the normal distribution approximation are:

$$3.465 = 3.678 - 1.645(0.017)^{\frac{1}{2}} < \delta < 3.678 + 1.645(0.017)^{\frac{1}{2}} = 3.89$$

The corresponding 90% interval for η results from exponentiating both sides of this interval:

$$31.99 = e^{3.465} < \eta < e^{3.89} = 48.91.$$

A 90% interval for ξ is:

$$0.224 = 0.387 - 1.645(9.855E - 3)^{1/2} < \xi < 0.387 + 1.645(9.855E - 3)^{1/2} = 0.551.$$

The corresponding interval for β results from taking the reciprocals of these two limits and interchanging the upper and lower since if $1/\beta < 0.551$, $\beta > 1/0.551$. The resultant interval is:

$$1.816 < \beta < 4.465.$$

To estimate the $100p$-th percentile of the z distribution, we use:

$$\hat{z}_p = \hat{\delta} + \hat{\xi} y_p \tag{5.66}$$

where the constant y_p is:

$$y_p \equiv \ln\ln\left(\frac{1}{1-p}\right). \tag{5.67}$$

For our example the estimate of the 10th percentile $z_{0.10}$ is:

$$\hat{z}_{0.10} = 3.678 - 2.25 \times 0.387 = 2.806.$$

The estimated 10th percentile of the Weibull distribution is:

$$\hat{x}_{0.10} = e^{2.806} = 16.56.$$

This agrees with what was found previously using direct ML estimation for the Weibull distribution. To find the variance of $\hat{z}_{0.10}$ we recognize that in general z_p is a linear combination of two random variables so its variance may be written:

$$var(\hat{z}_p) = var(\hat{\delta}) + y_p^2 \, var(\hat{\xi}) + 2y_p \, cov(\hat{\delta}, \hat{\xi}). \tag{5.68}$$

For the example,

$$var(\hat{z}_{0.10}) = 0.017 + (-2.25)^2 \times 9.855E - 3 + 2 \times (-2.25) \times (-4.046E - 3) = 0.085.$$

A 90% confidence interval on $z_{0.10}$ is:

$$2.327 = 2.806 - 1.645(0.085)^{1/2} < z_{0.10} < 2.806 + 1.645(0.085)^{1/2} = 3.285.$$

Exponentiating both sides gives the 90% interval for $x_{0.10}$:

$$10.25 < x_{0.10} < 26.72.$$

The intervals for the Weibull parameters η, β, and $x_{0.10}$ were recalculated using the local information matrix with Harter and Moore's matrix components. The table below summarizes the asymptotic approximate intervals computed with these two information matrices and the exact values computed previously:

Method	η	β	$x_{0.10}$
Observed matrix	31.99, 48.91	1.816, 4.465	10.25, 26.72
Local matrix	31.99, 48.90	1.837, 4.344	10.39, 26.36
Exact	30.9, 51.17	1.41, 3.51	7.25, 23.2

We see that for this example the difference between using the local matrix and the observed matrix is negligible, but both differ substantially from the exact limits.

The exact confidence limits for η are somewhat wider than both the asymptotic limits. The exact limits for β and $x_{0.10}$ are distinctly asymmetric, suggesting that the distributions have not yet converged to normality at a sample size of 10.

A Mathcad module titled asymptoticxample.xmcd that performs these calculations is shown below:

Mathcad Module for ML Estimation of the Shape and Scale Parameters and the Tenth Percentile of the Two-Parameter Weibull Distribution and Computation of Confidence Limits Based on Asymptotic Normality

The vector x contains the ordered data. n is the sample size and r is the number of failures in a type I or type II censored sample:

$$\mathbf{x} := \begin{pmatrix} 14.01 \\ 15.38 \\ 20.94 \\ 29.44 \\ 31.15 \\ 36.72 \\ 40.32 \\ 48.61 \\ 56.42 \\ 56.97 \end{pmatrix}$$

$$n := 10 \quad r := 10.$$

ML shape parameter estimate—initial guess

$$\beta := 1$$

$$f(\beta) := \frac{1}{\beta} + \frac{1}{r} \times \sum_{i=1}^{r} \ln(x_i) - \frac{\sum_{i=1}^{n} \left[(x_i)^\beta \times \ln(x_i) \right]}{\sum_{i=1}^{n} (x_i)^\beta}.$$

$$\beta := \text{root}(f(\beta), \beta) \qquad \beta = 2.582.$$

ML scale parameter estimate

$$\eta := \left[\frac{\sum_{i=1}^{n} (x_i)^\beta}{r} \right]^{\frac{1}{\beta}} \qquad \eta = 39.557.$$

ML estimate of $x_{0.10}$

$$x_{0.10} := \ln\left(\frac{1}{0.9} \right)^{\frac{1}{\beta}} \times \eta \quad x_{0.10} = 16.548$$

$$x_{0.50} := \ln(2)^{\frac{1}{\beta}} \times \eta \qquad x_{0.50} = 34.323.$$

Compute ML estimates of parameters of extreme value distribution

$$\delta := \ln(\eta) \quad \xi := \frac{1}{\beta}.$$

Transform data to standardized extreme value random variables

$$y := \frac{\ln(x) - \delta}{\xi}.$$

Compute the local information matrix

$$\mathbf{I} := \frac{1}{\xi^2} \begin{bmatrix} r & \sum_{i=1}^{n} (y_i \times e^{y_i}) \\ \sum_{i=1}^{n} (y_i \times e^{y_i}) & r + \sum_{i=1}^{n} \left[(y_i)^2 \times e^{y_i} \right] \end{bmatrix}.$$

Compute the estimated covariance matrix by inverting the local information matrix

$$\mathbf{Cov} := \mathbf{I}^{-1} \quad \mathbf{Cov} = \begin{pmatrix} 0.017 & -4.046 \times 10^{-3} \\ -4.046 \times 10^{-3} & 9.855 \times 10^{-3} \end{pmatrix}.$$

Compute 90% intervals using normality

$$\delta\text{upper} := \delta + 1.645(\text{Cov}_{1,1})^{0.5} \quad \delta\text{upper} = 3.89$$
$$\delta\text{lower} := \delta - 1.645(\text{Cov}_{1,1})^{0.5} \quad \delta\text{lower} = 3.465$$

$$\eta\text{upper} := e^{\delta\text{upper}} \quad \eta\text{upper} = 48.914$$
$$\eta\text{lower} := e^{\delta\text{lower}} \quad \eta\text{lower} = 31.99$$

$$\xi\text{upper} := \xi + 1.645(\text{Cov}_{2,2})^{0.5} \quad \xi\text{upper} = 0.551$$
$$\xi\text{lower} := \xi - 1.645(\text{Cov}_{2,2})^{0.5} \quad \xi\text{lower} = 0.224$$

$$\beta\text{upper} := \frac{1}{\xi\text{lower}} \quad \beta\text{upper} = 4.465$$

$$\beta\text{lower} := \frac{1}{\xi\text{upper}} \quad \beta\text{lower} = 1.816$$

Compute the variance of the log of 10th percentile

$$varz10 := \text{Cov}_{1,1} + 2 \times \ln\left(\ln(1.11111)\right) \times \text{Cov}_{1,2} + \left(\ln\left(\ln(1.11111)\right)\right)^2 \times \text{Cov}_{2,2}.$$

$$z10\text{upper} := \ln(x_{0.10}) + 1.645varz10^{0.5} \quad z10\text{upper} = 3.285$$
$$z10\text{lower} := \ln(x_{0.10}) + -1.645varz10^{0.5} \quad z10\text{lower} = 2.327$$

$$x10\text{upper} := e^{z10\text{upper}} \quad x10\text{upper} = 26.715$$
$$x10\text{lower} := e^{z10\text{lower}} \quad x10\text{lower} = 10.25$$

As a further example we consider the uncensored life test data obtained on 23 ball bearings and reported by Lieblein and Zelen (1956). These data have been considered by many writers including Lawless, Meeker and Escobar, and Leemis (Leemis 1995). The lives in millions of revolutions are in ascending order: 17.88, 28.92, 33.00, 41.52, 42.12, 45.60, 48.48, 51.84, 51.96, 54.12, 55.56, 67.80, 68.64, 68.88, 84.12, 93.12, 98.64, 105.12, 105.84, 127.92, 128.04, and 173.40. We will examine these data in the uncensored form, and after artificially time censoring it at 60.0 following Jeng and Meeker. To simulate type I censoring at $t_0 = 60$, the data values in the sample in excess of 60 are changed to 60. The number of actual failures prior to 60 is $r = 11$.

The ML estimates of η, β, and $x_{0.10}$ were computed to be 81.85, 2.103, and 28.07, respectively in the uncensored case and 68.68, 3.082, and 33.10 with censoring. To compute exact 90% confidence limits for η, β, and $x_{0.10}$ the 5% and 95% points of the distributions $of\ u(r, n, p)$ and $v(r, n)$ were obtained by simulation using the program Pivotal.exe. for $n = 23$, $r = 11$ and $r = 23$, and $p = 0.10$ and 0.632. These percentage points and the medians are tabled below:

Pivotal Function	5%	50%	95%
$u(11, 23, 0.10)$	−0.684	0.137	1.556
$u(11, 23, 0.632)$	−1.053	−0.1054	0.423
$v(11, 23)$	0.711	1.116	1.910
$u(23, 23, 0.10)$	−0.632	0.0801	1.107
$u(23, 23, 0.632)$	−0.393	−0.0146	0.375
$v(23, 23)$	0.8034	1.041	1.413

The exact and asymptotic 90% confidence limits are tabled below for the uncensored case. The observed information matrix was used in computing the asymptotic limits.

Method	η	β	$x_{0.10}$
Asymptotic	68.8, 97.3	1.67, 2.83	19.4, 40.67
Exact	68.9, 97.3	1.49, 2.62	16.6, 37.9

The agreement is quite good for η. As with the previous example, the asymptotic lower limits for β and $x_{0.10}$ are not small enough and the symmetry assumed in the normal approximation is not yet valid at $n = 23$.

The same comparison for the type I censored data is:

Method	η	β	$x_{0.10}$
Asymptotic	57.13, 82.58	2.109, 5.726	24.64, 44.46
Exact for type II	59.87, 96.65	1.614, 4.334	19.97, 41.32

The limits computed using the exact values for a type II censored test should not be far off in this case since a type II censored test with $r = 11$ would have concluded at a life close to $t_0 = 60$ anyway. The method based on the likelihood ratio cited in the Jeng and Meeker paper gives limits of (19.74, 43.26) for $x_{0.10}$ and (1.667, 5.00) for β and are in good agreement with the exact values at the lower end of the interval. The likelihood ratio method requires a numerical search technique.

Lawless cites a set of data due to Freireich (1963) on the remission time in weeks for a set of 21 leukemia patients treated with a drug called 6-MP. The remission time was observed for $r = 9$ patients and the other 12 left the study prior to remission after varying lengths of time. The censoring pattern in this data is therefore quite complex. The data are as follows:

Observed remission times: 6, 6, 6, 7, 10, 13, 16, 22, 23

Censoring times: 6, 9, 10, 11, 17, 19, 20, 25, 32, 32, 34, 35.

The ML estimates of η, β, and $x_{0.50}$ are 33.77, 25.7, and 1.354. The exact confidence interval computation is not applicable in this case because the data do not conform to the type II censoring model. Nevertheless, we have calculated 95% limits on that basis as well as asymptotic results with the observed information matrix. The results are tabled below:

Method	η	β	$x_{0.50}$
Asymptotic	21.5, 52.9	0.876, 2.98	15.8, 42.0
Exact type II	22.5, 128.6	0.546, 2.12	12.3, 73.9

For $x_{0.50}$ and β the results for the exact type II calculation are more conservative at the important lower end of the intervals.

Clearly unless the samples are quite large, possibly on the order of 50–100, the asymptotic approach is problematic. If at the planning stage of a life test it is found to be possible to avoid type I censoring by using another strategy, the experimenter is well advised to do so. If one has type I censored data, it might be useful in helping to assess the results to use the type II exact computation as a "second opinion" even though it is not strictly applicable.

REFERENCES

Bartholomew, D. 1963. The sampling distribution of an estimate arising in life testing. *Technometrics* 5(3): 361–374.

Benard, A. and E.C. Bos-Levenbach. 1953. The plotting of observations on probability paper (in Dutch). *Statistica Neerlandica* 7: 163–173.

Blom, G. 1958. *Statistical Estimates and Transformed Beta-Variables*. John Wiley and Sons, New York, NY.

Cohen, A.C. 1965. Maximum likelihood estimation in the Weibull distribution based on complete and on censored samples. *Technometrics* 7(4): 579–588.

Epstein, B. and M. Sobel. 1953. Life testing. *Journal of the American Statistical Association* 48(263): 486–502.

Fisher, R.A. 1934. Two new properties of mathematical likelihood. *Proceedings of the Royal Society A* 144: 285–307.

Freireich, E.J.E.A. 1963. The effect of 6-mercaptopurine on the duration of steroid induced remissions in acute leukemia. *Blood* 21: 699–716.

Genschel, U. and W.Q. Meeker. 2010. A comparison of maximum likelihood and median rank regression for weibull estimation. *Quality Engineering* 22(4): 236–255.

Gibbons, D.I. and L.C. Vance. 1981. A simulation study of estimators for the 2-parameter weibull distribution. *IEEE Transactions on Reliability* 30: 61–66.

Harter, H.L. and A.H. Moore. 1965. Point and interval estimators based on order statistics for rthe scale parameter of a Weibull population with known shape parameter. *Technometrics* 7(3): 405–422.

Harter, H.L. and A.H. Moore. 1967. Asymptotic variances and covariances of maximum likelihood estimates, from censored samples, of the paramers of the weibull and gamma distributions. *Annals of Mathematical Statistics* 38: 557–570.

Jeng, S.L. and W.Q. Meeker. 2000. Comparisons of approximate confidence interval procedures for type i censored data. *Technometrics* 42(2): 135–148.

Johnson, L.G. 1964. *The Statistical Treatment of Fatigue Experiments.* Elsevier, New York, NY.

Kaplan, E.L. and P. Meier. 1958. Nonparametric estimation from incomplete observations. *Journal of the American Statistical Association* 53: 457–481.

Kapur, K.C. and L.R. Lamberson. 1977. *Reliability in Engineering Design.* Wiley, New York, NY.

Lawless, J.F. 2002. *Statistical Models and Methods for Lifetime Data.* Wiley, New York, NY.

Leemis, L.M. 1995. *Reliability.* Prentice Hall, Englewood Cliffs, NJ. 07632.

Lieblein, J. and M. Zelen. 1956. Statistical investigation of the fatigue life of deep-groove ball bearings. *Journal of Research of the National Bureau of Standards* 57(5): 273–316.

McCool, J.I. 1970a. Evaluating Weibull endurance data by the method of maximum likelihood. *Tribology Transactions* 13(3): 189–202.

McCool, J.I. 1970b. Inference on Weibull percentiles and shape parameter from maximum likelihood estimates. *IEEE Transactions on Reliability* R-19: 2–9.

McCool, J.I. 1974. Inferential techniques for Weibull populations. *Aerospace Research Laboratories Report* ARL TR 74-0180.

McCool, J.I. 1998. Inference on the Weibull location parameter. *Journal of Quality Technology* 30(2): 119–126.

Meeker, W.Q. and L.A. Escobar. 1998. *Statistical Methods for Reliability Data.* Wiley, New York, NY.

Menon, M.V. 1963. Estimation of the shape and scale parameters of the Weibull distribution. *Technometrics* 5(2): 175–182.

Nelson, W. 1969. Hazard plotting for incomplete failure data. *Journal of Quality Technology* 1(1): 27–52.

Nelson, W. 1982. *Applied Life Data Analysis.* John Wiley & Sons, New York.

Phan, L.D. and J.I. Mccool. 2009. Exact confidence intervals for Weibull parameters and percentiles. *Proceedings of the Institution of Mechanical Engineers, Part O: Journal of Risk and Reliability* 223: 387–394.

Rausand, M. and A. Høyland. 2004. *System Reliability Theory: Models, Statistical Methods, and Applications.* 2nd ed., John Wiley and Sons, Inc., Hoboken, NJ.

Romanowski, E. 1998. The validity and accuracy of hazard plotting for multiply censored Weibull data. Thesis (M Eng), Pennsylvania State University, Great Valley.

Thoman, D.R., L.J. Bain, and C.E. Antle. 1969. Inferences on the parameters of the Weibull distribution. *Technometrics* 11(3): 445–460.

EXERCISES

1. If a sample of size 10 is drawn from the Weibull population $W(100, 2)$ compute the expected value of the first-order statistic.

2. The following sample was randomly drawn from $W(100, 2)$. Estimate η, β, and $x_{0.10}$ using Menon's method of estimation.

14.6375	66.9203	68.4248	69.2899	71.8208
77.4557	94.8249	100.317	101.453	104.603

3. The following is a random sample of size $n = 10$ type II censored at $r = 5$ and drawn from an exponential distribution with mean $\theta = 100$ hours. Compute two-sided 95% confidence limits for θ. Estimate the reliability at life = 25 hours.

7.764,84	18.1968	26.9440	37.7441	52.7651
52.7651	52.7651	52.7651	52.7651	52.7651

4. An acceptance test of an exponential sample is passed if $n = 20$ items survive $t_0 = 100$ hours. Compute a lower 95% confidence limit for the mean θ.

5. For the sample of problem 2, compute the raw maximum likelihood estimate of the scale parameter η assuming the shape parameter is known to be $\beta = 2$. Compute the bias corrected value of η.

6. For the sample in problem 2, use weib.exe or other software to estimate β, η, and $x_{0.10}$ using the method of maximum likelihood. Using appropriate pivotal quantity percentage points in Table 5.4, compute a 90% confidence interval and a median unbiased estimate for the shape parameter.

7. For an exponential distribution with mean θ, how long must 15 specimens run without failure to demonstrate with 90% confidence that that θ exceeds 80 hours?

CHAPTER 6

Sample Size Selection, Hypothesis Testing, and Goodness of Fit

6.1 PRECISION MEASURE FOR MAXIMUM LIKELIHOOD (ML) ESTIMATES

There are two reasons for conducting a life test, one being to test a hypothesis about one or more parameters of a distribution and the second being to form an estimate of one or more parameters to add to a growing body of knowledge and serve as a reference in future, yet unformulated, decision making.

In the second situation McCool (1999) proposed using the ratio of the upper to lower ends of a confidence interval (with some suitable associated confidence level) as a measure of the precision with which the parameter has been determined by a type II censored life test, and to select the sample size that yields a value of this ratio that the experimenter feels is needed. He suggested a 90% interval on the basis that it is wide enough to be useful but narrow enough to avoid using the less reliable extreme percentiles of a distribution determined by Monte Carlo simulation.

Equation 5.47 gives the expression for a two-sided 90% confidence interval for β. The ratio R of the upper to lower ends of this interval is,

$$R = \frac{v_{0.95}(r, n)}{v_{0.05}(r, n)}. \tag{6.1}$$

The precision measure R is independent of the data and varies with the sample size parameters n and r. Values of R for various combinations of r and n are given in Table 5.4. It is instructive to scan the R column in the table. When r is a small fraction of n, R is large, indicating that the shape parameter is best estimated if at least half the sample has failed. If one were to require that $R < 2$ at least 20 failures would be required.

Using the Weibull Distribution: Reliability, Modeling, and Inference, First Edition. John I. McCool.
© 2012 John Wiley & Sons, Inc. Published 2012 by John Wiley & Sons, Inc.

An alternate measure of the precision with which the shape parameter is determined by a sample of a given size was proposed in Phan and McCool (2009). It is based on the difference d between the upper and lower 90% confidence limits.

$$d = UCL - LCL = \frac{\hat{\beta}}{v_{0.05}} - \frac{\hat{\beta}}{v_{0.95}}.$$

Expressed this way the criterion is not very useful for test planning purposes since its magnitude depends on the shape parameter estimate $\hat{\beta}$. However, we may scale it by dividing by the true value β to give the relative error:

$$D = \frac{d}{\beta} = \frac{\hat{\beta}}{\beta} \left[\frac{1}{v_{0.05}} - \frac{1}{v_{0.95}} \right] = v \left[\frac{1}{v_{0.05}} - \frac{1}{v_{0.95}} \right].$$

The bracketed term is a constant and a function of n and r and v is the pivotal quantity $\frac{\hat{\beta}}{\beta}$. The median value of D designated $D_{0.50}$ depends only on n and r and seems to be a reasonable measure of the precision of the shape parameter determination resulting from a sample of size n with r failures. It is expressible as:

$$D_{0.50} = v_{0.50} \left[\frac{1}{v_{0.05}} - \frac{1}{v_{0.95}} \right]. \tag{6.2}$$

From Equation 5.49, the ratio of the upper to lower ends of a 90% confidence interval for x_p is:

$$R = \exp \left(\frac{u_{0.95}(r, n, p) - u_{0.05}(r, n, p)}{\hat{\beta}} \right) \tag{6.3}$$

Unlike the ratio for shape parameter estimation, this ratio is not a constant but depends upon the ML estimate $\hat{\beta}$ which is a random variable and varies from sample to sample. The median value of R, designated $R_{0.50}$, is accordingly proposed as a criterion and is computed by substituting the median value of $\hat{\beta}$, namely, $\beta v_{0.50}(r,n)$ into this expression. The result is:

$$R_{0.50} = \exp \left(\frac{u_{0.9}(r, n, p) - u_{0.05}(r, n, p)}{\beta v_{0.50}(r, n)} \right). \tag{6.4}$$

Raising both sides of Equation 6.4 to the power β gives,

$$R_{0.50}^{\beta} = \exp \left(\frac{u_{0.95}(r, n, p) - u_{0.05}(r, n, p)}{v_{0.50}(r, n)} \right). \tag{6.5}$$

In general the right-hand side of Equation 6.5 depends on sample size (r and n), and the quantile p under investigation. In practice one prescribes a value of $R_{0.50}$ and an assumed value of β and scans tables of the distribution $u(r, n, p)$ until values satisfying Equation 6.5 are found. Values of $R_{0.50}^{\beta}$ based on a 90% confidence interval for $x_{0.10}$ are listed in Table 5.4 in Chapter 5 for various values of r and n. Comparable tables may readily be developed for other values of p using the software called Pivotal.exe described in Chapter 7.

Example
Find the sample size required for determining the 10th quantile $x_{0.10}$ of a two-parameter Weibull distribution so that the median ratio of upper and lower 90% confidence intervals is less than 3.5. Assume a shape parameter $\beta = 1.3$.

$$R_{0.50}^{\beta} = 3.5^{1.3} = 5.1.$$

From Table 5.4 one sees that a sample size of 30 with at least 20 failures is needed.

Although the values of r and n in Table 5.4 are typical of the sample sizes used in bearing life tests, the values of $R_{0.50}^{\beta}$ are disappointingly large. They serve to discourage doing a life test if, for example, one could only afford a sample of size $n = 5$. It has been shown, however (McCool, 1979), that when a number of tests are run representing small differences in operating or design conditions (e.g., lubricant type, load, cage design) it is possible to combine the results to compute refined estimates of the shape parameter and the percentiles having greatly improved precision. The validity of the approach depends upon the shape parameters being equal among the test group populations. A way of testing the reasonableness of this assumption has been developed. When the assumption of equal population shape parameters can be justified, it has been found, for example, that with five groups having $n = r = 30$, $R_{0.50}^{\beta} = 2.33$. As seen in Table 5.4 in a single sample with $n = r = 30$ has $R_{0.50}^{\beta} = 4.22$. Since it is a frequent practice to test multiple groups it is recommended that experimenters exploit the improved precision that will result from pooling the results of a number of life test groups. The analysis of multiple groups of type II censored data from the two-parameter Weibull distribution is discussed in Chapter 8.

6.2 INTERVAL ESTIMATES FROM MENON'S METHOD OF ESTIMATION

As mentioned previously it has been found recently (Phan and McCool, 2009) that the same pivotal quantities of the ML estimates are pivotal for Menon's estimates as well, and therefore once the distributions are determined by simulation, one may set confidence intervals and form median unbiased estimates in the same manner. Since Menon's method is not applicable to censored samples, the notation for the pivotal functions reduces to $v(n)$ and $u(n, p)$. Software for simulating the distribution of these pivotal functions has been written in Microsoft Visual Basic. The executable version, titled Menon.exe, may be downloaded from http://www.personal.psu.edu/mpt. The program will prompt the user to specify values for the sample size, the percentile of interest, and the number of simulation samples to be used. It then computes the Menon shape and scale parameter estimates and pivotal functions u and v for each simulated sample and sorts them over the set of simulated samples. For 10,000 simulated samples of size 30 the computation time is nearly instantaneous on a contemporary personal computer. Since the precision of estimated percentage points $u(n, p)$ varies with both n and p, it is advisable to run the software several times to gauge the variability of the percentage points of interest.

The following percentage points of $v(10)$ and $u(10, 0.10)$ have been determined by running the software using 100,000 simulated samples to illustrate the calculations on the data sample we have been analyzing.

$v_{0.05}(10) = 0.663$	$u_{0.05}(10, 0.10) = -0.956$
$v_{0.50}(10) = 1.093$	$u_{0.50}(10, 0.10) = 0.167$
$v_{0.95}(10) = 1.861$	$u_{0.95}(10, 0.10) = 2.190$

Recall that for this sample, Menon's estimates of the shape parameter and $x_{0.10}$ were 2.539 and 16.04, respectively. The median unbiased estimate of the shape parameter is:

$$\tilde{\beta}' = \frac{2.539}{1.093} = 2.32.$$

The median unbiased estimate based on the ML estimate was 2.15.

Ninety percent confidence limits for the shape parameter are:

$$1.364 = \frac{2.539}{1.861} < \beta < \frac{2.539}{0.663} = 3.829.$$

The corresponding interval based on the ML estimates was (1.41, 3.51). and hence was tighter than the limits based on Menon's estimates. This was true at all sample sizes examined. Table 6.1 shows the value of the ratio R of the upper to lower limits of a 90% interval on β, as a function of sample size for Menon's method and the ML method. Also shown are the comparative values of the criterion based on differences of the upper and lower limits $D_{0.50}$.

The table shows the clear superiority of the method of ML with respect to estimating the Weibull shape parameter using either measure of precision. For

Table 6.1   Ratio and Median Differences of Upper and Lower 90% Limits on the Weibull Shape Parameter: Menon and ML

n		5	10	15	20	25	30	35	40	50	60	80	100
R	ML	4.11	2.50	2.05	1.84	1.71	1.62	1.55	1.52	1.45	1.40	1.34	1.30
	Me	4.57	2.79	2.28	2.09	1.93	1.83	1.74	1.70	1.61	1.54	1.45	1.40
$D_{0.50}$	ML	1.387	0.878	0.703	0.614	0.528	0.482	0.441	0.414	0.368	0.336	0.289	0.259
	Me	1.547	1.059	0.874	0.739	0.666	0.618	0.571	0.542	0.473	0.436	0.384	0.341

From the *Journal of Risk and Reliability* 223(O4), copyright 2009. Reprinted by permission of The Institute of Mechanical Engineers.

example, using ML, a sample of size 20 gives roughly the same precision as a sample of size 30 if Menon's method is used, regardless of the precision measure. A sample of size 13 gives an R value by Menon's method equal to 2.45. This is roughly equivalent to the precision of a sample of size 10 using ML estimation. With the $D_{0.50}$ criterion it takes a sample of size of 15 for Menon's method to be equivalent to a sample of size 10 using ML estimation.

The median unbiased estimate of the 10th percentile is:

$$\tilde{x}'_{0.10} = 16.04 \times \exp\left[-\left(-\frac{0.1670}{2.539}\right)\right] = 17.1.$$

This compares to 15.4 for the computation based on ML estimates.

The 90% confidence limits for $x_{0.10}$ are computed as follows:

$$6.77 = 16.04 \times \exp\left[-\frac{2.190}{2.539}\right] < x_{0.10} < 16.04 \times \exp\left[-\left(-\frac{0.956}{2.539}\right)\right] = 23.4.$$

This compares to the slightly tighter ML interval of (7.25, 23.2).

Figure 6.1 shows the precision measure $R^{\beta}_{0.50}$ plotted against p for $n = 10$ and $n = 30$ for both Menon and ML estimation. To within graphical accuracy, the precision of Menon's estimator is comparable to ML for $0.4 < p < 0.7$. If interest centers on the median or mean of the distribution, Menon's method is quite satisfactory. Such might be the case if the random variable is strength rather than lifetime.

6.3   HYPOTHESIS TESTING—SINGLE SAMPLES

Hypothesis testing is an application of statistical inference closely related to interval estimation. A statistical hypothesis is a mathematical statement about the parameters of one or more probability distributions. In testing a statistical hypothesis, two such statements are set down. One is called the null hypothesis, designated H_0, representing a "baseline" or "business as usual" condition. The other is called the alternative hypothesis. It is designated H_1 and represents

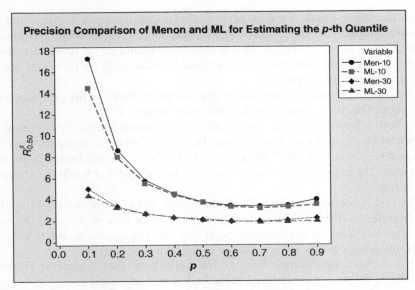

Figure 6.1 Precision comparison of ML and Menon as a function of p. From the *Journal of Risk and Reliability* 223(O4), copyright 2009. Reprinted by permission of The Institute of Mechanical Engineers.

the type of departure from the baseline that the experimenter is concerned with detecting.

For example, one may be interested in testing whether a new treatment or process has had an effect upon $x_{0.10}$ life. It is assumed that in the absence of any such effect the $x_{0.10}$ life should have the value $(x_{0.10})_0$ known from theory or from historical data.

The null hypothesis in this case is written as,

$$H_0 : x_{0.10} = (x_{0.10})_0.$$

The appropriate alternative hypothesis will be any one of the following three depending upon the purpose of the experiment.

$$H_1 : x_{0.10} < (x_{0.10})_0$$
$$H_1 : x_{0.10} > (x_{0.10})_0$$
$$H_1 : x_{0.10} \neq (x_{0.10})_0$$

The first alternative applies when the experimenter is concerned only with finding out whether $x_{0.10}$ is lower than the prescribed standard value and is not concerned about whether it is greater.

Such a case might arise, for example, if one has introduced a manufacturing change that lowers costs. It is then prudent to test whether the change has been detrimental to fatigue life. If the change has actually improved fatigue life, that is all to the good.

On the other hand, if a new process that increases costs is purported to improve fatigue life, one would be inclined to test the second alternative above, that is, $H_1: x_{0.10} > (x_{0.10})_0$. In this case there would be no reason to adopt the new process unless it could be established conclusively that it brings about an improvement in fatigue life.

The third (two-sided) alternative applies when there is no preconceived idea about the likely direction of an effect on fatigue life.

To test a hypothesis one must find a test statistic whose distribution is known when the null hypothesis is true. From this distribution one determines a critical region within which the statistic will lie with specified small probability α when the hypothesis is true and with higher probability when the alternative is true.

The test procedure consists of calculating the test statistic and rejecting the null hypothesis if the test statistic falls in the critical region. α is thus the risk that one is taking of erroneously rejecting the hypothesis when it is true. Rejecting a hypothesis when it is true is called a type I error.

To test, with significance level α, the hypothesis that a general Weibull quantile x_p assumes a designated value $(x_p)_0$, one computes,

$$\hat{\beta} \ln\left(\frac{\hat{x}_p}{(x_p)_0}\right). \tag{6.6}$$

where $\hat{\beta}$ and $\hat{x}_p$ are the uncorrected ML estimates of β and x_p from a sample of size n with r failures.

The hypothesis is accepted if $\hat{\beta} \ln\left[\dfrac{\hat{x}_p}{(x_p)_0}\right]$ does not fall in a critical region, the size of which depends upon n and r and also upon what alternative to the null hypothesis is being considered.

Likewise, to test the hypothesis that the Weibull shape parameter assumes a specified value β_0, one computes $\hat{\beta} / \beta_0$.

The acceptance criteria for hypothesis tests on a percentile and on the shape parameter are summarized in Table 6.2.

Table 6.2   Acceptance Regions for Tests on a Weibull Percentile

Alternate Hypothesis	Accept If
$x_p < (x_p)_0$	$\hat{\beta} \ln\left[\dfrac{\hat{x}_p}{(x_p)_0}\right] > u_\alpha(r, n, p)$
$x_p > (x_p)_0$	$\hat{\beta} \ln\left[\dfrac{\hat{x}_p}{(x_p)_0}\right] < u_{1-\alpha}(r, n, p)$
$x_p \neq (x_p)_0$	$u_{\frac{\alpha}{2}}(r, n, p) < \hat{\beta} \ln\left[\dfrac{\hat{x}_p}{(x_p)_0}\right] < u_{1-\frac{\alpha}{2}}(r, n, p)$

Table 6.2 (*Continued*)

Alternate Hypothesis	Accept If
$\beta < \beta_0$	$\dfrac{\hat{\beta}}{\beta_0} > v_\alpha(r,n)$
$\beta > \beta_0$	$\dfrac{\hat{\beta}}{\beta_0} < v_{1-\alpha}(r,n)$
$\beta \neq \beta_0$	$v_{\alpha/2}(r,n) < \dfrac{\hat{\beta}}{\beta_0} < v_{1-\alpha/2}(r,n)$

Example

A materials scientist claims to have a new finishing process that will increase the flexure strength of a ceramic material. To test this claim a group of 30 specimens so finished will be fractured in a three-point bending test under conditions for which the historical value of the scale parameter is 1000 MPa.

1. Set up the critical region for a 10% level significance test of this claim. The null hypothesis is,

$$H_0 : x_{0.632} = \eta_0 = (x_{0.632})_0 = 1000.0$$

The alternative hypothesis is

$$H_1 : x_{0.632} > (x_{0.632})_0$$

The significance level is $\alpha = 0.10$.
The null hypothesis of no improvement is accepted if

$$\hat{\beta} \ln\left(\frac{\hat{\eta}}{1000}\right) < u_{0.90}(30, 30, 0.632)$$

Running the Pivotal.exe software described in Chapter 7, it is determined that

$$u_{0.90}(30, 30, 0.632) = 0.256$$

2. Assume the flexure strength test is run and the raw ML estimates are calculated to be $\hat{\eta} = 1210$ MPa and $\hat{\beta} = 5.10$. To test the above hypothesis, compute:

$$\hat{\beta} \ln\left(\frac{\hat{\eta}}{1000}\right) = 5.10 \ln\left(\frac{1210}{1000}\right) = 0.972$$

Since $0.972 > 0.256$, the scientist's claim is accepted.

Example

For a shape parameter $\beta = 1.0$ the Weibull distribution reduces to the exponential distribution and the hazard function becomes a constant.

It is desired to test the hypothesis that bearing fatigue life is exponential rather than Weibull using an uncensored life test of 30 bearings. Set up the acceptance region for this test using a 5% level of significance.

The null hypothesis is,

$$H_0 : \beta = \beta_0 = 1.0.$$

The alternative is

$$H_1 : \beta \neq 1.0.$$

The null hypothesis will be accepted if $\hat{\beta} / \beta_0$ falls in the following interval,

$$v_{0.05}(r, n) < \frac{\hat{\beta}}{\beta_0} < v_{0.95}(r, n).$$

Running Pivotal.exe, the acceptance region is found to be:

$$0.788 < \frac{\hat{\beta}}{\beta_0} < 1.402.$$

Since the hypothesized value $\beta_0 = 1.0$, the exponential hypothesis will not be rejected if $\hat{\beta}$ is between 0.788 and 1.402. It is important to remember that failure to reject a hypothesis does not prove that the hypothesis is true. It simply means that the data do not offer evidence to dispute the claim embodied in the null hypothesis.

6.4   OPERATING CHARACTERISTIC (OC) CURVES FOR ONE-SIDED TESTS OF THE WEIBULL SHAPE PARAMETER

The hypothesis tests discussed in the previous section are constructed so as to accept the null hypothesis when it is true with a probability $1 - \alpha$, since, by definition, α is the probability of incorrectly rejecting H_0. Obviously, if the true parameter value were close to but not equal to the value specified by the null hypothesis, the probability of accepting the null hypothesis would still be close to $1 - \alpha$. As the true value departs more and more from the null hypothesized value, the probability of accepting the null hypothesis decreases. A plot of the probability P_a of accepting the null hypothesis against the true parameter value is known as the operating characteristic (OC) curve of the test and is a function of sample size. It is very useful for informing an experimenter whether his or her test has sufficient discriminating power and hence whether the sample size and amount of censoring selected are adequate for the purpose of the test.

OC curves may be constructed for the one-sided hypothesis tests given in Section 6.3.

Consider the test of the hypothesis $\beta = \beta_0$ against the alternative $\beta < \beta_0$. Assume the true value of $\beta = \beta_1$ and is, in fact, less than β_0. Erroneously accepting a hypothesis as true when it is false is termed a type II error in the statistical literature. This hypothesis will be erroneously accepted if,

$$\frac{\hat{\beta}}{\beta_0} > v_\alpha(r, n).$$

The probability that this is so, given that $\beta = \beta_1$ is formally expressible as:

$$P_a = \text{Prob}\left[\hat{\beta}/\beta_0 > v_\alpha(r, n) \mid \beta = \beta_1\right]. \tag{6.7}$$

Given that $\beta = \beta_1$, then $\hat{\beta}/\beta_1$ will follow the distribution of $v(r, n)$. The probability of acceptance may then be rewritten as:

$$P_a = \text{Prob}\left[\frac{\hat{\beta}}{\beta_1} = v(r, n) > \frac{\beta_0}{\beta_1} \cdot v_\alpha(r, n)\right]. \tag{6.8}$$

Since the probability that $v(r, n)$ exceeds $\beta_0/\beta_1 \cdot v_\alpha(r, n)$ is equal to P_a, then $\beta_0/\beta_1 \cdot v_\alpha(r, n)$ must equal the $100(1 - P_a)$th percentile of $v(r, n)$. Comparable reasoning applies to the test against the alternative $\beta > \beta_0$.

For the test of the hypothesis that $\beta = \beta_0$, the probability of acceptance, P_a, is related to the true value β_1 as summarized below.

For the alternative $\beta < \beta_0$ the value of β_1 is computed as:

$$\beta_1 = \frac{\beta_0 v_\alpha(r, n)}{v_{1-P_a(r, n)}}$$

For the alternative $\beta > \beta_0$ the value of β_1 is computed as:

$$\beta_1 = \frac{\beta_0 v_{1-\alpha}(r, n)}{v_{P_a(r, n)}}$$

Example

Determine several points on the OC curve for the hypothesis test that $\beta = 2.0$ against the alternative $\beta < 2.0$ at the 5% level of significance ($\alpha = 0.05$) using a type II censored life test with $r = 15, n = 30$.

Some percentage points of $v(15, 30)$ are shown in column 1 of Table 6.3. Using $v_{1-\alpha} = v_{0.95} = 1.703$, one divides $\beta_0 v_{1-\alpha}(r, n) = 2.0 \times 1.703$ by each of the percentage points in Table 6.3 to give β_1. These values are shown in column 2. The value of P_a associated with each β_1 value thus calculated is the complement of the corresponding quantile of the values in column 1. These values of P_a are written in column 3. A plot of P_a against β_1 forms the OC curve for the test and is shown in Figure 6.2. Additional points beyond those in Table 6.3 have been calculated to give the curve a smooth appearance.

Table 6.3 Computation of the OC Curve for a Test on the Weibull Shape Parameter, H_0: $\beta = 2.0$, H_1: $\beta < 2.0$, $n = 30$, $r = 15$, $\alpha = 0.05$

Selected Quantiles of v (15, 30)	β_1	P_a
$v_{0.05} = 0.7406$	4.60	0.05
$v_{0.10} = 0.8026$	4.24	0.10
$v_{0.25} = 0.9228$	3.69	0.25
$v_{0.50} = 1.090$	3.12	0.50
$v_{0.75} = 1.293$	2.63	0.75
$v_{0.90} = 1.525$	2.23	0.90
$v_{0.95} = 1.703$	2.0	0.95

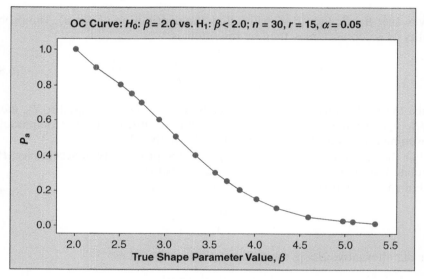

Figure 6.2 P_a versus β for testing H_0: $\beta = 2.0$ versus $\beta < 2.0$; $n = 30$, $r = 15$, $\alpha = 0.05$.

According to Table 6.3, the true shape parameter value would need to be greater than 4.60 before the probability of accepting $\beta = 2.0$ drops to below 5%.

If this amount of discrimination is unacceptable to an experimenter he should choose a larger sample size. For this example, assume that the experimenter wished to have $P_a = 0.05$ for $\beta_1 = 3.0$. He must therefore find a sample size n and number of failures r for which:

$$\frac{v_{0.10}(r, n)}{v_{0.90}(r, n)} = \frac{\beta_1}{\beta_0} = \frac{3.0}{2.0} = 1.50.$$

Table 6.4 shows values of $v_{0.95}/v_{0.05}$ for uncensored sample sizes ranging from 20 to 50 in increments of 5. It is seen that an uncensored sample size between 40 and 45 will meet the experimenter's objectives.

Table 6.4   Ratio of $v_{0.95}/v_{0.05}$ for Sample Size Selection

$n = r$	$v_{0.95}(r, n)$	$v_{0.05}(r, n)$	$\dfrac{v_{0.95}}{v_{0.05}}$
20	1.450	0.798	1.81
25	1.374	0.8061	1.70
30	1.330	0.8206	1.62
35	1.300	0.8332	1.56
40	1.279	0.8410	1.52
45	1.251	0.8470	1.48
50	1.236	0.8990	1.45

The power of a hypothesis test is defined as the probability of correctly rejecting a hypothesis when it is false. The power curve is the complement of the OC curve; that is, it is a plot of $1 - P_a$ against the set of alternative parameter values.

6.5   OC CURVES FOR ONE-SIDED TESTS ON A WEIBULL PERCENTILE

For the test of the hypothesis $x_p = (x_p)_0$, the probability of acceptance varies with the true value of x_p, say $(x_p)_1$. Consider the alternative H_1: $(x_p)_1 < (x_p)_0$. The null hypothesis will be accepted if:

$$\hat{\beta} \ln\left(\frac{\hat{x}_p}{(x_p)_0}\right) > u_a(r, n, p). \tag{6.9}$$

Add and subtract $\hat{\beta} \ln\left(\dfrac{(x_p)_1}{(x_p)_0}\right)$ to the left-hand side. After some rearrangement, the acceptance criterion becomes,

$$\hat{\beta} \ln\left(\frac{\hat{x}_p}{(x_p)_1}\right) + \hat{\beta} \ln\left(\frac{(x_p)_1}{(x_p)_0}\right) > u_a(r, n, p). \tag{6.10}$$

The leftmost term is now $u(r,n,p)$ since $(x_p)_1$ is the true value of x_p. After multiplying and dividing the second term by β and introducing $v = \hat{\beta}/\beta$, the acceptance criterion can be rewritten as:

$$u(r, n, p) + v \ln\left(\frac{(x_p)_1}{(x_p)_0}\right)^{\beta} > u_a(r, n, p). \tag{6.11}$$

The probability P_a that the null hypothesis will be accepted is therefore the probability that

$$P_a = \text{Prob}\left[s \equiv \frac{u_\alpha - u}{v} < \ln\left(\frac{(x_p)_1}{(x_p)_0} \right)^\beta \right]. \tag{6.12}$$

In this expression the dependence on r, n, and p have been suppressed for simplicity of notation. The random variable s, which depends on α, r, n, and p, is defined as:

$$s(\alpha, r, n, p) \equiv \frac{u_\alpha(r, n, p) - u(r, n, p)}{v}. \tag{6.13}$$

The expression for P_a establishes that $\ln\left(\frac{(x_p)_1}{(x_p)_0} \right)^\beta$ is the $100P_a$-th percentage point of the distribution of s. From this, one may solve for $(x_p)_1$ as summarized in Table 6.5 for the two one-sided alternative hypotheses.

For the alternative $(x_p)_1 < (x_p)_0$,

$$\left[\frac{(x_p)_1}{(x_p)_0} \right]^\beta = \exp[s_{p\alpha}(\alpha, r, n, p)].$$

For the alternative $(x_p)_1 > (x_p)_0$,

$$\left[\frac{(x_p)_1}{(x_p)_0} \right]^\beta = \exp[s_{1-p\alpha}(1-\alpha, r, n, p)].$$

The distribution of $s(\alpha, r, n, p)$ must be determined by Monte Carlo simulation. Percentage points of $s(\alpha, r, n, p)$ are given in McCool (1974) for 17 combinations of n and r, $\alpha = 0.10$ and 0.90, and $p = 0.10$ and 0.50. Other cases can be determined using a combination of the software described in the next chapter and a spreadsheet like Excel or a statistical package. This process is described and illustrated in Section 7.4 of Chapter 7. It is noted that the OC curve of a test on a Weibull quantile depends on the true but unknown value of β. In practice one must use a reasonable value obtained from experience for the purpose of selecting sample sizes. To be conservative one should err on the side of selecting smaller values of β.

Table 6.5   Relation between Acceptance Probability P_a and Alternative Values of the $100p$-th Percentile

Alternative Hypothesis	Relationship between P_a and $(x_p)_1$
$(x_p)_1 < (x_p)_0$	$(x_P)_1 = (x_P)_0 \exp[s_{Pa}(\alpha, r, n, p)/\beta]$ or $[(x_P)_1/(x_P)_0]^\beta = \exp[s_{Pa}(\alpha, r, n, p)]$
$(x_p)_1 > (x_p)_0$	$(x_P)_1 = (x_P)_0 \exp[s_{1-Pa}(1-\alpha, r, n, p)/\beta]$ or $[(x_P)_1/(x_P)_0]^\beta = \exp[s_{1-Pa}(1-\alpha, r, n, p)]$

Table 6.6 gives 15 percentage points of $s(0.10,10,10,0.10)$. These are the values needed for determining the OC curve for a one-sided 10% level test of the hypothesis that $x_{0.10} = (x_{0.10})_0$ against the one-sided alternative that $x_{0.10} < (x_{0.10})_0$ using an uncensored sample of size $n = r = 10$.

The probability of acceptance is shown plotted in Figure 6.3.

Table 6.6   Percentage Points of $s(0.10,10,10,0.10)$ and Computation of P_a

$100p \% = P_a$	$s(0.10,10,10,0.10)$	$\left(\dfrac{(x_{0.10})_1}{(x_{0.10})_0}\right)^{\beta} = \exp[s_{Pa}]$
1	−1.943	0.143
2	−1.832	0.160
5	−1.652	0.192
10	−1.487	0.226
20	−1.268	0.281
30	−1.104	0.332
40	−0.9606	0.383
50	−0.820	0.440
60	−0.6607	0.517
70	−0.5015	0.606
80	−0.2924	0.747
90	0.000	1
95	0.2316	1.260
98	0.5360	1.71
99	0.7421	2.10

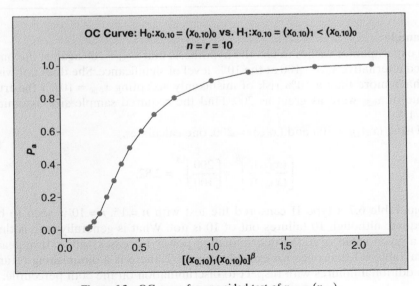

Figure 6.3   OC curve for one-sided test of $x_{0.10} = (x_{0.10})_0$.

Finding the sample size to achieve a required level of discrimination will now be discussed. Column 3 of Table 6.7 gives values of the critical value of u for conducting the test of H_0: $x_{0.50} = (x_{0.50})_0$ against the alternative H_1: $x_{0.50} < (x_{0.50})_0$ with $\alpha = 0.10$ for 15 pairs of r and n values. Column 4 gives the value of the 90th percentile of $s(0.90,r,n,0.50)$. The last column gives the ratio of the true to the hypothesized $x_{0.50}$ for which $P_a = 0.10$, raised to the power β.

Table 6.7   Values of $u_{0.90}(r,n,0.50)$ and $\left[\dfrac{(x_{0.50})_1}{(x_{0.50})_0}\right]^\beta$ for $P_a = 10\%$; H_1:$(x_{0.50})_1 < (x_{0.50})_0$

r	n	$u_{0.90}(r,n,0.50)$	$s_{0.90}(0.90,r,n,0.50)$	$\left[\dfrac{(x_{0.50})_1}{(x_{0.50})_0}\right]^\beta$
5	5	−1.07	−1.96	7.10
5	10	−0.546	−1.25	3.49
10	10	−0.603	−1.19	3.29
5	15	−0.429	−1.21	3.35
10	15	−0.444	−0.944	2.57
15	15	−0.462	−0.932	2.54
5	20	−0.420	−1.31	3.71
10	20	−0.367	−0.849	2.34
15	20	−0.382	−0.791	2.21
20	20	−0.392	−0.786	2.19
5	30	−0.459	−1.51	4.52
10	30	−0.329	−0.845	2.33
15	30	−0.298	−0.677	1.96
20	30	−0.292	−0.622	1.86
30	30	−0.305	−0.616	1.85

Example

An experimenter wishes to test the hypothesis that $x_{0.50} = 100$ against the one-sided alternative $x_{0.50} > 100$ at the 10% level of significance. She does not wish to have more than a 10% risk of mistakenly accepting $x_{0.50} = 100$ if the true value of $x_{0.50}$ were as great as 200. Find the required sample size assuming $\beta = 1.5$.

Using $(x_{0.50})_0 = 100$ and $(x_{0.50})_1 = 200$, one calculates,

$$\left[\frac{(x_{0.50})_1}{(x_{0.50})_0}\right]^\beta = \left[\frac{200}{100}\right]^{1.5} = 2.82.$$

From Table 6.7 a type II censored life test with $n = 15$, $r = 10$ is seen to be adequate although 10 failures out of 10 is not. What is generally true is that for a given value of n, the discriminating power increases with r. It appears from Table 6.7 that once r/n exceeds about 1/2, there is a diminishing return on additional failures with respect to discrimination on the 50th percentile.

Selecting between life test sample sizes that give nearly equivalent protection can be made on the basis of other considerations such as the relative cost of specimens and test time.

Assuming the experimenter chooses to test $n = 15$ specimens until $r = 10$ failures occur, the test procedure will be to form the ML estimates $\hat{x}_{0.10}$ and $\hat{\beta}$ and accept $x_{0.50} = 100$ if:

$$\hat{\beta} \ln\left(\frac{\hat{x}_{0.50}}{100}\right) < -0.444.$$

6.6 GOODNESS OF FIT

6.6.1 Completely Specified Distribution

A probability plot is the first line of defense against adopting the wrong model for a set of data. Minitab has a distribution identification module that creates plots of the user's data on up to 10 different grids corresponding to various alternative distributions. Obvious curvature in the plotted data may be taken as evidence against a given distribution. However, there are many formal tests of goodness of fit, with a rich literature surrounding them. The book edited by D'Agostino and Stephens (1986) is a good guide to the subject and the literature prior to 1986.

The foremost test for goodness of fit is the chi-square test covered in most introductory texts on statistics. This test is generally considered appropriate only for large samples and is not recommended for the range of sample sizes typically used in life testing.

More useful are the class of goodness-of-fit tests referred to as empirical distribution function (EDF) tests. The EDF based on a complete ordered sample $x_{(1)} < x_{(2)} < x_{(n)}$ is defined as:

$$F_n(x) \equiv \frac{i}{n}; \; x_{(i)} < x < x_{(i+1)}.$$

$F_n(x)$ is a staircase function having jumps of size $1/n$ at each successive ordered observed value of the random variable. The way to measure large discrepancies between $F_n(x)$ and a proposed cumulative distribution function (CDF) $F(x)$ is what differs among the various tests of the EDF class. The oldest EDF test is the Kolmogorov–Smirnov (K–S) test. (Kolmogorov, 1933; Smirnov, 1939). It is based on the absolute value of the largest difference D between $F_n(x)$ and some proposed completely specified CDF $F_0(x)$ over the range of x values in the sample.

The null hypothesis in this test is,

$$H_0 : F(x) = F_0(x), \tag{6.14}$$

and the alternative is

$$H_1 : F(x) \neq F_0(x). \tag{6.15}$$

This test is applicable when $F_0(x)$ is completely specified, that is, both its form and its parameter values are set forth. An example might be $F_0(x) = W(\eta = 56.46, \beta = 1.3)$. The alternative hypothesis is that $F(x)$ is not equal to this specific distribution; that is, $F(x)$ is some other distribution, possibly even a member of the same family as the hypothesized distribution. Because of the generality of this alternative, some refer to tests of this type as "omnibus" tests. This is as opposed to the case where the hypothesized distribution is pitted against a specific alternative. We will consider a test of this type in Section 6.7 where the hypothesized distribution is Weibull and the alternative is lognormal and vice versa. In Section 6.3 we considered testing the null hypothesis that a data sample is exponentially distributed against the alternative that it is Weibull.

The value of D is readily computed in three stages. First compute D^+ as:

$$D^+ = \max\left[\frac{i}{n} - F_0(x_{(i)})\right]. \tag{6.16}$$

Next compute D^- as:

$$D^- = \max\left[F_0(x_{(i)}) - \frac{(i-1)}{n}\right]. \tag{6.17}$$

The final value D is then:

$$D = \max(D^+, D^-). \tag{6.18}$$

A closely related variant of the K–S test is due to Kuiper (1960), who proposed the test statistic V, computed as:

$$V = D^+ + D^-. \tag{6.19}$$

One-sided tests of whether the distribution exceeds or is smaller than the hypothesized distribution may be based on the distribution of D^+ or D^-. Most usually the two-sided test based on D is conducted. If $F_0(x)$ is the true distribution, then the distribution of D is the same no matter what the form of the true distribution is. In this sense the K–S test is said to be a nonparametric test. It does not depend on the parametric form of any specific distribution. Large values of D argue against the hypothesized distribution. Upper percentage points are taken to define the acceptance region for the test. Critical values of D for various values of the type I error α are given compactly by Stephens as a constant D_α, which depends upon the desired type I error rate and which is adjusted for sample size by dividing by the term $\left(\sqrt{n}+0.12+\dfrac{0.11}{\sqrt{n}}\right)$ (Stephens, 1974). He lists $D_{0.05}$ as 1.358 so for a sample of size $n = 10$, the hypothesized distribution should be rejected at the 5% level of significance if D is found to exceed $1.358/3.317 = 0.409$. As noted, since the hypothesized distribution is completely specified, the critical values of D do not depend on the form of the hypothesized distribution, be it normal, Weibull, or any other.

Example

As an example application of the K–S test for a completely specified distribution we will compute the K–S test statistic for the data of Table 5.1, which was generated by simulation from the Weibull population having $\eta = 56.46$ and $\beta = 1.3$. We will test to see if the data are consistent with this distribution.

Table 6.8 Computation of the Kolmogorov–Smirnov Test Statistic

$x_{(i)}$	$i/10$	$F(x_{(i)}) = 1 - \exp -\left[\dfrac{x_{(i)}}{56.46}\right]^{1.3}$	$i/10 - F(x_{(i)})$	$F(x_{(i)}) - (i-1)/10$
14.01	0.1	0.150702	−0.050702	0.150702
15.38	0.2	0.168404	0.031596	0.068404
20.94	0.3	0.240752	0.059248	0.040752
29.44	0.4	0.348775	0.051225	0.048775
31.15	0.5	0.369703	0.130297	−0.030297
36.72	0.6	0.435392	0.164608	−0.064608
40.32	0.7	0.475613	0.224387	−0.124387
48.61	0.8	0.560955	0.239045	−0.139045
56.42	0.9	0.631782	0.268218	−0.168218
56.97	1.0	0.636421	0.363579	−0.263579
			$D^+ = 0.363579$	$D^- = 0.150702$

The sorted data are given in column 1 of Table 6.8 and the EDF is given in column 2. Column 3 shows the computed value of $F(x)$ at each observed sample value. Column 4 gives the computed values of $\dfrac{i}{n} - F_0(x_{(i)})$ and column 5 the computed values of $F_0(x_{(i)}) - \dfrac{(i-1)}{n}$. The values of D^+ and D^- are listed in the last row of columns 4 and 5, respectively. The test statistic D is then:

$$D = \max\left(D^+ = 0.363579, D^- = 0.150702\right) = 0.363579.$$

Since D is less than 0.409, we cannot reject that the data came from this specified distribution at the 5% level of significance.

For type I and type II censored samples, modified expressions for the K–S statistic and the associated tables of critical values are given in chapter 4 of D'Agostino and Stephens (1986).

6.6.2 Distribution Parameters Not Specified

It will be rare for an experimenter to wish to test his or her data against a completely specified distribution. A more likely question to ask of the data is whether they fit a specified form, regardless of the parameter values. An example is "do my data follow a Weibull distribution?" This is called a "composite" hypothesis by many writers. The natural thing to do in this case is to first estimate the parameters, substitute the estimates into the hypothesized distribution form and calculate the test statistic in the usual way. Unfortunately the critical values will change when this is done and their values will be specific to the form of the hypothesized distribution and the method used to estimate the parameters. Lilliefors (1967) was the first to publish tables of the critical values of the K–S statistic for use in testing whether an uncensored sample is drawn from an exponential population using the ML estimate of its mean. Critical values for uncensored samples from the two parameter Weibull distribution with parameters estimated by the method of ML have been computed by Littell et al. (1979) and Chandra et al. (1981). Evans, Johnson, and Greene (EJG; Evans et al., 1989) have extended these tables up to sample sizes as large as 400. They found that the critical values D_α for $\alpha = 0.01$. 0.05 and 0.10 are well approximated by the following functions of sample size:

$$D_{0.10} = 0.82645983 - \left(\frac{0.199103}{\sqrt{n}} \right) \tag{6.20}$$

$$D_{0.05} = 0.89820336 - \left(\frac{0.221577}{\sqrt{n}} \right) \tag{6.21}$$

$$D_{0.01} = 1.04550210 - \left(\frac{0.282595}{\sqrt{n}} \right). \tag{6.22}$$

Repeating the calculation in Table 6.8 but using the estimated parameters $\hat{\eta} = 39.56$ and $\hat{\beta} = 2.58$ the test statistic D is computed to be 0.124158. The 5% level critical value for a sample of size $n = 10$ is computed from the expression above for $D_{0.05}$ to be 0.828. Since $0.124 < 0.828$ one cannot reject the hypothesis that the data are Weibull distributed.

Competing with the K–S test are members of the class of tests based on the squared difference between $F_n(x)$ and $F_0(x)$ over the range of x and weighted in various ways. These include the Cramer–von Mises W^2, Watson's U^2, and the Anderson–Darling (AD) A^2 tests.

Littell et al. (1979) have studied five goodness-of-fit tests of the null hypothesis that $F(x)$ is of Weibull form based on ML-estimated Weibull parameters using simulated uncensored samples randomly drawn from six non-Weibull distributions. The proportion of times a test correctly rejects that the data came from a Weibull population is a measure of the power of the test. On this basis Littell et al. recommend A^2 and U^2 as the best overall choices. Wozniak (1994) used simulation to assess the power of five EDF-based tests including D, A^2, and W^2 using parameters estimated by ML as well as by two other techniques. She recommends W^2 and A^2 as more powerful than the other three.

Evans, Johnson, and Greene evaluated three goodness-of-fit tests namely, D, A^2, and another non-EDF–based test employing correlation that they call Rwe, for testing the two-parameter Weibull. They used complete samples and parameters estimated by the ML method. They concluded that A^2 was most powerful when tested against four alternative distributions. Weibull probability plots computed using the Minitab software include the value of A^2 computed using the ML estimates.

The computing formula for A^2 in uncensored samples is:

$$A^2 = -\frac{1}{n}\sum_{i=1}^{n}(2i-1)\left[\ln U_{(i)} + \ln U_{(n+1-i)}\right] - n \qquad (6.23)$$

where

$$U_{(i)} = 1 - \exp\left[-\left(\frac{x_{(i)}}{\hat{\eta}}\right)^{\hat{\beta}}\right]. \qquad (6.24)$$

Pettitt and Stephens (1974) proposed the following alternative form when the sample is type II censored at the r-th order statistic:

$$A^2 = -\frac{1}{n}\sum_{i=1}^{r}(2i-1)\left[\ln U_{(i)} - \ln(1-U_{(i)})\right]$$
$$-2\sum_{i=1}^{r}\ln(1-U_{(i)}) - \frac{1}{n\left[(r-n)^2\ln(1-U_{(r)}) - r^2\ln U_{(r)} + n^2 U_{(r)}\right]}. \qquad (6.25)$$

In the completely specified case when η and β are the hypothesized true parameter values, the quantities $U_{(i)}$ become order statistics from the uniform distribution when the hypothesis is true. When the parameters are estimated, however, they are no longer uniform order statistics and hence simulation must be used to find the distribution of A^2. Critical values A_{α}^2 valid when the distribution is completely specified as given by Stephens are independent of n provided $n > 5$. The critical values in Table 6.9 are extracted from Stephens' (1974) tables.

When the parameters are estimated by the ML method, the critical values do depend on sample size. The values in Table 6.10, valid for complete samples of size n, are excerpted from Evans et al. (1989).

Table 6.9   Critical Values of AD Statistic Completely Specified Uncensored Samples

A_α^2	1.933	2.492	3.857
α	0.10	0.05	0.01

Table 6.10   EJG Critical Values of A^2 for Weibull Parameters Estimated by ML Uncensored Samples

n	$\alpha = 0.10$	$\alpha = 0.05$	$\alpha = 0.01$
10	0.6171	0.7277	0.9878
15	0.6222	0.7372	1.0065
20	0.6265	0.7433	1.012
25	0.6236	0.7458	1.0179
30	0.6302	0.7489	1.0235
40	0.6320	0.7550	1.0362
50	0.6336	0.7559	1.0405

Example

Table 6.11 again shows the ordered, uncensored sample data that we have been using in all of our examples. Column 3 lists the values of $U_{(i)}$ computed as:

$$U_{(i)} = 1 - \exp\left[\frac{x_{(i)}}{56.46}\right]^{1.3}.$$

The next column gives $U_{(n-i+1)}$. This column is just column 3 sorted in descending order. Row i of column 4 contains $-\frac{1}{n}(2i-1)\{\ln U_{(i)} + \ln(1 - U_{(n-i+1)})\}$.

Table 6.11   Sample Calculation of AD Statistic Uncensored Completely Specified Case

Order i	$X_{(i)}$	$U_{(i)}$	$U_{(n-i+1)}$	$-(2i-1)[\ln U_{(i)} + \ln(1 - U_{(n+i-1)})]/10$
1	14.01	0.150702	0.636421	0.2904208913
2	15.38	0.168404	0.631782	0.8341410734
3	20.94	0.240752	0.560955	1.123569658
4	29.44	0.348775	0.475613	1.189197975
5	31.15	0.369703	0.435392	1.410010676
6	36.72	0.435392	0.369703	1.422380007
7	40.32	0.475613	0.348775	1.523665365
8	48.61	0.560955	0.240752	1.280313832
9	56.42	0.631782	0.168404	1.094153467
10	56.97	0.636421	0.150702	1.168956486
				11.33680943

The test statistic A^2 is the sum of the values in the last column minus the sample size $n = 10$ or 1.337. This is not significant at the 5% level since $A^2 < 2.492$ the 5% critical value given by Stephens for a completely specified distribution.

Repeating the calculation using the estimated parameter values $\hat{\beta} = 2.581$ and $\hat{\eta} = 39.56$ gives an A^2 value of 0.261, which is less than the 5% critical value, 0.7277, from the EJG table excerpt listed in Table 6.10, applicable when the parameters are estimated using the ML method.

The results show that we cannot reject either that the data are from a Weibull population or that the data come from the specific population $W(56.46, 1.3)$.

6.6.3 Censored Samples

Most of the work on the censored case applies when the hypothesized distribution is completely specified. Barr and Davidson (1973) give critical values of the K–S statistic applicable to type I and type II censoring when the distribution is completely specified. Guilbaud (1988) outlines a K–S approach to the multiply censored case based on the Kaplan–Meier product limit estimate of the distribution function. Hollander and Proschan (1979) suggest another approach to a multiply censored completely specified goodness-of-fit test. Koziol (1980) considered versions of the K–S and W^2 statistics for multiply censored data.

Michael and Schucany (1984) propose a method for modifying singly right censored data so that uncensored completely specified techniques can be applied. Stephens (D'Agostino and Stephens, 1986) offers a modified form of the A^2 computing formulas for singly type I or type II right censoring and gives critical values for the completely specified hypothesis as a function of sample size n, censoring fraction r/n, and type I error probability α.

For the composite case, Aho, Bain et al. provide tables of critical values for the K–S, Cramer–von Mises, and Kuiper tests for singly right censored data (Aho et al., 1983, 1985).

Wozniak and Li (1990) offer tables of critical values for a 5% level test using ML parameter estimates with single right censoring.

6.6.4 The Program ADStat

Recently, Mehta (2010) developed software for computing the null distribution of the AD statistic for censored and uncensored samples when the null distribution is the two-parameter Weibull distribution with the parameters either specified or estimated by the ML method. The software is written in Visual Basic and has been dubbed ADStat. It may be downloaded from the

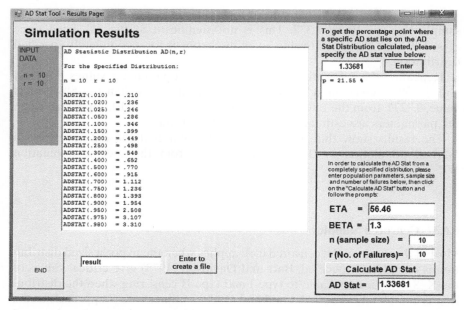

Figure 6.4   Output of ADStat for the completely specified case $n = r = 10$.

author's website along with a Readme file containing operating instructions. The software is based on simulating 10,000 type II censored samples with user-specified values of n and r. The 10,000 values are drawn from the Weibull population $W(1,1)$. For the completely specified case the AD statistic is computed for each of the 10,000 samples using $\beta = \eta = 1$. For the unspecified or composite case the AD statistic is computed after estimating β and η by the ML method for each of the 10,000 samples. In either case the 10,000 values of the AD statistic are sorted and 20 percentage points are displayed. The software is also capable of computing the AD statistic for a user's data sample.

Figure 6.4 shows the program output for the data in Table 5.1 when the Weibull population is completely specified as $W(56.46,1.3)$. The value of the AD statistic is calculated to be 1.33681 as shown on the bottom right-hand side of Figure 6.4. This value agrees with the value found in the manual calculation in Table 6.11. Entering this value in the text box on the upper right shows that the p value is 0.2155 so the hypothesis that the data were drawn from the population $W(56.46,1.3)$ cannot be rejected. Note that the 95th percentile of the distribution of the AD statistic is 2.508, consistent with the value 2.492 proposed by Stephens and given in Table 6.9.

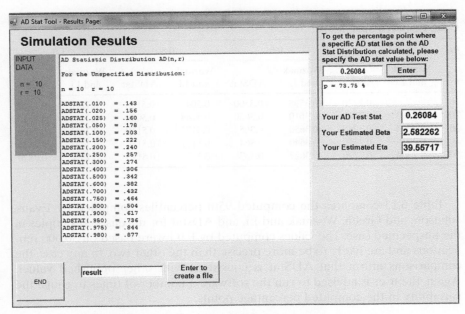

Figure 6.5 Output of ADStat for the unspecified case $n = r = 10$.

Running the program again with the same data sample but using the unspecified option results in the screen shown as Figure 6.5. The ML estimates of the scale and shape parameters and the computed AD statistic are shown on the right of the screen. Again, entering the AD test statistic in the text box on the upper right of the screen gives the p value of 0.7375. The hypothesis that the data follow a two-parameter distribution cannot be rejected. The AD statistic agrees with the value computed manually with the estimated parameters substituted for the population values in Table 6.11. The 95th percentile of the distribution of the AD statistic, 0.736, agrees well with the value 0.7277 determined by EJG and given in Table 6.10.

Table 6.12 shows some comparisons between the program output and the published values of Wozniak and Li for the unspecified case for various values of n and the censoring fraction r/n. Wozniak and Li also used 10,000 simulated samples. The differences exhibited are consistent with the variability found in repeat runs of ADStat. A user of ADStat is advised to run it several times to gauge the variability in the computed percentage points, particularly if the sample size is small and there is appreciable censoring.

Table 6.12 Comparison with 95th Percentile Values of Wozniak and Li, Unspecified Case, Censored Samples

r/n	0.4		0.6		0.8		1.0	
n	Wozniak and Li	ADStat	Wozniak and Li	ADStat	Wozniak and Li	ADStat	Wozniak and Li	ADStat
10		0.2220	0.3598	0.3500	0.5043	0.5130	0.7276	0.7260
20	0.2393	0.2310	0.3610	0.3630	0.5144	0.5060	0.7411	0.7340
30	0.2373	0.2280	0.3650	0.365	0.5172	0.5110	0.7461	0.7400
40	0.2316	0.2370	0.3640	0.364	0.5131	0.5188	0.7441	0.7530
50	0.2364	0.2320	0.3627	0.367	0.5227	0.5113	0.7394	0.7520

Table 6.13 compares the computed 95th percentiles computed by Evans, Johnson, and Green, Wozniak and Li, and ADstat for uncensored samples in the unspecified case. The values computed by EJG were based on 50,000 replications and are likely to be more precise than the other two. In any case, the comparisons affirm that ADStat is consistent with other published values. Again, the user is advised to run the software a number of times to gauge the variability in the computed percentage points.

Table 6.13 Comparison with 95th Percentile Values of Evans, Johnson, and Green Unspecified Case Uncensored Samples

n	Evans, Johnson, and Green	Wozniak and Li	ADStat
10	0.7277	0.7276	0.7260
20	0.7433	0.7411	0.7340
30	0.7489	0.7461	0.7400
40	0.7550	0.7441	0.7530
50	0.7559	0.7394	0.7520

6.7 LOGNORMAL VERSUS WEIBULL

With a small data sample it is difficult to distinguish between a Weibull distribution and a lognormal distribution. Both are employed as life distribution models, but the consequences of erroneously choosing between them can be grave. In the lognormal case the natural logarithm of the random variable follows a normal distribution with a mean μ and standard deviation σ. The natural logarithm of a Weibull-distributed random variable follows the extreme value distribution with mean and variance related to the Weibull parameters as follows:

$$\mu = \ln \eta - \frac{\gamma}{\beta} \tag{6.26}$$

$$\sigma = \frac{\pi}{\beta\sqrt{6}} \qquad (6.27)$$

where $\gamma = 0.5772$ is Euler's constant. For a Weibull random variable having $\eta = 56.46$ and $\beta = 1.3$, the mean and standard deviation of the logarithm of this Weibull variable is $\mu = 3.59$ and $\sigma = 0.987$. A normal random sample of size 10 having this mean and standard deviation was generated and then exponentiated to give the corresponding lognormal sample. These data plotted on a Weibull grid are shown in Figure 6.6.

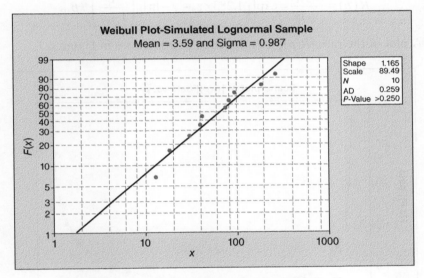

Figure 6.6 Lognormal sample plotted on Weibull grid.

The data show no systematic departure from a straight line and the A^2 statistic of 0.259 is well below the critical value at the 10% level for a sample of size 10 with estimated parameters.

It is instructive to compare the percentiles of a Weibull distribution with the percentiles computed under the lognormal model having the same mean and standard deviation as the Weibull.

It is recalled that for the Weibull model the $100p$-th percentile expressed in terms of its parameters is:

$$(x_p)_W = \eta\left(\ln\left[\frac{1}{1-p}\right]\right)^{\frac{1}{\beta}}. \qquad (6.28)$$

The $100p$-th percentile for the lognormal is:

$$(x_p)_L = \exp[\mu + z_p \cdot \sigma] \qquad (6.29)$$

where z_p is the $100p$-th percentile of the standard normal distribution. Substituting for μ and σ in terms of the Weibull parameters and simplifying gives:

$$(x_p)_L = \eta \exp\left[\frac{\frac{z_p \pi}{\sqrt{6}} - \gamma}{\beta} \right]. \tag{6.30}$$

The ratio $R(p)$ of the lognormal to the Weibull percentile is:

$$R(p) = \frac{(x_p)_L}{(x_p)_W} = \exp\left[\left(1.283 \times z_p - \gamma - \ln\ln\left(\frac{1}{1-p} \right) \right) / \beta \right].$$

A plot of the ratio$[R(p)]^{\beta}$ as a function of p is shown in Figure 6.7.

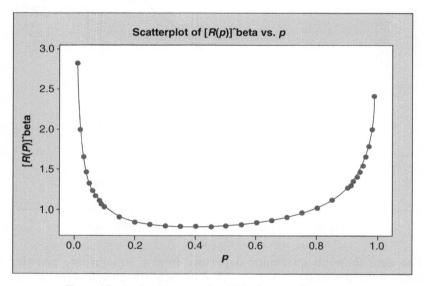

Figure 6.7   Ratio of lognormal to Weibull percentiles versus p.

The high and low percentiles for the lognormal are much larger than those for the corresponding Weibull model. The extremes become larger as the scale parameter β decreases. This plot teaches that if a Weibull model were mistakenly taken as lognormal the predicted values of the low percentage points, that is, the life associated with high reliability, will be overstated.

Dumonceaux and Antle used the ratio of the maximized likelihood (ML) under the two models as the basis for discriminating between them in uncensored samples. (Dumonceaux and Antle, 1973; Dumonceaux et al., 1973). Subsequently Antle and Klimko (1973) showed that the ratio of the ML estimates of the scale parameters of the distribution of the logarithms under the two models was simpler to compute and had comparable power.

Taking logarithms of the data, the ML estimate $\hat{\sigma}$ of the scale parameter is computed in terms of the logarithms of the data as:

$$\hat{\sigma} = \sqrt{\sum_{i=1}^{n} \frac{[z_i - \bar{z}]^2}{n}} = \left[\frac{n-1}{n}\right]^{\frac{1}{2}} s_z. \tag{6.31}$$

The uncorrected ML estimate of σ is computed with a denominator n and not the usual $n - 1$. The ordinary standard deviation must therefore be corrected by multiplying by the factor $\left[\dfrac{n-1}{n}\right]^{1/2}$.

The ML estimate of the scale parameter of the extreme value distribution is the reciprocal of the ML estimate of the Weibull shape parameter.

$$\hat{\xi} = \frac{1}{\hat{\beta}}. \tag{6.32}$$

To test H_0: Weibull versus H_1: Lognormal, the test statistic is $\dfrac{\hat{\xi}}{\hat{\sigma}} = \left(\hat{\sigma}\hat{\beta}\right)^{-1}$. On the other hand, if the null hypothesis is H_0: Lognormal versus H_1: Weibull, the test statistic is $\left(\hat{\sigma}\hat{\beta}\right)$.

The critical values of $\left(\hat{\sigma}\hat{\beta}\right)^{-1}$ for testing at significance levels $\alpha = 0.05$ and 0.10 are excerpted below for n ranging from 20 to 50 in increments of 10.

	20	30	40	50
$\alpha = 0.05$	0.9600	0.9315	0.9105	0.899
$\alpha = 0.10$	0.9254	0.9022	0.8847	0.8752

The critical values of $\left(\hat{\sigma}\hat{\beta}\right)$ are given below:

	20	30	40	50
$\alpha = 0.05$	1.262	1.216	1.187	1.166
$\alpha = 0.10$	1.212	1.177	1.152	1.135

The determination of which to choose as the null hypothesis depends upon which model corresponds to "business as usual." In the bearing industry the inclination would be to assume the Weibull distribution unless the data dictated otherwise and so the Weibull would be a null hypothesized distribution.

Another approach to distinguishing between a Weibull and a lognormal distribution due to Kappenman (1988) is framed as a selection problem rather than a hypothesis testing problem. The experimenter is viewed as having no basis for preferring one or the other distribution. A value r is calculated, and if $r > k$, choose the lognormal. If $r < k$ the Weibull is chosen. The quantity r is computed as follows:

$$r = \frac{A_3 - A_2}{A_2 - A_1}. \tag{6.33}$$

A_1 is the average of the lower 5% of the ordered logarithms and A_3 is the average of the upper 5%. A_2 is the average of what remains after the lower and upper 20% are discarded. Kappenman chooses $k = 0.7477$ for all sample sizes.

The basis for the test is the difference in shape of the extreme value distribution followed by the logarithms of Weibull data and the normal distribution followed by the logarithms of lognormal data. The normal distribution is symmetrical so the numerator and denominator of r will tend to be similar for lognormal data. The extreme value distribution is shown in Figure 6.8 for a location parameter of zero and a scale parameter of 1.0. The basic left skewed shape is the same regardless of the parameter values.

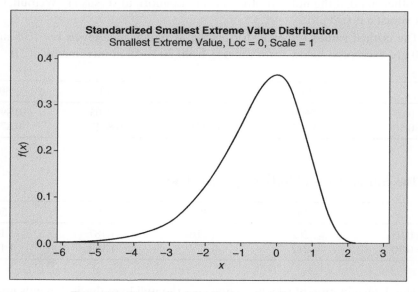

Figure 6.8 Standardized distribution of smallest extremes.

It is clear that because of this skewness, A_3-A_2 will be less than A_2-A_1 and therefore r will tend to be less than 1.0.

In computing the A values, Kappenman uses fractional values when the indicated fractions are not integers. For example, with a sample size of 30, A_1

is the average of the first $0.05 \times 30 = 1.5$ ordered logarithms. A_1 is then computed as:

$$A_1 = \frac{(z_{(1)} + 0.5z_{(2)})}{1.5}$$

Similarly A_3 would be calculated as:

$$A_3 = \frac{(z_{(30)} + 0.5z_{(29)})}{1.5}$$

And A_2 would be computed after discarding the $0.2 \times 30 = 6$ upper and lower order statistics:

$$A_2 = \frac{1}{18} \sum_{i=7}^{24} z_i.$$

Kappenman claims that in samples of size 50, 90% of the samples would be correctly identified as to whether they are lognormal or Weibull using his method. For samples of size 20 the corresponding percentage is 77% and for samples of size 30 it is 83%.

Example
A complete random sample of size 30 was generated from a normal distribution having $\mu = 3.59$ and $\sigma = 0.987$. This is the mean and standard deviation of the distribution of the logarithms of a Weibull population having $\eta = 56.46$ and $\beta = 1.3$. These values were exponentiated and the Weibull parameters were estimated even though this sample is lognormal. The value of the AD statistic A^2 was computed to be 0.753, which is larger than 0.7489, the largest of the three computed critical values in Table 6.13 for $n = 30$ and $\alpha = 0.05$. Thus, the AD test invites skepticism of the Weibull model for this data.

The Weibull parameters were estimated to be $\hat{\beta} = 1.064$ and $\hat{\eta} = 61.75$. The logarithms of the sorted data are tabled below:

1.4171126	2.8473605	3.138776	3.6917256	3.9947116	4.5138741
2.429432	2.8856396	3.1792441	3.7922064	4.2446643	4.615145
2.5617789	3.0469363	3.3558955	3.8552721	4.2881964	4.9614991
2.6524817	3.0654405	3.5946731	3.8881352	4.295375	5.5373569
2.8297682	3.1045878	3.6295689	3.9605876	4.4814342	5.7385071

The corrected value of $\hat{\sigma} = 0.925$. Taking the null hypothesis as the Weibull, the test statistic is:

$$\left(\hat{\sigma}\hat{\beta}\right)^{-1} = \frac{1}{1.064 \times 0.925} = 1.016.$$

This exceeds the critical value of 0.9315 from Antle and Klimko's table for $n = 30$ with $\alpha = 0.05$. Thus, the Weibull is rejected in favor of the lognormal. The values of A_1 and A_3 are:

$$A_1 = (1.4171126 + 0.5 \times 2.429432)/1.5 = 1.754$$

$$A_3 = \frac{5.7385071 + 0.5 \times 5.5373569}{1.5} = 5.479504.$$

A_2 is calculated to be 3.36168. The value of Kappenman's r is then:

$$r = \frac{5.4795 - 3.3617}{3.3617 - 1.754} = 1.317.$$

Since r exceeds $k = 0.7477$, the lognormal is chosen. The tests are in agreement that this data sample is not Weibull.

Suppose the null hypothesis was that the sample came from a lognormal distribution. The test statistic in this case is $(\hat{\sigma}\hat{\beta}) = 1/0.9858 = 1.014$. From the second set of Antle and Klimko's tables above, the critical value for $n = 30$ and $\alpha = 0.05$ is 1.216. Since $1.014 < 1.216$, we cannot reject the null hypothesis that the data are lognormally distributed.

The tests pitting the Weibull and lognormal against each other are more powerful than the AD test in this case. This is not surprising since the AD test is pitting the Weibull against all other distributions and not just against a single distribution.

REFERENCES

Aho, M., L. Bain, et al. 1983. Goodness-of-fit tests for the Weibull distribution with unknown parameters and censored sampling. *Journal of Statistical Computation and Simulation* 18(1): 59–69.

Aho, M., L. Bain, et al. 1985. Goodness-of-fit tests for the Weibull distribution with unknown parameters and heavy censoring. *Journal of Statistical Computation and Simulation* 21(3): 213–225.

Antle, C. and L. Klimko. 1973. Choice of Model for Reliability Studies and Related Topics III, AeroSpace Research Laboratories, United States Air Force.

Barr, D. and T. Davidson. 1973. A Kolmogorov-Smirnov test for censored samples. *Technometrics* 15(4): 739–757.

Chandra, M., N. Singpurwalla, et al. 1981. Kolmogorov statistics for tests of fit for the extreme value and Weibull distributions. *Journal of the American Statistical Association* 76(375): 729–731.

D'Agostino, R.B. and M.A. Stephens. 1986. *Goodness-of-Fit Techniques*. Marcel Dekker Inc., New York.

Dumonceaux, R. and C. Antle. 1973. Discrimination between the log-normal and the Weibull distributions. *Technometrics* 15(4): 923–926.

Dumonceaux, R., C. Antle, et al. 1973. Likelihood ratio test for discrimination between two models with unknown location and scale parameters. *Technometrics* 15(1): 19–27.

Evans, J.W., R.A. Johnson, et al. 1989a. Two- and Three-Parameter Weibull Goodness-of-Fit Tests. Research Paper FPL-RP-493, US Forest Products Laboratory, Madison, WI.

Guilbaud, O. 1988. Exact Kolmogorov-type tests for left-truncated and/or right-censored data. *Journal of the American Statistical Association* 83(401): 213–221.

Hollander, M. and F. Proschan. 1979. Testing to determine the underlying distribution using randomly censored data. *Biometrics* 35(2): 393–401.

Kappenman, R. 1988. A simple method for choosing between the lognormal and Weibull models. *Statistics & Probability Letters* 7(2): 123–126.

Kolmogorov, A. 1933. Sulla Determinazione Empirica di una Legge di Distribuzione. *Giornale dell'Istituto Italiano degli Attuari* 4: 83–91.

Koziol, J. 1980. Goodness-of-fit tests for randomly censored data. *Biometrika* 67(3): 693–696.

Kuiper, N.H. 1960. Tests concerning random points on a circle. *Proceedings of the Koninklijke Nederlandse Akademie van Wetenschappen. Series A* 63: 38–47.

Lilliefors, H. 1967. On the Kolmogorov-Smirnov test for normality with mean and variance unknown. *Journal of the American Statistical Association* 62(318): 399–402.

Littell, R., J. Clave, et al. 1979. Goodness-of-fit tests for the two parameter Weibull distribution. *Communications in Statistics-Simulation and Computation* 8(3): 257–269.

McCool, J. 1999. Life test sample size selection under a Weibull failure model Paper No. 01-2860. International Off-Highway & Powerplant Congress and Exposition. Indianapolis IN, SAE International.

McCool, J.I. 1974. Inferential techniques for Weibull populations. Aerospace Research Laboratories Report ARL TR: 74-0180.

McCool, J.I. 1979. Analysis of single classification experiments based on censored samples for the two-parameter Weibull distribution. *Journal of Statistical Planning and Inference* 3: 39–68.

Mehta, N. 2010. Support Software for Assessing if a Data Set Follows the Two Parameter Weibull Distribution. Malvern PA, Penn State Great Valley School of Graduate Professional Studies. Master of Systems Engineering.

Michael, J. and W. Schucany. 1984. Analysis of data from censored samples. *Contract* 14: 82-K-0207.

Pettitt, A.N. and M.A. Stephens. 1974. Studies of goodness of fit and some comparisons. *Journal of American Statistical Association* 69(347): 730–737.

Phan, L.D. and J.I. McCool. 2009. Exact confidence intervals for Weibull parameters and percentiles. *Proceedings of the IMechE O: Journal of Risk and Reliability* 223: 387–394.

Smirnov, N. 1939. Sur les Ecarts de la Courbe de Distribution Empirique (Russian, French Summary). *Matematicheskii Sbornik* 48(6): 3–26.

Stephens, M.A. 1974. EDF statistics for goodness of fit and some comparisons. *Journal of the American Statistical Association* 69(347): 730–737.

Wozniak, P. 1994. Power of goodness of fit tests for the two-parameter Weibull distribution with estimated parameters. *Journal of Statistical Computation and Simulation* 50(3): 153–161.

Wozniak, P. and X. Li. 1990. Goodness of fit for the two parameter Weibull distribution with estimated parameters. *Journal of Statistical Computation and Simulation* 34(2): 133–143.

EXERCISES

1. For a test of H_0: $\beta = 2.0$, against the alternative H_1: $\beta < 2.0$, find the value of β corresponding to $P_a = 0.10$ when using $\alpha = 0.10$, with an uncensored sample size of $n = 20$.

2. For an uncensored sample of size 10 and $\alpha = 0.10$ in a test of the hypothesis H_0: $x_{0.10} = 100$ against the alternative H_1: $x_{0.10} < 100$, find the value of $x_{0.10}$ for which $P_a = 0.10$. Assume the true value of β is 2.0.

3. For the same hypothesis and shape parameter as in exercise 2, determine the sample size for which $P_a = 0.10$ when $x_{0.10} = (x_{0.10})_1 = 56.9$.

4. The following uncensored sample of size 10, repeated from exercise 2 in Chapter 5, was randomly drawn from $W(100,2)$.

14.6375	66.9203	68.4248	69.2899	71.8208
77.4557	94.8249	100.317	101.453	104.603

Compute the ML estimate of β and use it to test $H_0 = 1.0$ against the two-sided alternative using $\alpha = 0.05$.

5. Use the data sample in exercise 4 and compute the AD statistic to assess whether the sample came from the completely specified distribution $W(100,2)$.

   a. Using the ML estimates of η and β, compute the AD test statistic to assess whether the sample was drawn from a Weibull distribution irrespective of the parameter values.

   b. Using ADStat, determine the critical value for the AD statistics in the two cases using a 10% level of significance.

6. Using the sample in problem 4, test whether the sample comes from a Weibull population against the alternative that it comes from a lognormal population, using Antle and Klimko's test.

CHAPTER 7

The Program Pivotal.exe

As mentioned in Section 5.6.3, a software program named Pivotal.exe has been developed for computing the distribution of pivotal functions needed for the construction of confidence intervals and hypothesis tests related to the shape parameter and percentiles of the two-parameter Weibull distribution based on complete or type II censored samples. The program is general enough to include the pivotal functions applicable to series systems of m identical components so that inferences may be made about the system characteristics in terms of the characteristics of the independent identical components comprising the system. This generalization is discussed in Section 7.2. The distribution of the pivotal functions discussed previously are obtained for the special case that $m = 1$. This generalization to the series framework can be further exploited for analyzing sudden death tests as we will show in Section 7.6. Another program option lets the user name, and write to a text file, the values of the maximum likelihood (ML) shape and scale parameters computed from 10,000 type II censored Weibull samples of user-specified sample size n and number of failures r. These values are generated from populations having true η and β values of 1. They are used within the program to construct and sort 10,000 values of the pivotal quantity $u(r, n, p, m)$ as discussed below. The distribution of the pivotal quantity $v(r, n)$ is directly obtained by sorting the 10,000 values of the ML estimates of the shape parameter since the true β is equal to 1. These output values can be imported into a spreadsheet or statistical package such as Minitab and modified to permit the computation of prediction intervals, confidence limits on reliability, and operating characteristic (OC) curves for the hypothesis tests discussed in Section 6.3. These further applications will be described and illustrated in Sections 7.3 to 7.5.

A zipped copy of the software and a readme file may be downloaded from the author's website. The input screen appears as follows:

Using the Weibull Distribution: Reliability, Modeling, and Inference, First Edition. John I. McCool.
© 2012 John Wiley & Sons, Inc. Published 2012 by John Wiley & Sons, Inc.

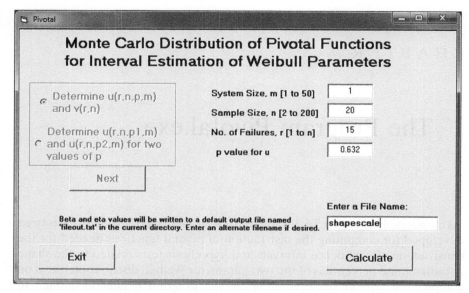

Figure 7.1   Input screen for Pivotal.exe for $m = 1$, $n = 20$, $r = 15$, and $p = 0.632$.

This screen shows the input for a single component system ($m = 1$) for a type II censored sample of size 20 with $r = 15$. For this run, the scale parameter, which is the 0.632 quantile, was of interest so $p = 0.632$ is input. The program simulates 10,000 samples of the specified size and censoring amount using population parameters $\beta = \eta = 1$. An output file named "shapescale" is specified in the case illustrated. The file name is the user's choice. A text file named shapescale.txt will be written to the directory where the program resides. An alternate path may be specified if convenient. This file will contain the 10,000 values of the estimated scale and shape parameters and may be processed further to obtain other useful results as discussed subsequently.

A short table excerpt of the output as captured in an Excel spreadsheet is shown in Table 7.1.

The results screen shown in Figure 7.2 lists 20 standard values of the percentage points of $u(15, 20, 0.632)$ and $v(15, 20)$ corresponding to the input shown in Figure 7.1. If some other nonstandard percentage point is needed, it may be entered in the dialog box on the upper right of the screen. In this case, 0.55 is entered and the resultant quantiles of u and v are shown in Figure 7.2. The user may write the result screen to a text file for record keeping or inclusion in a document. In Figure 7.2 this file was given the name pivotalpercentpoints.txt.

Table 7.1   Sample Text File Output of Pivotal.exe

$\hat{\beta}$	$\hat{\eta}$
0.801503	0.783556
1.339761	1.291463
1.234625	0.989331
0.757292	1.144527
1.885939	0.982971
1.071219	1.033692
1.347778	0.800993
1.14115	0.953514
0.862079	0.947985
1.821859	0.834359

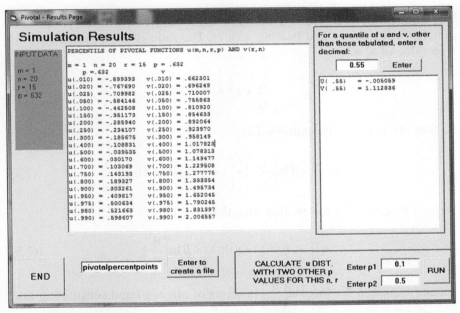

Figure 7.2   Output screen corresponding to the input screen of Figure 7.1.

As shown by Equation 7.6 in section 7.1, the pivotal function needed for inference on a quantile x_q is related to the pivotal function for quantile x_p. The software exploits this relationship. Having computed 10,000 values of v and u for some p, it is easy to recalculate u for another quantile by computing this new function of the u's and v's. This is the basis for the further option to use the same simulation results for a rapid computation of $u(r, n, p, m)$ for two additional values of p. For example, $p_1 = 0.1$ and $p_2 = 0.5$ in the case shown in Figure 7.2.

We now consider the relationship between the pivotal function for the $100p$-th percentile x_p and that for some other percentile, say, x_q.

7.1 RELATIONSHIP AMONG QUANTILES

Recall that the $100p$-th percentile is related to the Weibull parameters as:

$$x_p = k_p^{1/\beta}\eta \tag{7.1}$$

where

$$k_p \equiv -\ln(1-p). \tag{7.2}$$

The estimate of x_p is given by the same expression but with the parameters denoted as estimates and bearing a caret. The ratio of the estimated to the true value of x_p is:

$$\frac{\hat{x}_p}{x_p} = k_p^{\frac{1}{\hat{\beta}}-\frac{1}{\beta}} \cdot \frac{\hat{\eta}}{\eta}. \tag{7.3}$$

So that $u(r, n, p)$ can be expressed as:

$$u(r, n, p) = \hat{\beta}\ln\left(\frac{\hat{x}_p}{x_p}\right) = \left(1-\frac{\hat{\beta}}{\beta}\right)\ln k_p + \hat{\beta}\ln\left(\frac{\hat{\eta}}{\eta}\right). \tag{7.4}$$

Writing Equation 7.4 for another quantile x_q:

$$u(r, n, q) = (1-v)\ln k_q + \hat{\beta}\ln\left(\frac{\hat{\eta}}{\eta}\right). \tag{7.5}$$

Solving Equation 7.4 for $\hat{\beta}\ln\left(\frac{\hat{\eta}}{\eta}\right)$ in terms of $u(r, n, p)$ and substituting into Equation 7.5 gives, after rearrangement,

$$u(r, n, q) = (1-v)\ln\left(\frac{k_q}{k_p}\right) + u(r, n, p). \tag{7.6}$$

The pivotal function needed for inference on quantile x_q is thus related to the pivotal function for quantile x_p. As noted, the software exploits this relationship. Having computed 10,000 values of v and u for some p, it is easy to recalculate u for another q by computing this new function of the u's and v's.

7.2   SERIES SYSTEMS

Consider a series system of m components each of which has an independent lifetime distribution that follows $W(\eta, \beta)$. The life, y, of the system is the minimum of the component lives $x_1, x_2, \ldots, x_m$ and was shown in Section 3.2 to follow the distribution $W(m^{-1/\beta}\eta, \beta)$.

The p-th quantile y_p of the series system is thus expressible in terms of the p-th quantile x_p of a component:

$$y_p = m^{-1/\beta} \cdot x_p. \tag{7.7}$$

Similarly, from an ML estimate $\hat{x}_p$ of the p-th quantile of the component, one may estimate y_p as:

$$\hat{y}_p = m^{-1/\hat{\beta}}\hat{x}_p. \tag{7.8}$$

Thus for the series system,

$$u(r, n, p, m) = \hat{\beta}\ln\left(\frac{\hat{y}_p}{y_p}\right) = \hat{\beta}\left[\ln\left(m^{-1/\hat{\beta}}\hat{x}_p\right) - \ln\left(m^{-1/\beta}x_p\right)\right]. \tag{7.9}$$

Rearranging,

$$u(r, n, p, m) = (v - 1)\ln(m) + u(r, n, p). \tag{7.10}$$

Equation 7.10 defines a generalized version of $u(r, n, p)$ to account for system size m. For $m = 1$, $u(r, n, p, m)$ reduces to $u(r, n, p)$. We will continue to use the notation $u(r, n, p)$ when it is clear from the context that $m = 1$. The random variable $u(r, n, p, m)$ is a function of pivotal functions and so, is itself, pivotal. Analogous to Equation 5.51, percentage points of $u(r, n, p, m)$ may be used to set confidence limits on the p-th quantile of a series system when that system's estimate of the p-th quantile is computed from the corresponding quantile of the component by multiplying by $m^{-1/\hat{\beta}}$.

Comparing Equations 7.6 and 7.10, it is seen that $u(r, n, p, m)$ may alternately be computed as $u(r, n, q)$ after determining q so as to satisfy:

$$m = \frac{k_p}{k_q}. \tag{7.11}$$

Using the defining relations for k_p and k_q and simplifying gives:

$$q = 1 - (1 - p)^{1/m}. \tag{7.12}$$

Example

A housing contains $m = 6$ equally loaded, identical radial ball bearings. A type II censored life test with $n = 20$ and $r = 15$ was conducted on the bearings. The ML estimates of the $x_{0.50}$ life and shape parameter were 65.2 million revolutions and 1.2, respectively. Set 90% confidence limits on the median housing life.

The estimated median of the housing life is:

$$\hat{y}_{0.50} = 6^{-1/1.2} \times 65.2 = 20.54.$$

Running Pivotal.exe with $n = 20$, $r = 15$, and $m = 6$, the following percentage points were obtained:

$$u_{0.05}(15, 20, 0.50, 6) = -0.670 \quad u_{0.50}(15, 20, 0.50, 6) = 0.113$$
$$u_{0.95}(15, 20, 0.50, 6) = 1.363.$$

Using these values, the confidence limits are computed in the usual way:

$$6.60 = 20.54e^{-1.363/1.2} < y_{0.50} < 20.54e^{0.670/1.2} = 35.90.$$

The median unbiased estimate of $y_{0.50}$ is:

$$\tilde{y}_{0.50} = 20.54e^{-\frac{0.113}{1.2}} = 18.69$$

As noted, the percentage points used in this calculation could have been obtained with $m = 1$ if the software had been run with the percentile of interest taken to be x_q with $q = 1 - (1 - 0.5)^{1/6} = 0.109$. It might at first be thought that one could compute a confidence interval for $y_{0.50}$ by first computing a confidence interval with the same confidence coefficient for $x_{0.50}$ and then multiplying both ends of the interval by $m^{-1/\hat{\beta}}$. This is incorrect and understates the variability in the system percentile. The interval that would be computed in this manner in the present example is $(9.68, 22.2)$ and is considerably tighter than the correct interval.

7.3 CONFIDENCE LIMITS ON RELIABILITY

The reliability at life x under the two-parameter Weibull model is:

$$R(x) = \exp\left[-\left(\frac{x}{\eta}\right)^{\beta}\right]. \tag{7.13}$$

An estimate of the reliability at life x may be computed by substituting the ML estimates for the parameters.

$$\hat{R}(x) = \exp\left[-\left(\frac{x}{\hat{\eta}}\right)^{\hat{\beta}}\right]. \tag{7.14}$$

Solving Equation 7.13 for $\ln x$ in terms of the true reliability R results in:

$$\ln x = \frac{1}{\beta}\ln\ln\left(\frac{1}{R}\right) + \ln\eta. \tag{7.15}$$

Taking logarithms twice transforms the expression in Equation 7.14 to:

$$\ln\ln\left(\frac{1}{\hat{R}}\right) = \hat{\beta}\ln x + \hat{\beta}\ln\hat{\eta}. \tag{7.16}$$

Now substituting the expression for $\ln x$ in Equation 7.15 into Equation 7.16 and simplifying results in:

$$\ln\ln\left(\frac{1}{\hat{R}}\right) = v\ln\ln\left(\frac{1}{R}\right) + \hat{\beta}\ln\left(\frac{\hat{\eta}}{\eta}\right). \tag{7.17}$$

The right-hand side of Equation 7.17 involves the pivotal functions v, $\hat{\beta}\ln\left(\dfrac{\hat{\eta}}{\eta}\right)$ and the true reliability value R. For a specified value of true reliability R, one can simulate the distribution of the right-hand side. These values can then be transformed to give values of $\hat{R}$. This distribution of $\hat{R}$ will be contingent on the true value of R. Repeating for other values of R one will generate the family of distributions we might designate $f(\hat{R} \mid R)$. A graph of suitable upper and lower limits of $\hat{R} \mid R$ can then be plotted and the plots used to generate confidence intervals for R based on an observed value of $\hat{R}$. Using the output option of the program Pivotal.exe and importing the output into the first two columns of a spreadsheet, the right-hand side of the expression above can be computed more simply as follows since the population values of the Weibull parameters are unity:

$$\ln\ln\left(\frac{1}{\hat{R}}\right) = \hat{\beta}\left[\ln\ln\left(\frac{1}{R}\right) + \ln\hat{\eta}\right]. \tag{7.18}$$

With $\hat{\beta}$ and $\hat{\eta}$ in the first two columns of a spreadsheet and a chosen value of true reliability R, one can compute the right-hand side and insert it into column 3. The values in column 3 are then exponentiated twice to give 10,000 values of $\dfrac{1}{R}$. Finally the reciprocal of these values is taken and sorted to yield the distribution of $\hat{R} \mid R$ for whatever value of R was used in the computation.

The 500th value from the bottom is an estimate of the 5% point of the distribution of $\hat{R} \mid R$. Likewise the 9500th sorted value estimates the 95th percentile. The table below shows the value of these percentiles as R varies from 10% to 95% for a type II censored sample of size $n = 20$ with $r = 15$. The sequence of computations can be formed into a macro. Only the value of the true reliability R varies between computations.

R	$\hat{R}_{0.05}$	$\hat{R}_{0.95}$
10	1	24.5
15	3.02	31.2
20	6.23	37.6
30	15.1	49.6
40	25.8	60.7
50	36.9	70.7
60	47.9	79.9
70	58.8	87.8
80	69.9	93.9
90	82.1	98.1
95	89.3	99.4

A plot of these percentage points against R is shown in Figure 7.3.

Now suppose that an experimenter has conducted a life test with $n = 20$ and $r = 15$, computed ML estimates of the Weibull parameters, and used them

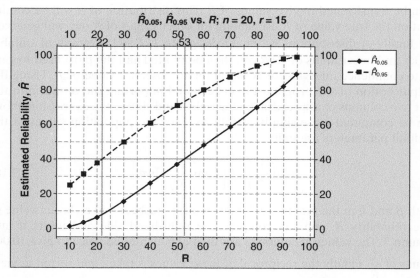

Figure 7.3 Plot of upper and lower 5% points of estimated reliability versus R.

in the reliability expression to compute the reliability at some life x. Suppose that the estimated life was $\hat{R} = 0.40$. The next step is to locate 0.40 on the ordinate of Figure 7.3 and determine the values of R where a horizontal line drawn at $\hat{R} = 0.40$ intersects the two curves. These points determine a 90% confidence interval for R, which, based on reading the graph, appears to be roughly (0.22, 0.53).

7.4 USING Pivotal.exe FOR OC CURVE CALCULATIONS

It is recalled that to test the hypothesis about a percentile x_p:

$$H_0 : x_p = (x_p)_0. \tag{7.19}$$

Against the one-sided alternative:

$$H_1 : x_p < (x_p)_0. \tag{7.20}$$

The null hypothesis is accepted at the $100\alpha\%$ significance level if:

$$\hat{\beta} \ln\left(\frac{\hat{x}_p}{(x_p)_0}\right) > u_\alpha(r, n, p). \tag{7.21}$$

The probability P_a of accepting the null hypothesis when the alternative is true and $x_p = (x_p)_1$ was shown to be expressible as:

$$P_a = \text{Prob}\left[s < \ln\left[\frac{(x_p)_1}{(x_p)_0}\right]^\beta\right]. \tag{7.22}$$

This expression means that $\ln\left[\frac{(x_p)_1}{(x_p)_0}\right]$ is the $100P_a$-th percentile of the random variable s, where s is defined as:

$$s \equiv \frac{u_\alpha(r, n, p) - u(r, n, p)}{v}. \tag{7.23}$$

Finding the percentage points of s by simulation gives the value of $\ln\left[\frac{(x_p)_1}{(x_p)_0}\right]^\beta$ and hence assuming a value for β, of $(x_p)_1$ associated with the value of P_a equal to each percentage point. We will now discuss how to compute the points on a plot of P_a against $(x_p)_1$ or, equivalently, against $\left[\frac{(x_p)_1}{(x_p)_0}\right]$.

We begin with a spreadsheet containing 10,000 values of $\hat{\beta}$ in column 1 and $\hat{\eta}$ in column 2 for some n and r of interest. Since these values were generated from a Weibull population having $\eta = \beta = 1.0$, the first column is the same as $v(r, n)$. The true value of x_p is:

$$x_p = \eta (k_p)^{1/\beta} = k_p \qquad (7.24)$$

The estimated value of x_p row for row of the spreadsheet is computed as:

$$\hat{x}_p = \hat{\eta}(k_p)^{\frac{1}{\hat{\beta}}} \qquad (7.25)$$

The value of $u(r, n, p)$ row for row may therefore be computed as:

$$u(r, n, p) = \hat{\beta} \ln\left(\frac{\hat{x}_p}{x_p}\right) = \hat{\beta} \ln\left(\frac{\hat{\eta}(k_p)^{\frac{1}{\hat{\beta}}}}{k_p}\right) = \hat{\beta}\left[\ln\hat{\eta} + \left(\frac{1}{\hat{\beta}} - 1\right)\ln(k_p)\right] \qquad (7.26)$$

The random variable s may now be computed row for row remembering that $v = \hat{\beta}$:

$$s = \frac{u_\alpha - \hat{\beta}\left[\ln\hat{\eta} + \left(\frac{1}{\hat{\beta}} - 1\right)\ln(k_p)\right]}{\hat{\beta}} = u_\alpha / \hat{\beta} - \ln\hat{\eta} + \left(1 - \frac{1}{\hat{\beta}}\right)\ln k_p. \qquad (7.27)$$

Given u_α and k_p the value of s may be computed row by row and inserted in column 3. Sorting the values in this column provides estimates of the percentiles s_p of s where $p = i/10{,}000$ and i refers to the i-th ordered value of s. Inserting the integers from 1 to 10,000 in column 4 and dividing them by 10,000 puts the P_a values in column 4 associated with each sorted value of s. The final step is to exponentiate the sorted values of s to give the ratio $\left[\frac{(x_p)_1}{(x_p)_0}\right]^\beta$.

Example

We use the 10,000 values of $\hat{\beta}$ and $\hat{\eta}$ produced from a run of the software using $n = 20$, $r = 15$. We will test H_0: $x_{0.10} = 10$ against the alternative H_1: $x_{0.10} < 10$ at the 10% level of significance. From a run of Pivotal.exe we find that $u_{0.10}(15, 20, 0.10) = -0.541$. $k_p = k_{0.10} = -\ln(1 - 0.1) = 0.10536$. The value of s is therefore computed as:

$$s = -\frac{0.541}{\hat{\beta}} - \ln\hat{\eta} + \left(1 - \frac{1}{\hat{\beta}}\right) \times 0.10536.$$

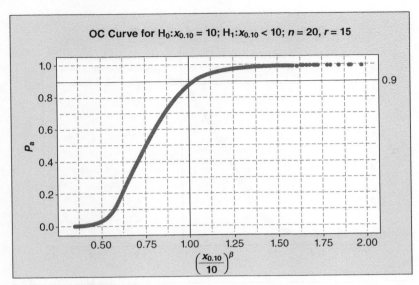

Figure 7.4 Operating characteristic curve for a one-sided test of $x_{0.10}$, $n = 20$, $r = 15$.

Computing s row by row and then sorting, and exponentiating gives $\left[\dfrac{(x_{0.10})_1}{10}\right]^{\beta}$ for each value of P_a. Figure 7.4 gives the OC curve in dimensionless form. With an assumed value of β the plot can be rescaled to display $(x_{0.10})_1$ on the abscissa if necessary.

Since the type I error was set at $\alpha = 0.10$, the curve has a P_a value of 0.90 corresponding to an abscissa value of 1.0.

The OC curves for a test on the shape parameter are particularly easy to calculate. Consider H_0: $\beta = \beta_0$ against the alternative $\beta < \beta_0$ using a significance level α. From Section 6.4 we see that when the ratio of the true to the hypothesized value of β is $\beta_1/\beta_0 = v_{\alpha}/v_p$, then P_a, the probability of accepting the null hypothesis, is the complement of the corresponding percentage point of v (i.e., $P_a = 1 - p$).

Example

Continuing with the $n = 20$, $r = 15$ sample size example, one sorts the $\hat{\beta}$ (or v) values in column 1. The value $v_{0.10}(15, 20) = 0.8398$ is then divided by every row of the sorted sample. An adjacent column is used to store P_a computed as $1 - i/10{,}000$, where i is the order number of each row of the sorted values of v. The OC curve in terms of the ratio β_1/β_0 is shown in Figure 7.5.

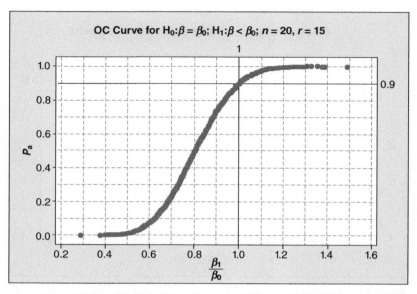

Figure 7.5 Operating characteristic curve for a one-sided test of the Weibull shape parameter.

7.5   PREDICTION INTERVALS

If the Weibull parameters are known, an interval may be calculated such that an observed future value of the random variable will fall within that interval with specified probability. For example, suppose that $\eta = 500$ and $\beta = 2.0$ and we calculate the 5% and 95% percentiles:

$$x_{0.05} = \eta(k_{0.05})^{1/\beta} = 500(0.05129)^{1/2.0} = 113.2$$

$$x_{0.95} = \eta(k_{0.95})^{1/\beta} = 500(2.996)^{1/2.0} = 865.4.$$

One may therefore predict that with a probability of 90% a value of x drawn from this distribution will fall between 113.2 and 865.4.

Suppose, however, the population parameters were not known but instead they were estimated using the ML estimates in a sample. Using these estimates as if they were population values as above results in what Meeker and Escobar (1998) call a naïve prediction interval. It does not account for the uncertainty due to the fact that the Weibull parameters are estimated and not known.

Exact prediction intervals may be computed. however, because, as will be shown, the following function is pivotal:

$$y = \left(\frac{x}{\hat{\eta}}\right)^{\hat{\beta}}. \tag{7.28}$$

In this expression x is an observation from a two-parameter Weibull distribution $W(\eta, \beta)$ and $\hat{\beta}$ and $\hat{\eta}$ are the ML estimates of the parameters from a type II censored sample of size n having r failures.

Remember that a Weibull random variable x can be generated in terms of a variable u that is uniformly distributed over the interval $(0, 1)$:

$$x = \eta(-\ln u)^{1/\beta}. \tag{7.29}$$

Substitute this value for x into the expression for y:

$$y = \left(\frac{\eta(-\ln u)^{1/\beta}}{\hat{\eta}}\right)^{\hat{\beta}} = \left(\frac{\eta}{\hat{\eta}}\right)^{\hat{\beta}} (-\ln u)^{\nu} = \exp\left[-\hat{\beta}\ln\left(\frac{\hat{\eta}}{\eta}\right)\right](-\ln u)^{\nu} \tag{7.30}$$

$$= \exp[-u(r, n, 0.632)](-\ln u)^{\nu}$$

This shows that y is a function of two pivotal quantities and a random value u from the distribution $U(0, 1)$. Thus, y is also a pivotal function. Its distribution does not depend on the parameters of the underlying Weibull population.

We may thus use the distribution of y to set a prediction interval on x. Consider again an uncensored sample of size $n = r = 20$. We have 10,000 values of the ML estimates of the parameters from a population for which $\eta = \beta = 1.0$ in two columns of a spreadsheet. We now add a third column consisting of 10,000 random values of x drawn from the same population with unit parameters. Now form 10,000 values of y by dividing the x value in each row by the scale parameter estimate in that row and raising the quotient to a power equal to the estimated shape parameter in that row. Sorting the 10,000 values of y, the 5th and 95th percentiles are found to be:

$$y_{0.05} = 0.0323 \qquad y_{0.95} = 3.941.$$

Suppose that in a type II censored sample with $n = 20$ and $r = 15$, the estimated parameters were computed and happened to be the same as the population values assumed above. That is $\hat{\beta} = 2.0$ and $\hat{\eta} = 500$. The probability statement may now be written:

$$0.0323 < \left[\frac{x}{500}\right]^{2.0} < 3.941.$$

Solving the inequalities for x gives the prediction interval:

$$89.9 < x < 992.6.$$

These limits are seen to be wider than the naïve interval $(113.2, 865.4)$. The same approach could be used to predict the smallest value in a future sample of size m. In this case the simulated x values should be drawn from the Weibull distribution followed by the smallest member of a sample of size m. When the population scale and shape parameters are 1.0 as they are in the output of the Pivotal program, the first member of a sample of size m follows $W(m^{-1}, 1)$. The procedure remains as above with this modification.

7.6   SUDDEN DEATH TESTS

The term sudden death is generally attributed to Leonard Johnson, who applied it to a strategy for conducting rolling bearing life tests which he claimed was more efficient in terms of test time than the comparable conventional life test (Johnson, 1964). In a sudden death test, one observes the first failures that occur in a set of g testers each of which contains m test elements that undergo simultaneous testing. As practiced in the bearing industry, the number of samples within each group was generally dictated by the number of physical test heads on a test machine, a number generally set by the designer of the test equipment without regard to statistical efficiency with respect to the purpose of the test. Sudden death tests were studied by McCool (1970), who showed how to compute confidence limits for a Weibull quantile using ML estimates computed from the results of a sudden death test conducted with a given group size m and number of groups g. The values of some percentage points of pivotal quantities needed for setting confidence intervals were tabulated for various values of m and g along with the expected relative test duration for a conventional test on $n = mg$ items tested until the $r = g$-th earliest failure occurs (McCool, 1974). Pascual and Meeker (1998) proposed a modified form of sudden death testing in which testing continues beyond the first failure in each group.

Consider once again a series system of m components each of which has an independent lifetime distribution that follows $W(\eta, \beta)$. The life, y, of the system is the minimum of the component lives $x_1, x_2, \ldots, x_m$. and follows the distribution $W(m^{-1/\beta}\eta, \beta)$. In Section 7.2 we based our confidence limits on system life on the result of a test of component life. In analyzing sudden death tests we do the opposite. We treat the g first failures in each group of m as the results of an uncensored test of size g. We may estimate the shape parameter β directly by the ML method since the first failure distribution is Weibull with shape parameter β. We may also estimate the p-th quantile of the distribution of y. As we have seen, the p-th quantile y_p of the series system is thus expressible in terms of the p-th quantile x_p of a component:

$$y_p = m^{-1/\beta} \cdot x_p. \tag{7.31}$$

From an ML estimate $\hat{y}_p$ of the p-th quantile of the series system, one may now estimate the p-th quantile x_p of the component as:

$$\hat{x}_p = m^{1/\hat{\beta}} \cdot \hat{y}_p. \tag{7.32}$$

Suppose y_p is estimated from a complete sample of size g so that $n = r = g$. Thus for the sudden death test the analog of Equation 7.9 is,

$$w(g, m, p) = \hat{\beta} \ln(\hat{x}_p / x_p) = \hat{\beta} \left[\ln\left(m^{\frac{1}{\beta}} \hat{y}_p \right) - \ln\left(m^{\frac{1}{\beta}} y_p \right) \right]. \qquad (7.33)$$

Rearranging gives:

$$w(g, m, p) = (1 - v)\ln m + u(g, g, p). \qquad (7.34)$$

Compare this expression to the Equation 7.10 for $u(r, n, p, m)$ in Section 7.2 above. Remembering that $\ln(1/m) = -\ln(m)$, it is apparent that $w(g, m, p) = u(g, g, 1/m, p)$. To get the percentage points of $w(g, m, p)$ one has only to run the Pivotal code with $n = r = g$, the number of groups in the test, and m equal to the reciprocal of the number m of test elements in each group.

Analogous to the analysis for conventional type II censored tests, percentage points of w may be used to set confidence limits on the p-th quantile of a component in terms of the p-th quantile of a system of m components estimated from an uncensored sample of g system lives.

$$\hat{x}_p \cdot \exp\left[-\frac{w_{\left(1 - \frac{\alpha}{2}\right)}}{\hat{\beta}} \right] < x_p < \hat{x}_p \cdot \exp\left[-\frac{w_{\left(\frac{\alpha}{2}\right)}}{\hat{\beta}} \right] \qquad (7.35)$$

A median unbiased estimate of x_p is computed as

$$\hat{x}'_p = \hat{x}_p \exp\left[-\frac{w_{0.50}}{\hat{\beta}} \right] \qquad (7.36)$$

Example
Eight samples of size four each were generated from the Weibull population W(100, 1.5). The simulated results, sorted within each of the $g = 8$ groups, are shown in Table 7.2.

Table 7.2 Simulated Sudden Death Test Data

Grp 1	Grp 2	Grp 3	Grp 4	Grp 5	Grp 6	Grp 7	Grp 8
88.2	23.5	121.7	32.1	28.4	27.2	63.2	16.3
98.2	41.0	150.0	33.4	104.7	85.6	151.5	50.8
143.1	51.7	193.5	94.0	132.8	114.9	254.7	160.1
154.2	113.6	222.2	98.3	161.2	117.2	257.8	181.1

A Weibull plot produced by the Minitab software using the first ordered lives in each group is shown in Figure 7.6.

The box on the right side of the plot gives the ML parameter estimates of the shape and scale parameters. The Anderson-Darling statistic is also shown and the associated p value of 0.203 indicates that the Weibull model cannot be rejected for this data at any meaningful level of significance.

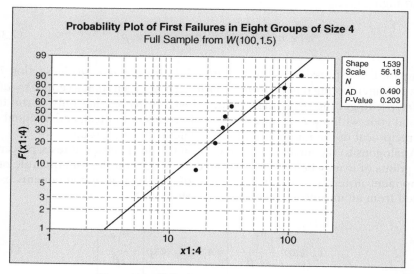

Figure 7.6   Weibull plot of group first failures.

The following percentage points were found from running the Pivotal program with $m = 1/4 = 0.25$ and $n = r = 8$ and $p = 0.632$.

α	$w_\alpha(8, 4, 0.632)$	$v_\alpha(8, 8)$
0.05	−1.60	0.721
0.50	−0.189	1.129
0.95	0.468	2.019

Using the 5th and 95th percentage points for $v = v(8, 8)$ gives the following confidence limits for the shape parameter:

$$0.762 = \frac{1.539}{2.019} < \beta < \frac{1.539}{0.721} = 2.134.$$

The interval includes the true value of $\beta = 1.5$.

The median unbiased estimate of the shape parameter is:

$$\hat{\beta}' = \frac{1.539}{1.129} = 1.36.$$

The estimate of the scale parameter of the distribution of first failures is seen from the Weibull plot:

$$\hat{y}_{0.632} = \hat{\eta} = 56.18.$$

The estimated scale parameter of x is now computed as:

$$\hat{x}_{0.632} = m^{1/\hat{\beta}} \hat{y}_{0.632} = 4^{1/1.539} \, 56.18 = 138.3.$$

A 90% confidence interval for $\eta = x_{0.632}$ is:

$$102.0 = 138.3 \exp\left(-\frac{0.468}{1.539}\right) < x_{0.632} < 138.3 \exp\left(-\frac{-1.60}{1.539}\right) = 391.1.$$

This interval fails to include the true value of $x_{0.632} = 100$ as 10% of such intervals will. The median unbiased estimate is:

$$\hat{x}'_{0.632} = 138.3 \exp\left(\frac{0.189}{1.539}\right) = 156.4.$$

Analogous to Equation 6.5, a precision measure may be computed as the median ratio of the upper to lower 90% confidence limits giving:

$$R^{\beta}_{0.50} = \exp\left[\frac{(w_{0.95}(g, m, p) - w_{0.05}(g, m, p))}{v_{0.50}(g, g)}\right]. \tag{7.37}$$

The precision measure for β is as before:

$$R = \frac{v_{0.95}(g, g)}{v_{0.05}(g, g)}. \tag{7.38}$$

This sudden death test had eight failures out of a total sample size of 32. A reasonable question is whether this test is any more precise than a conventional type II censored test where 32 items are tested until the first eight fail. A further question is how does the total time on test compare for the two strategies? For the sudden death test the measures of precision are:

$$R = \frac{2.019}{0.721} = 2.80$$

$$R^{\beta}_{0.50} = \exp\left[\frac{0.468 + 1.60}{1.129}\right] = 6.24.$$

The values needed for the conventional type II censored test evaluation are tabled below:

α	$u_\alpha(8, 32, 0.632)$	$v_\alpha(8, 32)$
0.05	−2.247	0.680
0.50	−0.278	1.182
0.95	0.551	2.351

$$R = \frac{2.351}{0.680} = 3.55$$

$$R_{0.50}^\beta = \exp\left[\frac{0.551 + 2.247}{1.182}\right] = 10.7.$$

In this case we see that the sudden death test is superior with respect to estimation of both the shape and scale parameters. Designing a sudden death test means selecting the value of m to best predict a certain percentile. A test that is optimum for predicting one percentile will not be optimum for predicting a different percentile.

It is possible to compute the optimum size of the subgroup m needed to estimate a given quantile with the greatest precision (McCool, 2009). Calculating $R_{0.50}^\beta$ for various values of p in complete samples of size g, it was found that there is a value p^* for which the p-th quantile is estimated with maximum precision and p^* increases with sample size g over the range of about 0.6 to 0.75.

7.7   DESIGN OF OPTIMAL SUDDEN DEATH TESTS

Two Weibull populations are shown in the probability plot sketched in Figure 7.7. One is the Weibull distribution about which interest is centered and the other is the distribution of smallest values in samples of size m drawn from that distribution.

Estimating $x_{0.10}$ of the target population is the same as estimating x_p in the distribution of first-order statistics.

It is readily shown that p is expressible as:

$$p = 1 - \exp\left\{m \ln\left[\frac{1}{1 - 0.10}\right]\right\} \tag{7.39}$$

We can therefore estimate $x_{0.10}$ for the parent distribution by estimating the $100p$-th percentile of the first failure distribution using the complete uncensored sample of size g from that distribution obtained from a sudden death test. The value of m that makes $p = p^*$ is then given by:

Weibull Plot Geometry

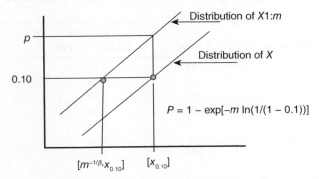

Figure 7.7 Weibull plot of a Weibull population and the associated distribution of the minimum of samples of size m. From Journal of Testing and Evaluation, 37(4), copyright 2009 ASTM International. Reprinted by permission of ASTM International.

$$m = \frac{\ln\left[\dfrac{1}{1-p^*}\right]}{\ln\left[\dfrac{1}{1-0.10}\right]}. \tag{7.40}$$

The m values for optimum estimation of $x_{0.10}$ for g varying from 3 to 20 were calculated in this way using p^* for each value of g. The optimum m values rounded to the nearest integer are given in Table 7.3. For each value of g and m the values of the logarithm of Equation 7.37 were computed by Monte Carlo sampling. Comparable values were computed from Equation 6.5 for the conventional type II censored test having the overall sample size $n = mg$ and total number of failures $r = g$. The comparison could only be made for g values up to and including 16 because the simulation software was limited to $n \leq 200$. It is seen that for every case calculated, the precision of the sudden death test having optimal subgroup size was superior to the corresponding conventional test.

As noted in Section 7.2, the first failures in each subgroup follow the Weibull distribution with scale parameter $m^{-1/\beta}\eta$. The first failure times y_i of each subgroup are the order statistics of a sample of size g drawn from that distribution, that is, from $W\left(m^{-\frac{1}{\beta}}\eta, \beta\right)$. The total time on test for all specimens in a sudden death test is thus:

$$T_s = m \times \sum_{i=1}^{g} y_{i:g}. \tag{7.41}$$

Table 7.3   Comparison of the Precision in Estimating $x_{0.10}$ with Optimum Sudden Death Tests and Corresponding Conventional Type II Censored Tests Having $n = gm$ and $r = g$

$g = r$	Optimal m (Rounded)	$N = gm$	Precision Sudden Death	Precision Conventional
3	9	27	3.11	3.48
4	10	40	2.28	2.41
5	10	50	1.88	2.00
6	11	66	1.66	1.76
7	11	77	1.44	1.54
8	11	88	1.33	1.41
9	12	108	1.23	1.32
10	12	120	1.17	1.22
11	12	132	1.09	1.17
12	12	144	1.03	1.07
13	12	156	0.988	1.03
14	12	168	0.935	0.989
15	12	180	0.917	0.948
16	12	192	0.891	0.907
17	12	204	0.857	–
18	12	216	0.818	–
19	12	228	0.794	–
20	12	240	0.780	–

where $y_{j:n}$ is the i-th order statistic in a sample of size n. Because the sample is complete, the sum of the ordered sample is the same as the sum of a random sample, and hence the expected value of the sum is $gE(y)$. The expected total test time is therefore:

$$E(T_s) = m^{1-\frac{1}{\beta}} g \eta \Gamma\left(1 + \frac{1}{\beta}\right). \tag{7.42}$$

For the conventional type II censored test the total time on test is:

$$T_r = \sum_{i=1}^{g} x_{i:gm} + (gm - g)x_{g:gm} \tag{7.43}$$

where x refers to the distribution $W(\eta, \beta)$.

Harter (1964) has tabled the expected value of Weibull order statistics for sample sizes of up to 40 and with $\beta = 0.5(0.5)2.0$. This limit permits the evaluation of the total expected test times for just the first two cases in Table 7.2, that is, the cases with $g = 3$ and $m = 9$ and $g = 4$ and $m = 10$.

The expected test times scaled by η are given in Table 7.4 for these two cases and with these shape parameter values.

Table 7.4 Comparison of Expected Test Times for Conventional and Sudden Death Tests for the Various Shape Parameter Values

Shape Parameter β	$g = 3, m = 9$ and $r = 3$, $n = 27$		$g = 4, m = 10$ and $r = 4$ and $n = 40$	
	$E(T_s)/\eta$	$E(T_r)/\eta$	$E(T_s)/\eta$	$E(T_r)/\eta$
0.5	0.667	0.456	0.8	0.527
1.0	3	3	4	4
1.5	5.633	5.988	7.78	8.342
2.0	7.976	8.58	11.21	13.717

The sudden death test takes a greater total test time than the corresponding conventional test for $\beta < 1$ and less time for $\beta > 1$. A comparable evaluation for many arbitrary sudden death sample sizes shows that this seems to be generally true for any sudden death test.

For $\beta = 1$ the times are equal and are shown by McCool (2009) to have a normalized value equal to the group size g. Thus, provided the hazard rate is increasing ($\beta > 1$), sudden death testing is more efficient than the corresponding conventional test with respect to test time. If the objective of the testing is to estimate a specific quantile as precisely as possible, the subgroup size m may be chosen so that the sudden death test achieves as great or greater precision than the equivalent conventional test. When the value of m is arbitrarily selected, the sudden death test may or may not have greater precision than the corresponding conventional test.

REFERENCES

Harter, H. 1964. Expected Values of Exponential, Weibull and Gamma Order Statistics, US Airforce Aerospace Research Laboratory.

Johnson, L.G. 1964. *The Statistical Treatment of Fatigue Experiments*. Elsevier, New York, NY.

McCool, J.I. 1970. Inference on Weibull percentiles and shape parameter from sudden death tests using maximum likelihood estimates. *IEEE Transactions on Reliability* R-19: 177–179.

McCool, J.I. 1974. Analysis of sudden death tests of bearing endurance. *ASLE Transactions* 17: 8–13.

McCool, J.I. 2009. Design of sudden death tests for estimation of a Weibull percentile. *Journal of Testing and Evaluation* 37(4): 311–315.

Meeker, W.Q. and L.A. Escobar. 1998. *Statistical Methods for Reliability Data*. Wiley, New York, NY.

Pascual, F.G. and W.Q. Meeker. 1998. The modified sudden death test: planning life tests with a limited number of test positions. *Journal of Testing and Evaluation* 26(5): 434–443.

EXERCISES

1. A series system contains $m = 3$ identical components. An uncensored life test on a sample of $n = 10$ components resulted in the following set of observed lifetimes in hours:

14.6375	66.9203	68.4248	69.2899	71.8208
77.4557	94.8249	100.317	101.453	104.603

   (To avoid redundant effort, this is the same data set that appears in exercises in Chapters 5 and 6.) Use Pivotal.exe and compute 90% confidence limits on the median ($y_{0.50}$) of the system life.

2. An uncensored sample of size 20 resulted in the raw ML estimates $\hat{\eta} = 85$ hours and $\hat{\beta} = 2.1$. Using Figure 7.3, graphically determine the 90% confidence interval for the reliability at life 61.7 hours.

3. For the same data sample referred to in problem 2, compute a 90% interval for a single future value from the same population.

4. Using Monte Carlo simulation, generate $g = 5$ uncensored samples of size $m = 4$ from $W(100, 1.5)$. Construct an uncensored sample of size 5 using the first-order statistics in each of the five samples. Compute the ML estimates of the scale and shape parameters using this sample of first-order statistics and then compute the raw ML estimate of the scale parameter of the population from which the samples were drawn. Run Pivotal.exe using appropriate values of m, n, and r and compute a 90% interval and a median unbiased estimate for the population scale parameter.

CHAPTER 8

Inference from Multiple Samples

In experimental investigations involving life tests or material strength, there are often one or more factors of concern, and tests are frequently run at several levels of those factors. This might include perhaps a set of lubricants, different design variations, heat treatment recipes, or chemical compositions. In any case, a project may involve a number of related life tests. Typically these tests are analyzed individually and then the results are compared in the hope of detecting differences ascribable to the factors under investigation. In contrast, in designed experiments with a response variable that is assumed to be normally distributed, it is the custom to analyze the data as a set using an estimate of the variability that results from pooling the variability in each of the individual tests. This is justified by the assumption that the factor or factors under investigation affect the mean but not the standard deviation of the response variable and that the standard deviations in the tests at each factor level differ only due to sampling error. Careful experimenters may use one of several hypothesis tests of the validity of this assumption, which is often referred to as the *homogeneity of variance*, before proceeding. If the test fails to reject the assumption that the variance is the same at all levels of the factor or factors, experimenters will then assume that it is true.

8.1 MULTIPLE WEIBULL SAMPLES

In life testing with a two-parameter Weibull response, it is often tacitly assumed that the scale parameter may be affected by the factor levels but that the shape parameter is unaffected. This assumption is explicit in rolling bearing life distribution theory wherein the effect of an increased load is assumed to decrease the scale parameter while leaving the shape parameter unaffected. Although it is generally assumed to be true that the shape parameter estimates differ

Using the Weibull Distribution: Reliability, Modeling, and Inference, First Edition. John I. McCool.
© 2012 John Wiley & Sons, Inc. Published 2012 by John Wiley & Sons, Inc.

only due to sampling error, this fact is seldom exploited to give an improved estimate of the underlying common shape parameter. We will show that pooling the information about the shape parameter in sets of tests will increase the precision with which the shape parameter can be estimated. Furthermore, the assumption need not be made blindly. Its validity may be tested. A shape parameter that is the same at every factor level does not result in homogeneous variance for the Weibull since the variance depends on both the shape and scale parameters. If the scale parameters vary with factor level so will the variance even if the shape parameter does not. The homogeneity of shape parameter assumption is therefore not equivalent to the homogeneity of variance assumption.

8.2   TESTING THE HOMOGENEITY OF SHAPE PARAMETERS

Consider a set of k uncensored or type II censored tests. We assume the sample size n and number of failures r is the same for each test. This assumption is made for manageability in the calculation of tables and is not a restriction in theory. If the shape parameter is estimated by the method of maximum likelihood (ML) for each sample, the k estimates may be sorted from low to high. Under the assumption of a common shape parameter β, the k ML shape parameter estimates could, in principle, be divided by β to give an ordered sample of k values of the pivotal function $v(r, n)$. The ratio w of the largest value of v to the smallest is a pivotal quantity since it is the ratio of two-order statistics in a sample of k values of a pivotal function. Thus, we have,

$$w(r, n, k) = \frac{v_{(k)}}{v_{(1)}} = \frac{\hat{\beta}_{\max} / \beta}{\hat{\beta}_{\min} / \beta} = \frac{\hat{\beta}_{\max}}{\hat{\beta}_{\min}}, \qquad (8.1)$$

and the unknown true value of β cancels. The distribution of $w(r, n, k)$ under the null hypothesis that the k groups have a common shape parameter may be estimated by Monte Carlo sampling by generating sets of k samples all from the same Weibull population, estimating β for each sample and forming the ratio of the largest to the smallest value within each set.

In practice, a difference in the population shape parameters among the k groups will tend to manifest itself in a greater ratio $\dfrac{\hat{\beta}_{\max}}{\hat{\beta}_{\min}}$ than if the groups all had the same population shape parameter value. We may therefore use an upper percentile value of the w distribution as the critical value in a test for the homogeneity of shape parameter values. Table 8.1 lists the 90th percentile of the w distribution for various n, r and k. These values were determined in simulations of 10,000 sets of size k for each combination of n and r.

Table 8.1 Values of $w_{0.90}(r, n, k)$ for Various r, n and k

n	r	$k = 2$	$k = 3$	$k = 4$	$k = 5$	$k = 10$
5	3	5.45	8.73	11.0	13.4	22.2
5	5	2.77	3.59	4.23	4.69	6.56
10	3	6.04	9.93	12.5	15.4	26.8
10	5	3.21	4.35	5.16	5.69	7.98
10	10	1.87	2.23	2.47	2.61	3.16
15	5	3.20	4.48	5.28	5.90	8.39
15	10	2.02	2.44	2.69	2.90	3.56
15	15	1.65	1.87	2.05	2.16	2.45
20	5	3.31	4.54	5.28	6.24	9.05
20	10	2.11	2.50	2.80	3.00	3.69
20	15	1.72	1.96	2.15	2.30	2.70
20	20	1.52	1.70	1.80	1.90	2.14
30	5	3.28	4.47	5.38	6.11	8.95
30	10	2.11	2.54	2.90	3.10	3.88
30	15	1.78	2.05	2.27	2.40	2.82
30	20	1.61	1.82	1.95	2.06	2.40
30	30	1.41	1.53	1.61	1.67	1.84

From Analysis of Sets of Two-Parameter Weibull Data Arising in Rolling Contact Endurance Data in ASTM Technical Publication 771, copyright 1982. Reproduced by permission of ASTM International.

A software program called Multi-Weibull described in Section 8.8 can be used to compute, via simulation, 16 percentiles of the distribution of $w(r, n, k)$ for any choices of r, n and k.

Example

The data in Table 8.2 are the ordered rolling contact fatigue lives measured in millions of revolutions in uncensored samples of size 10 using specimens made from each of five different types of steel. The data were originally reported by Brown and Potts (1977), who give the full details of the test conditions and materials. Since those details are not relevant to our present illustrative purpose, we refer to the materials simply using the letters A–E. The last two rows of the table contain the ML estimates of the 10th percentiles and the shape parameter for each group of 10 tests.

To test the hypothesis that the population shape parameters are equal, we compute the ratio of the largest to smallest ML estimates of the shape parameter:

$$w = \frac{\hat{\beta}_{max}}{\hat{\beta}_{min}} = \frac{3.65}{1.94} = 1.88.$$

The critical value for $\alpha = 0.10$ is found in Table 8.1 for $k = 5$, $n = r = 10$, to be 2.61. Since $1.88 < 2.61$, there is no cause to reject the supposition that the data

Table 8.2   Rolling Contact Fatigue Data for Five Steel Compositions 10^6 Stress Cycles

	A	B	C	D	E
	3.03	3.19	3.46	5.88	6.43
	5.53	4.26	5.22	6.74	9.97
	5.60	4.47	5.69	6.90	10.39
	9.30	4.53	6.54	6.98	13.55
	9.92	4.67	9.16	7.21	14.45
	12.51	4.69	9.40	8.14	14.72
	12.95	5.78	10.19	8.59	16.81
	15.21	6.79	10.71	9.80	18.39
	16.04	9.37	12.58	12.28	20.84
	16.84	12.75	13.41	25.46	21.51
$\hat{x}_{0.10}$	5.06	2.60	4.72	3.49	8.83
$\hat{\beta}$	2.59	2.32	3.13	1.94	3.65

have a common shape parameter value. The observed ratio is, in fact, quite typical, since the median value of w for $k = 5, n = r = 10$ was found to be 1.84.

8.3   ESTIMATING THE COMMON SHAPE PARAMETER

Having accepted the equality of the population shape parameters but allowing that the scale parameters may differ from among the k sampled populations, the model for the i-th population is:

$$F(x) = 1 - \exp\left[-\left(\frac{x}{\eta_i}\right)^{\beta}\right]; \quad i = 1 \ldots k. \tag{8.2}$$

Where η_i denotes the scale parameter for the i-th sample. The likelihood function for multiple samples is the product over the k samples of the likelihood function as previously expressed for a single sample. If the shape parameters are taken to be the same for each sample, the ML estimate of the common shape parameter, designated $\hat{\beta}_1$, is the solution of the following nonlinear equation written for the general case where the values of n and r vary with the sample (cf. Schafer and Sheffield, 1976; McCool, 1979).

$$\sum_{i=1}^{k} \frac{r_i}{\hat{\beta}_1} + \sum_{i=1}^{k} \sum_{j=1}^{r_i} \ln(x_{ij}) - \sum_{i=1}^{k} r_i \left[\sum_{j=1}^{n_i} x_{ij}^{\hat{\beta}_1} \ln(x_{ij}) \right] / \sum_{j=1}^{n_i} x_{ij}^{\hat{\beta}_1} = 0. \tag{8.3}$$

It is assumed that within each sample the r_i failures are listed first followed by the censored lives. The lives need not be ordered within the failed and censored sets. x_{ij} denotes the j-th listed value within the i-th sample. The value of $\hat{\beta}_1$ found as the solution of this equation is the valid ML estimate of the shape parameter irrespective of the nature of the censoring method. However, computation of confidence limits and tests of hypotheses discussed subsequently are valid only when the data are uncensored or type II censored.

The data from all of the samples contribute to the determination of this estimate of β. The effect is that the estimate is much more precisely determined than any one of the estimates from individual samples within the set.

Having computed $\hat{\beta}_1$, the scale parameter and $100p$-th percentile may be computed for the i-th sample as:

$$\hat{\eta}_i = \left[\sum_{j=1}^{n_i} x_{ij}^{\hat{\beta}_1} / r_i \right]^{1/\hat{\beta}_1} \tag{8.4}$$

$$(\hat{x}_p)_i = [k_p]^{1/\hat{\beta}_1} \cdot \hat{\eta}_i. \tag{8.5}$$

The scale parameter within each sample is computed using just the lives in the sample but with the shape parameter estimate based on the data in all of the groups. The estimate $\hat{\beta}_1$ acts as a sort of average of the shape parameters within the groups. It is unaffected by any differences in the scale parameters among the groups. Multiplying the data within any group by a constant leaves the value of $\hat{\beta}_1$ unchanged.

8.3.1 Interval Estimation of the Common Shape Parameter

It has been proven that when the samples are type II censored, the following function is pivotal when the k population shape parameters are the same (McCool, 1979):

$$v_1 = \frac{\hat{\beta}_1}{\beta}. \tag{8.6}$$

This function is a generalization of the pivotal function $v(r, n)$ introduced for single samples in Chapter 5. The arguments have been suppressed for convenience but in full generality v_1 could be expressed as $v_1(r, n, k)$. The distribution of v_1 applicable when the sample size and censoring number are the same for each group were determined by Monte Carlo simulation for various choices of n, r and k. Table 8.3 contains values of the 5th, 50th, and 95th percentiles of v_1 for a range of values of n and r with k varying from 1 to 5. The table contains percentage points of several other random variables, which will be discussed subsequently. These percentage points of v_1 are used in the same way as the single sample version in computing confidence limits and correcting bias.

Table 8.3 Selected Percentage Points as a Function of Sample Size

(a)

n	r	k	$\hat{\beta}_1/\hat{\beta}_0$ 0.90	0.95	$-2\ln\lambda$ 0.90	0.95	v_1 0.05	0.50	0.95	u_1 0.05	0.50	0.95	t_1 0.90	0.95	c	$-2c\ln\lambda$ 0.95	$\chi^2(k-1)$ 0.95
5	3	2	1.799	2.144	5.357	7.235	0.7618	1.505	3.833	-1.175	0.6861	5.325	2.833	3.752	0.514	3.717	3.81
5	3	3	1.788	2.078	7.913	10.24	0.8431	1.471	3.022	-1.135	0.6111	3.864	3.294	4.192	0.574	5.878	5.99
5	3	4	1.769	2.009	10.29	12.80	0.8915	1.462	2.673	-1.131	0.6124	3.417	3.605	4.342	0.596	7.673	7.81
5	3	5	1.744	1.925	12.34	15.07	0.9331	1.446	2.485	-1.112	0.6216	3.073	3.733	4.512	0.623	9.384	9.49
5	3	10	1.644	1.747	21.78	24.88	1.038	1.43	2.030	-1.084	0.5889	2.439	4.158	4.707	0.667	16.59	16.9
5	5	2	1.259	1.366	3.794	5.290	0.7773	1.195	2.056	-0.9715	0.3875	2.750	1.441	1.802	0.715	3.785	3.81
5	5	3	1.273	1.361	5.984	7.804	0.8265	1.184	1.816	-0.8715	0.3473	2.255	1.742	2.111	0.758	5.914	5.99
5	5	4	1.275	1.353	8.076	10.08	0.8609	1.178	1.703	-0.8321	0.341	1.980	1.950	2.285	0.776	7.819	7.81
5	5	5	1.267	1.339	9.843	12.10	0.8888	1.177	1.621	-0.7907	0.347	1.810	2.082	2.392	0.788	9.538	9.49
5	5	10	1.246	1.284	17.94	20.72	0.9545	1.167	1.453	-0.7314	0.3336	1.471	2.409	2.702	0.818	16.95	16.9
10	5	2	1.363	1.508	4.142	5.782	0.7677	1.256	2.326	-0.8691	0.2856	2.424	1.576	1.999	0.662	3.826	3.81
10	5	3	1.372	1.491	6.443	8.372	0.8281	1.242	2.039	-0.8526	0.2727	1.959	1.900	2.296	0.708	5.93	5.99
10	5	4	1.367	1.463	8.467	10.51	0.8679	1.234	1.877	-0.8303	0.2761	1.729	2.084	2.426	0.739	7.765	7.81
10	5	5	1.365	1.438	10.39	12.45	0.8904	1.23	1.768	-0.7987	0.2794	1.587	2.209	2.547	0.757	9.420	9.49
10	5	10	1.328	1.378	18.84	21.58	0.9669	1.219	1.575	-0.7994	0.2733	1.361	2.549	2.823	0.782	16.88	16.9
10	10	2	1.105	1.150	3.165	4.472	0.8130	1.087	1.516	-0.7161	0.1587	1.444	0.8509	1.040	0.855	3.822	3.81
10	10	3	1.116	1.149	5.317	6.834	0.8508	1.082	1.414	-0.6590	0.146	1.204	1.062	1.231	0.872	5.962	5.99
10	10	4	1.115	1.142	6.981	8.706	0.8794	1.077	1.355	-0.6117	0.1448	1.066	1.171	1.33	0.893	7.772	7.81
10	10	5	1.114	1.139	8.755	10.55	0.8939	1.076	1.319	-0.5883	0.1519	0.9899	1.256	1.406	0.895	9.442	9.49
10	10	10	1.105	1.121	16.16	18.61	0.9405	1.073	1.235	-0.5766	0.145	0.8384	1.468	1.598	0.905	16.84	16.9
15	5	2	1.374	1.535	4.090	5.770	0.768	1.269	2.414	-0.879	0.203	1.976	1.569	2.02	0.656	3.788	3.81
15	5	3	1.392	1.522	6.393	8.428	0.822	1.256	2.080	-0.876	0.190	1.645	1.899	2.308	0.71	5.987	5.99
15	5	4	1.392	1.491	8.579	10.626	0.857	1.246	1.927	-0.848	0.193	1.474	2.110	2.475	0.736	7.819	7.81

(b)

n	r	k	$\hat{\beta}_1/\hat{\beta}_0$ 0.90	0.95	$-2\ln\lambda$ 0.90	0.95	v_1 0.05	0.50	0.95	u_1 0.05	0.50	0.95	t_1 0.90	0.95	c	$-2c\ln\lambda$ 0.95	$\chi^2(k-1)$ 0.95
15	5	5	1.377	1.462	10.238	14.443	0.891	1.239	1.815	−0.864	0.189	1.377	2.206	2.548	0.753	10.889	9.49
15	10	2	1.131	1.190	3.260	4.653	0.799	1.106	1.611	−0.668	0.157	1.352	0.873	1.08	0.816	3.797	3.81
15	10	3	1.144	1.185	5.370	6.846	0.845	1.1	1.489	−0.613	0.145	1.152	1.08	1.254	0.858	5.873	5.99
15	10	4	1.145	1.181	7.224	8.962	0.866	1.097	1.412	−0.586	0.143	1.033	1.200	1.381	0.869	7.79	7.81
15	10	5	1.143	1.176	8.881	10.873	0.888	1.094	1.375	−0.573	0.138	0.987	1.289	1.457	0.877	9.531	9.49
15	15	2	1.066	1.096	3.101	4.436	0.836	1.056	1.369	−0.615	0.108	...	0.671	0.817	0.876	3.885	3.81
15	15	3	1.072	1.095	5.071	6.562	0.871	1.052	1.302	−0.559	0.098	0.904	0.827	0.949	0.909	5.964	5.99
15	15	4	1.074	1.091	6.843	8.497	0.890	1.049	1.258	−0.515	0.097	0.805	0.920	1.036	0.912	7.749	7.81
15	15	5	1.074	1.090	8.477	10.378	0.902	1.048	1.230	−0.497	0.095	0.772	0.987	1.101	0.917	9.514	9.49
20	5	2	1.382	1.546	4.110	5.749	0.761	1.278	2.428	−0.937	0.119	1.660	1.575	2.020	0.664	3.821	3.81
20	5	3	1.408	1.542	6.455	8.469	0.818	1.267	2.078	−0.912	0.127	1.403	1.922	2.348	0.708	5.996	5.99
20	5	4	1.398	1.495	8.487	10.505	0.864	1.251	1.916	−0.918	0.118	1.299	2.121	2.486	0.732	7.685	7.81
20	5	5	1.389	1.478	10.339	12.412	0.892	1.246	1.836	−0.937	0.119	1.226	2.233	2.580	0.745	9.251	9.49
20	10	2	1.139	1.197	3.220	4.587	0.796	1.109	1.650	−0.621	0.125	1.231	0.871	1.072	0.829	3.801	3.81
20	10	3	1.155	1.202	5.453	7.004	0.836	1.104	1.507	−0.587	0.125	1.035	1.098	1.282	0.852	5.968	5.99
20	10	4	1.155	1.195	7.233	9.120	0.866	1.100	1.446	−0.573	0.115	0.953	1.213	1.384	0.862	7.867	7.81
20	10	5	1.155	1.190	9.068	11.008	0.885	1.099	1.402	−0.550	0.117	0.915	1.304	1.475	0.865	9.520	9.49
20	15	2	1.079	1.112	3.088	4.297	0.826	1.064	1.425	−0.569	0.0982	1.013	0.672	0.809	0.887	3.813	3.81
20	15	3	1.088	1.114	5.102	6.583	0.857	1.062	1.337	−0.531	0.0988	0.874	0.830	0.964	0.899	5.920	5.99
20	15	4	1.088	1.111	6.862	8.591	0.882	1.058	1.289	−0.497	0.0914	0.800	0.922	1.057	0.907	7.789	7.81
20	15	5	1.089	1.108	8.603	10.560	0.896	1.059	1.263	−0.472	0.0955	0.743	0.999	1.120	0.907	9.583	9.49
20	20	2	1.048	1.068	3.012	4.194	0.855	1.040	1.300	−0.531	0.0723	0.874	0.566	0.678	0.917	3.848	3.81

(*Continued*)

Table 8.3 *(Continued)*

(c)

n	r	k	$\hat{\beta}_1/\hat{\beta}_0$ 0.90	0.95	$-2\ln\lambda$ 0.90	0.95	v_1 0.05	0.50	0.95	u_1 0.05	0.50	0.95	t_1 0.90	0.95	c	$-2c\ln\lambda$ 0.95	$\chi^2(k-1)$ 0.95
20	20	3	1.054	1.069	4.986	6.456	0.881	1.037	1.242	−0.486	0.0746	0.7422	0.702	0.803	0.918	5.926	5.99
20	20	4	1.054	1.068	6.740	8.460	0.898	1.037	1.208	−0.461	0.0715	0.665	0.785	0.883	0.924	7.821	7.81
20	20	5	1.053	1.065	8.428	10.231	0.911	1.036	1.187	−0.433	0.0714	0.623	0.836	0.941	0.927	9.491	9.49
25	5	2	1.383	1.545	4.003	5.634	0.764	1.287	2.427	−1.002	0.0776	1.407	1.560	1.200	0.662	3.729	3.81
25	5	3	1.400	1.522	6.365	8.142	0.822	1.264	2.102	−0.996	0.0786	1.246	1.910	2.308	0.707	5.754	5.99
25	5	4	1.402	1.498	8.518	10.401	0.866	1.256	1.952	−0.977	0.0740	1.144	2.114	2.469	0.730	7.596	7.81
25	5	5	1.390	1.481	10.287	12.419	0.894	1.251	1.846	−0.986	0.0719	1.098	2.231	2.562	0.749	9.306	9.49
25	10	2	1.143	1.199	3.211	4.485	0.791	1.117	1.652	−0.605	0.107	1.099	0.871	1.061	0.831	3.728	3.81
25	10	3	1.160	1.208	5.403	7.038	0.832	1.107	1.527	−0.586	0.0957	0.979	1.091	1.278	0.851	5.992	5.99
25	10	4	1.161	1.202	7.274	9.074	0.864	1.103	1.456	−0.575	0.0987	0.887	1.219	1.387	0.863	7.831	7.81
25	10	5	1.158	1.191	8.872	10.828	0.885	1.100	1.401	−0.557	0.0946	0.838	1.289	1.466	0.878	9.502	9.49
25	15	2	1.084	1.122	3.079	4.465	0.817	1.069	1.447	−0.551	0.0969	0.974	0.673	0.827	0.877	3.917	3.81
25	15	3	1.094	1.123	5.116	6.633	0.855	1.063	1.358	−0.501	0.0830	0.849	0.835	0.972	0.894	5.933	5.99
25	15	4	1.096	1.123	6.993	8.885	0.878	1.061	1.308	−0.477	0.0847	0.753	0.932	1.072	0.902	8.017	7.81
25	15	5	1.095	1.116	8.544	10.520	0.893	1.061	1.277	−0.458	0.0843	0.710	0.996	1.118	0.912	9.594	9.49
25	20	2	1.056	1.078	2.994	4.250	0.841	1.045	1.336	−0.519	0.0762	0.857	0.567	0.682	0.911	3.871	3.81
25	20	3	1.063	1.061	4.988	6.453	0.874	1.043	1.268	−0.469	0.0710	0.745	0.701	0.811	0.922	5.952	5.99
25	20	4	1.063	1.080	6.783	8.448	0.892	1.041	1.236	−0.435	0.0692	0.661	0.785	0.882	0.929	7.846	7.81
25	20	5	1.062	1.077	8.212	10.154	0.905	1.041	1.208	−0.426	0.0681	0.618	0.836	0.931	0.936	9.505	9.49
25	25	2	1.037	1.054	2.945	4.210	0.864	1.032	1.255	−0.494	0.0629	0.752	0.497	0.601	0.925	3.895	3.81
25	25	3	1.042	1.054	4.894	6.450	0.890	1.030	1.204	−0.445	0.0607	0.655	0.615	0.710	0.935	6.030	5.99
25	25	4	1.042	1.053	6.616	8.330	0.905	1.030	1.178	−0.409	0.0541	0.581	0.686	0.779	0.941	7.836	7.81
25	25	5	1.041	1.051	8.122	9.919	0.916	1.028	1.051	−0.387	0.0570	0.539	0.736	0.814	0.949	9.409	9.49

(d)

n	r	k	$\hat{\beta}_1/\hat{\beta}_0$ 0.90	$\hat{\beta}_1/\hat{\beta}_0$ 0.95	$-2\ln\lambda$ 0.90	$-2\ln\lambda$ 0.95	v_1 0.05	v_1 0.50	v_1 0.95	u_1 0.05	u_1 0.50	u_1 0.95	t_1 0.90	t_1 0.95	c	$-2c\ln\lambda$ 0.95	$\chi^2(k-1)$ 0.95
30	5	2	1.394	1.571	4.111	5.852	0.761	1.285	2.436	-1.066	0.0177	1.289	1.591	2.060	0.656	3.840	3.81
30	5	3	1.414	1.544	6.489	8.396	0.827	1.267	2.120	-1.070	0.0342	1.151	1.938	2.367	0.698	5.863	5.99
30	5	4	1.411	1.529	8.601	10.827	0.871	1.256	1.963	-1.065	0.0331	1.049	2.148	2.515	0.723	7.831	7.81
30	5	5	1.399	1.492	10.373	12.674	0.895	1.256	1.838	-1.062	0.0275	1.019	2.256	2.607	0.739	9.367	9.49
30	10	2	1.148	1.212	3.259	4.704	0.789	1.117	1.674	-0.587	0.0874	1.020	0.880	1.095	0.821	3.862	3.81
30	10	3	1.165	1.215	5.452	7.200	0.831	1.109	1.528	-0.576	0.0804	0.896	1.099	1.303	0.845	6.086	5.99
30	10	4	1.167	1.206	7.402	9.171	0.861	1.108	1.454	-0.570	0.0819	0.828	1.228	1.407	0.854	7.832	7.81
30	10	5	1.165	1.199	9.071	11.028	0.883	1.104	1.410	-0.570	0.0795	0.787	1.313	1.492	0.863	9.517	9.49
30	15	2	1.087	1.124	3.033	4.312	0.813	1.071	1.458	-0.513	0.0823	0.886	0.669	0.813	0.884	3.817	3.81
30	15	3	1.096	1.042	5.046	6.549	0.815	1.068	1.365	-0.583	0.0767	0.776	0.832	0.965	0.908	5.945	5.99
30	15	4	1.099	1.042	6.967	8.663	0.875	1.065	1.319	-0.470	0.0734	0.721	0.940	1.067	0.907	7.857	7.81
30	15	5	1.098	1.042	8.588	10.513	0.893	1.066	1.288	-0.460	0.0768	0.692	1.004	1.136	0.908	9.546	9.49
30	20	2	1.059	1.042	2.921	4.167	0.833	1.049	1.349	-0.490	0.0737	1.083	0.560	0.679	0.920	3.834	3.81
30	20	3	1.065	1.042	4.844	6.256	0.871	1.045	1.280	-0.442	0.0680	0.704	0.698	0.796	0.935	5.849	5.99
30	20	4	1.067	1.042	6.687	8.405	0.889	1.045	1.243	-0.430	0.0664	0.629	0.781	0.884	0.934	7.848	7.81
30	20	5	1.067	1.042	8.361	10.167	0.903	1.045	1.219	-0.410	0.0701	0.621	0.842	0.942	0.934	9.493	9.49
30	25	2	1.044	1.042	2.956	4.116	0.853	1.037	1.280	-0.481	0.0572	0.882	0.500	0.595	0.921	3.790	3.81
30	25	3	1.047	1.042	4.753	6.168	0.882	1.035	1.226	-0.425	0.0593	0.627	0.609	0.699	0.956	5.894	5.99
30	25	4	1.048	1.042	6.562	8.082	0.899	1.033	1.195	-0.396	0.0575	0.583	0.683	0.771	0.949	7.670	7.81
30	25	5	1.048	1.042	8.129	9.886	0.913	1.033	1.175	-0.380	0.0599	0.549	0.736	0.816	0.950	9.388	9.49
30	30	2	1.031	1.042	2.890	4.025	0.873	1.026	1.224	-0.458	0.0530	0.668	0.448	0.534	0.930	3.744	3.81
30	30	3	1.035	1.042	4.765	6.192	0.895	1.025	1.184	-0.406	0.0513	0.574	0.552	0.635	0.954	5.906	5.99
30	30	4	1.035	1.042	6.511	8.123	0.911	1.024	1.156	-0.387	0.0467	0.529	0.618	0.698	0.950	7.717	7.81
30	30	5	1.035	1.042	8.121	9.938	0.921	1.024	1.142	-0.368	0.0493	0.496	0.665	0.740	0.956	9.938	9.49

For the rolling contact fatigue data the ML estimate $\hat{\beta}_1$ was found to be 2.480. (A Mathcad module for performing this and other computations will be given subsequently.) Using the values of the 5th and 95th percentage points of v_1 for $n = r = 10$ and $k = 5$, a 90% confidence interval for the common shape parameter is computed as:

$$1.88 = \frac{2.480}{1.319} < \beta < \frac{2.480}{0.8939} = 2.77.$$

The ratio R of the upper to lower ends of this confidence interval is 1.47. This is superior to the ratio of 1.62 shown in Table 6.1 for an uncensored sample of size 30. The value of R for a single uncensored sample of size 10 is 2.49. The pooled shape parameter thus has the precision associated with a sample size larger than that of the individual samples.

The median unbiased estimate of the shape parameter is:

$$\beta_1 = \frac{2.480}{1.076} = 2.304.$$

8.4 INTERVAL ESTIMATION OF A PERCENTILE

Analogous to $u(r, n, p)$ from Chapter 5, we have $u_1(r, n, p, k)$ defined as:

$$u_1(r, n, p, k) = \hat{\beta}_1 \ln\left(\frac{\hat{x}_p}{x_p}\right).$$

The population value x_p and its estimate refer to any group of the k sets of data. Confidence intervals and median unbiased estimates for x_p may be computed for each group using percentage points of the u_1 distribution just as in the single sample case. The ML estimates for each group are computed using $\hat{\beta}_1$ as the shape parameter. The 5th, 50th, and 95th percentiles of u_1 for $p = 0.10$ are given in Table 8.3 for various n, r and k. For group A in the rolling contact fatigue data set given in Table 8.2, the 10th percentile estimate recomputed using $\hat{\beta}_1 = 2.48$ gives the revised estimate $\hat{x}_{0.10} = 4.838$. Based on the tabled 5th and 95th percentiles for $n = r = 10$, $k = 5$, 90% confidence intervals are computed as:

$$0.671 \cdot \hat{x}_{0.10} = \hat{x}_{0.10} \exp\left(-\frac{0.9899}{2.480}\right) < x_{0.10} < \hat{x}_{0.10} \exp\left(-\frac{-0.5883}{2.480}\right) = 1.27 \cdot \hat{x}_{0.10}.$$

The factors 0.671 and 1.27 apply to every group. So for group A:

$$3.25 = 0.671 \times 4.838 < x_{0.10} < 1.27 \times 4.838 = 6.14.$$

Table 8.4   Raw, Median Unbiased, and Interval Estimates of $x_{0.10}$ for Each Material

Group	$\hat{x}_{0.10}$	$\hat{x}'_{0.10}$	90% Limits
B	2.811	2.645	1.89–3.56
C	3.798	3.574	2.55–4.82
A	4.873	4.553	3.25–6.13
D	4.838	4.586	3.27–6.18
E	6.342	5.968	4.25–8.04

A median unbiased estimate is computed as:

$$\hat{x}'_{0.10} = \hat{x}_{0.10} \cdot \exp\left(\frac{-0.1519}{2.480}\right) = 0.941 \cdot \hat{x}_{0.10}.$$

For group A, we have $0.941 \times 4.838 = 4.550$. Table 8.4 gives the computed or raw ML estimate of the 10th percentile, its median unbiased estimate, and 90% confidence limits in ascending order of the estimated 10th percentile value.

This ranking of the steels (BCADE) differs from the ranking based on the single test estimates, (BDCAE) although both rankings agree on the best and worst steels.

The precision measure $R^{\beta}_{0.50}$ is computed as for single samples but using the multiple group analogs. Thus for the example:

$$R^{\beta}_{0.50} = \exp\left[\frac{0.9899 + 0.5883}{1.076}\right] = 4.34.$$

This precision is comparable to a complete sample of size 30 (4.22) and vastly superior to a single uncensored sample of size 10 (15.3).

Table 8.5 on the following pages gives the values of R and $R^{\beta}_{0.50}$ for estimation of $x_{0.10}$ for the sample sizes given in Table 8.3. For a given value of the number of tests in the set, k, Table 8.5 can be used to select n and r to give a desired precision of estimation.

Figure 8.1 shows R plotted against the number of failures r for $n = 30$ for a single sample ($k = 1$), and for a set of five ($k = 5$) samples.

For a single sample with $r = 5$, the precision is poor and improves dramatically as r is increased. When $k = 5$ groups are considered, the precision is reasonably good at $r = 5$ and improves further as r increases. The benefit of grouping samples, provided that the population shape parameters are equal, is quite evident.

The precision criterion $R^{\beta}_{0.50}$ for estimation of $x_{0.10}$ is shown in Figure 8.2.

Table 8.5   Precision Measures for Shape Parameter and 10th Percentile for Various Sample Sizes

k	n	r	R	$R_{0.50}^{\beta}$
1	5	3	10.7	900
2	5	3	5.03	75.1
3	5	3	3.58	29.9
4	5	3	3	22.4
5	5	3	2.66	19.1
10	5	3	1.96	11.7
1	5	5	4.14	92.4
2	5	5	2.65	22.5
3	5	5	2.2	14
4	5	5	1.98	10.9
5	5	5	1.82	9.11
10	5	5	1.52	6.6
1	10	5	5.06	36.7
2	10	5	3.03	13.8
3	10	5	2.46	9.62
4	10	5	2.16	7.96
5	10	5	1.99	6.96
10	10	5	1.63	5.88
1	10	10	2.49	15.3
2	10	10	1.86	7.3
3	10	10	1.66	5.59
4	10	10	1.54	4.75
5	10	10	1.48	4.34
10	10	10	1.31	3.57
1	15	5	5.78	18.2
2	15	5	3.14	9.49
3	15	5	2.53	7.44
4	15	5	2.25	6.45
5	15	5	2.04	6.1
1	15	10	2.72	11.4
2	15	10	2.02	6.21
3	15	10	1.76	4.98
4	15	10	1.63	4.37
5	15	10	1.55	4.16
1	15	15	2.03	8.41
2	15	15	1.64	4.88
3	15	15	1.49	4.02
4	15	15	1.41	3.52
5	15	15	1.36	3.36
1	20	5	5.45	13.8
2	20	5	3.19	7.63
3	20	5	2.54	6.22
4	20	5	2.22	5.88
5	20	5	2.06	5.67
1	20	10	2.83	8.89
2	20	10	2.07	5.31

Table 8.5 (*Continued*)

k	n	r	R	$R_{0.50}^{\beta}$
3	20	10	1.8	4.35
4	20	10	1.67	4
5	20	10	1.58	3.79
1	20	15	2.19	7.37
2	20	15	1.73	4.42
3	20	15	1.56	3.75
4	20	15	1.46	3.41
5	20	15	1.41	3.15
1	20	20	1.82	6.13
2	20	20	1.52	3.26
3	20	20	1.41	3.06
4	20	20	1.35	2.96
5	20	20	1.3	2.77
1	25	5	5.3	9.86
2	25	5	3.18	6.5
3	25	5	2.56	5.89
4	25	5	2.25	5.41
5	25	5	2.06	5.29
1	25	10	2.91	7.13
2	25	10	2.09	4.6
3	25	10	1.84	4.11
4	25	10	1.69	3.76
5	25	10	1.58	3.55
1	25	15	2.24	6.27
2	25	15	1.77	4.16
3	25	15	1.59	3.56
4	25	15	1.49	3.19
5	25	15	1.43	3.01
1	25	20	1.94	5.68
2	25	20	1.59	3.73
3	25	20	1.45	3.2
4	25	20	1.39	2.87
5	25	20	1.33	2.73
1	25	25	1.71	4.91
2	25	25	1.45	3.34
3	25	25	1.35	2.91
4	25	25	1.3	2.61
5	25	25	1.15	2.46
1	30	5	5.36	8.12
2	30	5	3.2	6.25
3	30	5	2.56	5.77
4	30	5	2.25	5.38
5	30	5	2.05	5.24
1	30	10	2.89	5.99
2	30	10	2.12	4.22
3	30	10	1.82	3.77

(*Continued*)

Table 8.5 *(Continued)*

k	n	r	R	$R_{0.50}^{\beta}$
4	30	10	1.69	3.53
5	30	10	1.6	3.42
1	30	15	2.26	5.37
2	30	15	1.79	3.89
3	30	15	1.67	3.57
4	30	15	1.51	3.06
5	30	15	1.44	2.95
1	30	20	1.98	5.04
2	30	20	1.62	4.48
3	30	20	1.47	2.99
4	30	20	1.4	2.75
5	30	20	1.35	2.68
1	30	25	1.79	4.74
2	30	25	1.5	3.72
3	30	25	1.39	2.76
4	30	25	1.33	2.58
5	30	25	1.29	2.46
1	30	30	1.62	4.22
2	30	30	1.4	3
3	30	30	1.32	2.6
4	30	30	1.27	2.45
5	30	30	1.24	2.33

From Analysis of Sets of Two-Parameter Weibull Data Arising in Rolling Contact Endurance Data in ASTM Technical Publication 771, copyright 1982. Reproduced by permission of ASTM International.

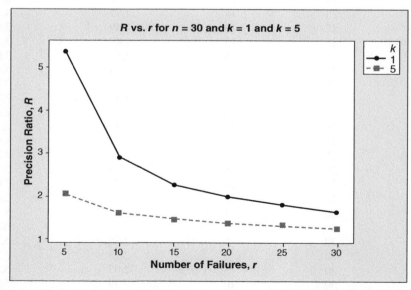

Figure 8.1 Precision measure for shape parameter in censored samples of size $n = 30$; $k = 1$ and $k = 5$.

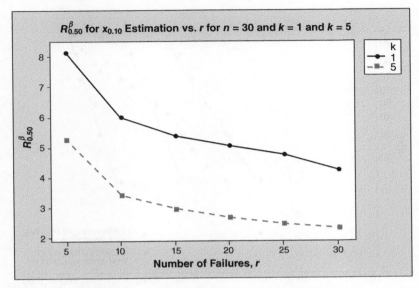

Figure 8.2 Precision measure for 10th percentile in censored samples of size $n = 30$; $k = 1$ and $k = 5$.

Figure 8.2 shows that although the precision is greatly improved using $k = 5$, the sensitivity to r appears to be about the same, as with $k = 1$; that is, the curves are roughly parallel.

8.5 TESTING WHETHER THE SCALE PARAMETERS ARE EQUAL

Having established that it is reasonable to assume that the populations from which the k samples are drawn have a common shape parameter, the next issue is whether the scale parameters differ. Formally the hypothesis is expressible as:

$$H_0 : \eta_1 = \eta_2 = \ldots = \eta_k = \eta \qquad (8.7)$$

The alternative hypothesis is that at least one of the shape parameters differs from the others. A test of this hypothesis is comparable to the one way analysis of variance (ANOVA) in normal distribution theory. In the one-way ANOVA, one tests for the equality of the population means having shown or assumed that the population variances do not differ. We consider two tests for the hypothesis that the scale parameters are equal. One of them we dub the shape parameter ratio (SPR) test. It has an intuitive graphical interpretation. The other is the classical likelihood ratio test.

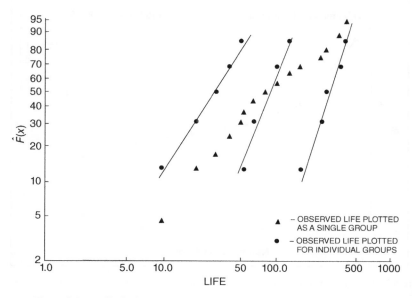

Figure 8.3   Weibull plots of three data samples separately and combined.

8.5.1   The SPR Test

Figure 8.3 shows three data samples plotted on a Weibull grid. The plots are nearly parallel, but it is clear from the lateral separation of the three plots that their scale parameters differ. Consider what happens graphically if the data are combined into a single sample. The group with the lowest scale parameter will tend to have the earliest lives in the combined sample and hence the lowest plotting positions on the vertical scale. The group with the highest scale parameter will tend to have the largest lives in the combined sample and the greatest plotting positions. It is seen that the plot of the combined sample has a lower slope and hence a lower shape parameter than each of the groups. The greater the scale parameter differences among the groups, the lower will be the shape parameter estimated from the combined data. Imposing the condition that both the scale and shape parameters are the same for each group is equivalent to assuming that the entire data set is a single sample from a single Weibull population. In this case the shape parameter estimate $\hat{\beta}_0$ is given as the solution of the following equation:

$$\sum_{i=1}^{k} \frac{r_i}{\hat{\beta}_0} + \sum_{i=1}^{k}\sum_{j=1}^{r_i} \ln(x_{ij}) - \sum_{i=1}^{k}\sum_{j=1}^{n_i} x_{ij}^{\hat{\beta}_0} \ln(x_{ij}) / \sum_{i=1}^{k}\sum_{j=1}^{n_i} x_{ij}^{\hat{\beta}_0} = 0. \qquad (8.8)$$

This is equivalent to the usual estimate from a single sample but expressed in double subscript notation. Having determined $\hat{\beta}_0$, the scale parameter estimate is:

$$\hat{\eta}_0 = \left[\sum_{i=1}^{k} \sum_{j=1}^{n_i} x_{ij}^{\hat{\beta}_0} \bigg/ \sum_{i=1}^{k} r_i \right]^{1/\hat{\beta}_0}. \tag{8.9}$$

The $100p$-th percentile is, as usual,

$$(\hat{x}_p)_0 = [k_p]^{1/\hat{\beta}_0} \times \hat{\eta}_0. \tag{8.10}$$

$\hat{\beta}_1$ is unaffected by scale parameter differences, but $\hat{\beta}_0$ is diminished by scale parameter differences. The two shape parameter estimates should be comparable when all the scale parameters are equal. When there are scale parameter differences, however, the ratio $\dfrac{\hat{\beta}_1}{\hat{\beta}_0}$ should increase. Large values of this ratio therefore argue against the equality of the scale parameters. This ratio is a pivotal function because $\hat{\beta}_0 / \beta$ is pivotal under the null hypothesis of equal shape parameters and $\hat{\beta}_1 / \beta$ is pivotal in any case. Thus, their quotient must be pivotal when the null hypothesis is true. Table 8.3 gives the 90th and 95th percentile values of $\dfrac{\hat{\beta}_1}{\hat{\beta}_0}$ for a number of n, r and k values computed using sets of k samples for which the shape and scale parameters were equal. The Mathcad module below shows the computation of the two shape parameter estimates for the data in Table 8.2. The computed ratio, 1.184, exceeds the upper 5% point of the shape parameter distribution, and thus we reject the hypothesis that the scale parameters are equal.

It has been shown (McCool, 1979) that the probability of accepting the shape parameters as equal when they are not can be expressed in terms of a single parameter ϕ_1 given by:

$$\phi_1 = \sum_{i=1}^{k} \beta \ln \eta_i \left(\frac{\eta_i^{\beta}}{\sum_{i=1}^{k} \eta_i^{\beta}} - \frac{1}{k} \right) \tag{8.11}$$

The function ϕ_1 is analogous to the noncentrality parameter of a fixed effect ANOVA. It has the following properties:

1. It is symmetric in η_i^{β} and therefore the value of ϕ_1 does not depend on the arbitrary numbering of the populations.
2. It is non-negative and reduces to zero when the η_i values are equal.
3. It is scale invariant; that is, its value is the same irrespective of the units in which η_i^{β} values are expressed. Also, since x_p is the same multiple of η_i for every sample, the value of ϕ_1 will not change if some percentile is used in lieu of η_i.

The probability of accepting the scale parameters as equal is a decreasing function of ϕ_1.

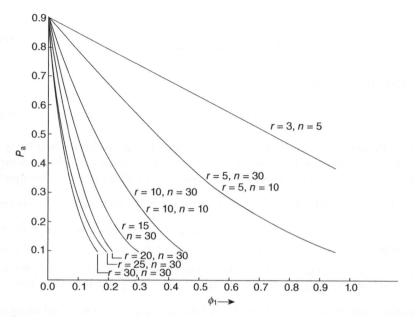

Figure 8.4 Acceptance probability as a function of ϕ_1 for various sample sizes. From Analysis of Sets of Two-Parameter Weibull Data Arising in Rolling Contact Endurance Data in ASTM Technical Publication 771, copyright 1982. Reproduced by permission of ASTM International.

Figure 8.4 shows P_a as a function of ϕ_1 for several pairs of n and r values with $k = 2$.

These operating characteristic curves depend primarily on r and only slightly on n. Because the curves for the same r are so close, only one has been plotted. To design a test with $k = 2$ such that $P_a = 0.10$ if $(x_{0.10})_1 = 3(x_{0.10})_2$ and using $\beta = 1.3$, we compute ϕ_1:

$$\phi_1 = 1.3\ln[3]\times\left[\frac{3^{1.3}}{3^{1.3}+1^{1.3}}-\frac{1}{2}\right]+1.3\ln[1]\times\left[\frac{1^{1.3}}{3^{1.3}+1^{1.3}}-\frac{1}{2}\right]=0.437.$$

From the Figure 8.4 it appears that to within graphical accuracy, $r = 10$ with any choice of $n > r$ will suffice.

8.5.2 Likelihood Ratio Test

The likelihood ratio is a classic way of testing statistical hypotheses (cf. Kendall and Stuart, 1961). In the present context the procedure is to compute the ratio λ of the likelihood function maximized under the hypothesis that the scale parameters are equal to the likelihood function maximized under the hypothesis that they may differ. The shape parameter is taken to be constant in either case. The numerator has two unknown parameters since all groups are assumed to have the same shape and scale parameters. The denominator has $k + 1$

parameters, a single shape parameter, and k scale parameters. When the hypothesis is true, that is, the scale parameters are the same, then in large samples, the quantity $-2\ln\lambda$ will follow the chi-square distribution with degrees of freedom equal to the difference in the number of parameters namely, $k+1-2=k-1$. In the general case of unequal sample sizes, the expression for $\ln\lambda$ may be written as:

$$-\left(\sum_{i=1}^{k} r_i\right)\ln\left(\frac{\hat{\beta}_1}{\hat{\beta}_0}\right)+\sum_{i=1}^{k}\sum_{j=1}^{n}\hat{\beta}_0\ln\left(\frac{x_{ij}}{\hat{\eta}_0}\right)-\sum_{i=1}^{k}\sum_{j=1}^{n}\hat{\beta}_1\ln\left(\frac{x_{ij}}{\hat{\eta}_i}\right) \tag{8.12}$$

Note that the leading term is just a monotonic function of the SPR test statistic $\dfrac{\hat{\beta}_1}{\hat{\beta}_0}$. The distribution of $-2\ln\lambda$ was also determined by Monte Carlo sampling, and the 90th and 95th percentile values are listed in Table 8.3. They do not compare well with the corresponding percentage points of the chi-square distribution with $k-1$ degrees of freedom. The expected value of a chi-square variable is equal to its degrees of freedom. Kendall has suggested that multiplying $-2\ln\lambda$ by a constant c determined so that the expected value of $-2c\ln\lambda$ is equal to $k-1$ gives a much improved fit to the asymptotic distribution. The value of c was accordingly computed as follows:

$$c=\frac{k-1}{\overline{-2\ln\lambda}}. \tag{8.13}$$

where $\overline{-2\ln\lambda}$ is the average value of $-2\ln\lambda$. Values of c are listed in Table 8.3 for each combination of n, r and k.

The last two columns of Table 8.3 show that when the 95th percentile of $-2\ln\lambda$ is multiplied by c it compares favorably with the 95th percentile of the chi-square distribution with $k-1$ degrees of freedom.

The probability of accepting the null hypothesis when it is false using the likelihood ratio test was found to depend on both ϕ_1 and, to a lesser extent, on an additional function ϕ_2 defined as follows:

$$\phi_2=k\ln\left(\frac{\sum_{i=1}^{k}\eta_i^{\beta}}{k}\right)-\sum_{i=1}^{k}\ln(\eta_i^{\beta}) \tag{8.14}$$

When the scale parameters are all equal, ϕ_2, like ϕ_1, is zero. When the scale parameters are not equal, $\phi_2>0$. For combinations of scale parameter values that result in the same values of these two functions, the probability of accepting the scale parameters as equal will be the same when using the likelihood ratio test.

A program called Weibest among the set of six disk operating system (DOS) programs on the author's website will read the data for each of k sets

Figure 8.5 Weibest output for data of Table 8.1.

of test results. It computes the ML estimates of the parameters and several percentiles for each test, and then for the whole set it computes the ML estimates $\hat{\beta}_0$ and $\hat{\beta}_1$ and their ratio, the test statistic $w = \dfrac{\hat{\beta}_{max}}{\hat{\beta}_{min}}$, and the likelihood ratio. A copy of the last screen of the output is shown as Figure 8.5 when it was run with the data of Table 8.1.

The Mathcad module below summarizes the computations for the same data set. It is titled WeibullGroups.xmcd. The results are in agreement with the Weibest output. Small differences are attributable to differences in convergence criteria in the iterative solutions for the shape parameter estimates. In both instances the value of $-2\ln\lambda$ for the rolling contact fatigue data is computed to be 17.963. This is well beyond the upper 95th percentile of 10.55 for $n = r = 10$ and $k = 5$. The c value for this case is 0.895 (Table 8.3). The corrected value, $-2c\ln\lambda = 16.077$, well exceeds the 95th percentile of the chi-square distribution with 4 degrees of freedom. It is in fact equal to the 99.7th percentile of the chi-square distribution and so is a value that would be exceeded by chance only 0.3% of the time. The scale parameters are clearly not equal.

The SPR and the likelihood ratio tests have been found to have comparable power; that is, they have a comparable probability of rejecting the hypothesis of equal scale parameters when they are not equal. The SPR test appeared to be superior when all but one of the scale parameters are equal. The likelihood ratio test was somewhat superior when several scale parameters differed.

Mathcad Module for Testing the Equality of k Shape Parameter Estimates
   ORIGIN = 1

This example is for $k = 5$ groups of size $n = 10$. The groups are uncensored so $r = n = 10$. The shape parameter β is estimated under the assumption that each group has a common shape parameter. The scale parameters and 10th percentile are estimated for each group using the common shape parameter estimate.

In the matrix $\mathbf{y}$, each column contains the lives in a group

$$y := \begin{bmatrix} 12.51 & 4.69 & 9.40 & 8.14 & 14.72 \\ 3.03 & 3.19 & 3.46 & 5.88 & 6.43 \\ 5.53 & 4.26 & 5.22 & 6.74 & 9.97 \\ 5.6 & 4.47 & 5.69 & 6.9 & 10.39 \\ 9.3 & 4.53 & 6.54 & 6.98 & 13.55 \\ 9.92 & 4.67 & 9.16 & 7.21 & 14.45 \\ 12.95 & 5.78 & 10.19 & 8.59 & 16.81 \\ 15.21 & 6.79 & 10.71 & 9.8 & 18.39 \\ 16.04 & 9.37 & 12.58 & 12.28 & 20.84 \\ 16.84 & 12.75 & 13.41 & 25.46 & 21.51 \end{bmatrix}$$

$$\mathbf{x} := y^T$$

Transpose the data matrix so rows represent different groups:

$$\mathbf{x} = \begin{bmatrix} 12.51 & 3.03 & 5.53 & 5.6 & 9.3 & 9.92 & 12.95 & 15.21 & 16.04 & 16.84 \\ 4.69 & 3.19 & 4.26 & 4.47 & 4.53 & 4.67 & 5.78 & 6.79 & 9.37 & 12.75 \\ 9.4 & 3.46 & 5.22 & 5.69 & 6.54 & 9.16 & 10.19 & 10.71 & 12.58 & 13.41 \\ 8.14 & 5.88 & 6.74 & 6.9 & 6.98 & 7.21 & 8.59 & 9.8 & 12.28 & 25.46 \\ 14.72 & 6.43 & 9.97 & 10.39 & 13.55 & 14.45 & 16.81 & 18.39 & 20.84 & 21.51 \end{bmatrix}$$

Set sample size parameters

$$n := 10$$
$$k := 5$$
$$r := 10$$

Define the ML function

$$f(\beta_1) := \left(\frac{1}{\beta_1}\right) + \left[\sum_{i=1}^{k}\sum_{j=1}^{r}\frac{\ln(x_{i,j})}{r \times k} - \sum_{i=1}^{k}\frac{\sum_{j=1}^{n}\left[(x_{i,\varphi})^{\beta_1} \times \ln(x_{i,j})\right]}{k \cdot \sum_{j=1}^{n}(x_{i,j})^{\beta_1}}\right].$$

Initial guess at the root of the ML function

$$\beta_1 := 2.$$

Solve

$$\beta_1 := \mathrm{root}(f(\beta_1), \beta_1).$$

Converged solution

$$\beta_1 = 2.48.$$

$$i := 1, 2 \ldots k$$

Compute the scale parameter estimates

$$\eta_i := \left[\sum_{j=1}^{n} \frac{(x_{i,j})^{\beta_1}}{r} \right]^{\frac{1}{\beta_1}}.$$

Compute the x_{10} estimates

$$x_{10_i} := 0.1053^{\frac{1}{\beta_1}} \times \eta_i.$$

$i =$	$x_{10_i} =$	$\eta_i =$
1	4.837	11.987
2	2.81	6.965
3	3.797	9.41
4	4.871	12.072
5	6.341	15.713

Compute the shape parameter estimate using all the data combined

$$g(\beta_0) := \left(\frac{1}{\beta_0} \right) \left[\sum_{i=1}^{k} \sum_{j=1}^{r} \frac{\ln(x_{i,j})}{r \times k} - \frac{\sum_{i=1}^{k} \sum_{j=1}^{r} \left[(x_{i,j})^{\beta_0} \times \ln(x_{i,j}) \right]}{\sum_{i=1}^{k} \sum_{j=1}^{r} (x_{i,j})^{\beta_0}} \right].$$

Guess at solution using combined data

$$\beta_0 := 2.$$

$$\beta_0 := \mathrm{root}(g(\beta_0), \beta_0).$$

$$\eta_0 := \left[\sum_{i=1}^{k} \sum_{j=1}^{r} \frac{(x_{i,j})^{\beta_0}}{r \times k} \right]^{\frac{1}{\beta_0}}.$$

Combined group shape and scale parameter estimates

$$\beta_0 = 2.094.$$

$$\eta_0 = 11.318.$$

Compute ratio of shape parameter estimates

$$z := \frac{\beta_1}{\beta_0}.$$

$$z = 1.184.$$

Z exceeds the upper 5% point of the null distribution (1.139 for $n = r = 10$, $k = 5$) so conclude that the scale parameters differ.

Compute minus the log of the likelihood ratio λ

$$\text{minusln}\lambda := (r \times k) \times \ln\left(\frac{\beta_1}{\beta_0}\right) - \sum_{i=1}^{k}\sum_{j=1}^{r}[(\beta_0 \times \ln(x_{i,j})) - \beta_0 \times \ln(\eta_0)] +$$

$$\sum_{i=1}^{k}\sum_{j=1}^{r}[(\beta_1 \times \ln(x_{i,j})) - \beta_1 \times \ln(\eta_i)].$$

$$2 \times \text{minusln}\lambda = 17.963$$

correction factor for chi-square distribution

$$c := 0.895$$

$$c \times 2 \times \text{minusln}\lambda = 16.077.$$

A chi-square distribution with 4 degrees of freedom has a probability of 0.003 of exceeding 16.077, supporting the conclusion that the scale parameters differ.

8.6 MULTIPLE COMPARISON TESTS FOR DIFFERENCES IN SCALE PARAMETERS

Once the SPR or likelihood ratio test indicates that there are differences between the scale parameters, interest centers on determining exactly which ones differ. A similar problem occurs in applications of the normal distribution. When the ANOVA indicates that the means differ, there are a number of procedures, often referred to as post hoc tests, to establish differences among the samples.

The procedure we are using is analogous to a test called Tukey's test (Tukey, 1949). It focuses on the largest and smallest scale parameter estimates in the set of k such estimates. Interior values that differ by more than a value that would be unusual for the largest and smallest values are *a fortiori* considered to differ.

The test statistic is:

$$t_1(r, n, k) \equiv \hat{\beta}_1 \ln\left(\frac{\hat{\eta}_{\max}}{\hat{\eta}_{\min}}\right). \qquad (8.15)$$

Since we are operating under the assumption that the samples have a common shape parameter, the percentiles for any sample are the same multiple of the scale parameter. Thus t_1 could have been written in terms of any percentile and would have the same distribution, that is,

$$t_1(r, n, k) \equiv \hat{\beta}_1 \ln\left(\frac{[\hat{x}_p]_{max}}{[\hat{x}_p]_{min}}\right).$$

That t_1 is a pivotal quantity under the hypothesis that the scale parameters are equal follows from forming u_1 for each of the k samples. Sorting the k values of u_1 and taking the difference between the largest and smallest leads to:

$$u_{max} - u_{min} = \hat{\beta}_1 \ln\left(\frac{\hat{\eta}_{max}}{\eta}\right) - \hat{\beta}_1 \ln\left(\frac{\hat{\eta}_{min}}{\eta}\right) = \hat{\beta}_1 \ln\left(\frac{\hat{\eta}_{max}}{\hat{\eta}_{min}}\right).$$

Differences of pivotal quantities are themselves pivotal. Under the hypothesis that every sample comes from the same population the scale parameters cancel.

Table 8.3 shows 90% and 95% points of the distribution of t_1. For the rolling contact fatigue data for which $n = r = 10$ and $k = 5$, the value of t_1 appropriate for a 10% significance level test is 1.26. If the value of t_1 exceeds 1.26 for any pair of estimates and not just the most extreme, we can declare that they differ as well since a critical value for a pair of interior samples would be less than 1.26. Knowing that $\hat{\beta}_1 = 2.48$, we can compute the smallest ratio that would be significant according to this criterion. Designating this value the least significant ratio (LSR), we have:

$$2.48\ln(LSR) = 1.26.$$

Solving for LSR,

$$LSR = \exp\left(\frac{1.26}{2.48}\right) = 1.66.$$

Thus, any two scale parameters, or percentiles, whose ratios exceed 1.66 may be declared to differ at a 10% experiment-wise level of significance. Ordinarily when a set of confidence interval statements are made, each having, say, a 90% chance of being true statements, the probability that *every* statement in the set is true is lower than 90%. In the present case, when we declare some set of scale parameters different, there is a 90% chance that the entire set of statements is correct. That is the meaning of the experiment-wise error significance level or error rate.

Taking the ratios of the 10th percentile estimates computed using the estimate of the common shape parameter, we can summarize which pairs differ, that is, which pairs are in a ratio in excess of 1.66. Pairs that differ significantly are noted in the two-way table below:

	B	C	A	D	E
B	*		Sig.	Sig.	Sig.
C		*			Sig
A			*		
D				*	
E					*

Another common way to summarize the results of a multiple comparison test of this type is to list the scale parameter values or just the group names in increasing order of estimated scale parameter value and draw a line under pairs which do not differ:

B   C   A   D   E

The results will often be only partially conclusive as in the present case. It would be desirable if, say, B and C formed a set and A, D, and E were a second set with no differences within each set but with every member of the first set different from every member of the second.

One might ask: How can B not differ from C and C not differ from D and yet B differs from D? The answer is that these comparisons are not transitive. The results only summarize what cannot be distinguished in view of the variability or noise in the data. The values of the 10th percentile estimates for B and C are within the observed noise limits. C, A, and D are also within the observed noise limits as are A, D, and E. At this point the results serve to show where further testing may be best conducted. If our desire is to increase the life, we might consider further tests on A, D, and E. We should not be surprised if on further testing the materials rank in a different sequence.

8.7 AN ALTERNATIVE MULTIPLE COMPARISON TEST FOR PERCENTILES

It sometimes happens that one does not have access to the raw data from an investigation but does have a summary in terms of ML estimates of the shape parameters and some percentile such as the scale parameters $x_{0.632}$ or the medians $x_{0.50}$. One also needs the knowledge that the tests are type II censored with the same values of n and r. The following random variable is readily shown to be a pivotal function when the k samples have a common shape parameter and also equal values of the $100p$-th percentile(McCool, 1975, 1978).

$$t(r, n, p, k) = \bar{\beta} \ln \left(\frac{[\hat{x}_p]_{max}}{[\hat{x}_p]_{min}} \right). \tag{8.16}$$

Here the parameters are estimated using just the data within each of the k samples. The average of the k estimated shape parameters is denoted $\bar{\beta}$. (This function will still be a pivotal function if the harmonic mean is used instead of the arithmetic mean of the shape parameters. The harmonic mean is the reciprocal of the average reciprocals of the shape parameters.)

The statistic t_1 defined in Equation 8.15 was independent of the percentage point p under consideration because the scale parameters were estimated using the same shape parameter estimate for each of the k samples. This caused the ratio $\dfrac{[\hat{x}_p]_{max}}{[\hat{x}_p]_{min}}$ to be the same no matter what p was at issue.

On the other hand, the distribution of t depends on the percentile p since the percentiles are not all estimated with the same shape parameter value. The critical values of t for a 10% level test of the equality of shape parameters are given in Table 8.6 for $p = 0.10$. Table 8.7 lists the corresponding values for $p = 0.50$. The software program Multi-Weibull.exe described in the next section may be used to generate critical values of $t(r, n, k, p)$ for any significance level and percentile p and a broad range of n, r and k.

Once the test for a common shape parameter has failed to reject that hypothesis, one may calculate the quantity t above. We take the data on the fatigue lives of $k = 5$ materials given in Table 8.5 as an example. The data were

Table 8.6 Values of $t_{0.90(r, n, k, p)}$ for $p = 0.10$ and Various n, r and k

n	r	$k = 2$	$k = 3$	$k = 4$	$k = 5$	$k = 10$
5	3	6.18	8.74	9.84	10.78	12.8
5	5	3.48	4.36	4.92	5.29	6.29
10	3	5.33	7.02	8.19	8.98	10.8
10	5	3.00	3.84	4.33	4.61	5.59
10	10	1.97	2.52	2.81	3.02	3.64
15	5	2.53	3.35	3.80	4.02	4.74
15	10	1.82	2.33	2.58	2.82	3.34
15	15	1.55	1.93	2.20	2.36	2.79
20	5	2.31	2.98	3.36	3.61	4.32
20	10	1.67	2.06	2.29	2.46	2.94
20	15	1.44	1.82	2.02	2.20	2.63
20	20	1.31	1.62	1.81	1.95	2.32
30	5	1.98	2.53	2.90	3.11	3.63
30	10	1.38	1.73	1.95	2.09	2.49
30	15	1.25	1.54	1.78	1.90	2.23
30	20	1.18	1.46	1.63	1.76	2.08
30	30	1.03	1.28	1.42	1.53	1.84

Table 8.7   Values of $t_{0.90}(r, n, k, p)$ for $p = 0.50$ and Various r, n and k

n	r	$k = 2$	$k = 3$	$k = 4$	$k = 5$	$k = 10$
5	3	3.85	5.25	5.88	6.44	7.85
5	5	1.85	2.29	2.57	2.80	3.33
10	3	4.71	6.42	7.53	8.41	9.83
10	5	1.81	2.32	2.63	2.83	3.35
10	10	1.01	1.30	1.44	1.57	1.87
15	5	2.05	2.67	3.02	3.31	3.93
15	10	0.942	1.18	1.31	1.44	1.71
15	15	0.798	0.994	1.11	1.21	1.45
20	5	2.36	3.01	3.50	3.78	4.61
20	10	0.967	1.18	1.35	1.42	1.73
20	15	0.728	0.914	1.00	1.07	1.30
20	20	0.657	0.837	0.933	1.01	1.21
30	5	2.76	3.65	4.12	4.42	5.34
30	10	1.06	1.34	1.51	1.62	1.93
30	15	0.708	0.898	1.01	1.08	1.27
30	20	0.583	0.744	0.817	0.882	1.05
30	30	0.531	0.649	0.727	0.788	0.933

already tested for a common shape parameter and that assumption was found to be plausible. The average shape parameter is calculated to be $\hat{\bar{\beta}} = 2.726$. The t statistic is calculated as:

$$t(10, 10, 5, 0.10) = 2.726 \ln\left(\frac{8.83}{2.60}\right) = 3.33.$$

This exceeds the critical value $t_{0.90} = 3.02$ from Table 8.6, so we may affirm that the largest and smallest values differ. As before, any interior estimates whose ratios exceed that allowed for the extreme ratio may be declared to differ. Again, denoting the smallest significant ratio as LSR, we may establish its value from the relation:

$$2.726 \ln(LSR) = 3.02.$$

From which the LSR is computed as:

$$LSR = \exp\left(\frac{3.02}{2.726}\right) = 3.03.$$

8.8   THE PROGRAM MULTI-WEIBULL.EXE

The software program called Multi-Weibull was written in the Visual Basic language to enable multiple comparisons for the Weibull shape parameter and percentiles. A zipped file containing the software and a readme file may be downloaded from http://www.personal.psu/mpt.

The software simulates from 100 to 10,000 sets of k type II censored Weibull samples, each comprising a sample size n with r failures. For each set the value of the ratio $w(r, n, k) = \hat{\beta}_{(k)} / \hat{\beta}_{(1)}$ of the largest to the smallest ML estimate of the shape parameters is formed and 16 percentiles of the distribution are printed to the output screen. The statistic $w(r, n, k)$ has been shown to be a pivotal function; that is, its distribution is independent of the true population value of the shape parameter. An upper percentile of the distribution of $w(r, n, k)$ may therefore be used to test the hypothesis that a set of data samples were drawn from populations having a common shape parameter.

A second function $t(r, n, k, p) = \bar{\beta} \ln[x_{p(k)} / x_{p(1)}]$ is also formed consisting of the product of the average shape parameter estimate and the logarithm of the ratio of the largest to smallest ML estimate of a quantile x_p. Given that the k samples are drawn from populations having a common shape parameter, an upper percentage point of the distribution of $t(r, n, k, p)$ may be used to test whether the populations also have a common value of a specified quantile.

The software accepts values of k ranging from 2 to 25, test group sample sizes n from 2 to 200, and number of failures ranging from 2 to n.

The introductory program screen is shown in Figure 8.6.

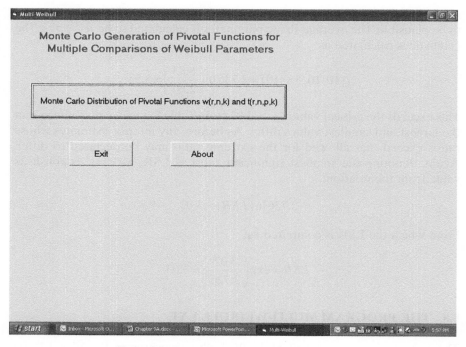

Figure 8.6   Introductory screen of Multi-Weibull.

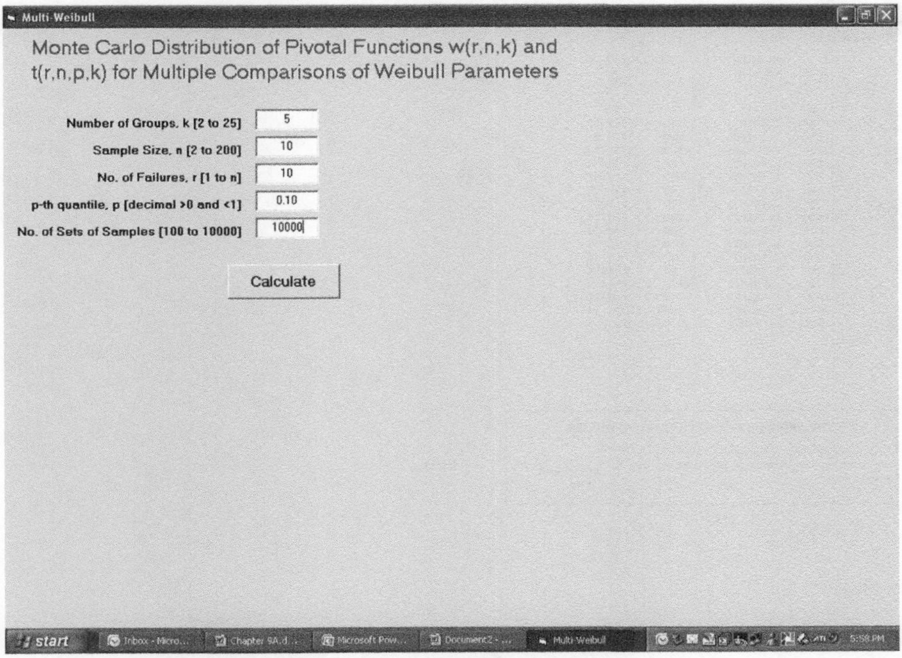

Figure 8.7   Input screen for program Multi-Weibull.

The following screen (Figure 8.7) appears when the user clicks the large button. The user then enters the number of groups, the sample size n, the number of failures r if the data is type II censored, the quantile of interest and the number of sets of data to simulate. As shown, the input screen is set for the case $n = r = 10$, $p = 0.10$, and $k = 5$.

Clicking the calculate button produces 16 percentage points of the distribution of $w(r, n, k) = \hat{\beta}_{(k)} / \hat{\beta}_{(1)}$ and $t(r, n, k, p) = \bar{\beta} \ln[\hat{x}_{p(k)} / \hat{x}_{p(1)}]$ as shown in Figure 8.8. These values may be written to a text file at the user's option. The values shown are for $k = 5$, $n = r = 10$, and $p = 0.10$.

As a further example of the use of MultiWeibull.exe, we use it to analyze a set of 13 test results analyzed using Pivotal.exe in conjunction with a spreadsheet in McCool (2000). The fracture stress in four point bending was recorded on 10 specimens of each of 13 different composite matrix resins, typical of the materials used in dental restorations. The materials differed with respect to chemistry, degree of cure, and exposure to soaking in water. Weibull plots and formal tests affirmed that the fracture data for all materials followed the two-parameter Weibull model. Table 8.8 lists the raw ML estimates of η and β for each material.

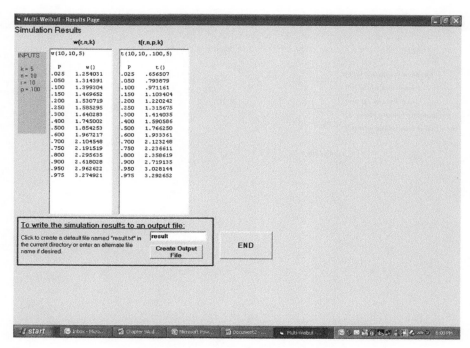

Figure 8.8 Output screen for program Multi-Weibull.

Table 8.8   Maximum Likelihood Estimates of Weibull Parameters

Material	$\hat{\beta}$	$\hat{\eta}$
A	10.58	146.3
B	8.24	48.81
C	11.9	158.95
D	10.07	134.47
E	9.58	173.7
F	19.49	127.9
G	11.16	216.79
H	11.75	144.87
I	16.13	73.1
J	7.45	132.73
K	13.4	82.9
L	20.85	118.05
M	16	147.4

Running Multi-Weibull with $n = r = 10$, $k = 13$, and $p = 0.632$ and using 10,000 samples gave the following results:

INPUT PARAMETERS
k = 13
n = 10
r = 10
p = .632

w(10, 10, 13)   t(10, 10,632, 13)

p	w_p	p	t_p
.025	1.696847	.025	.712227
.050	1.792500	.050	.782060
.100	1.908321	.100	.865702
.150	1.985398	.150	.928024
.200	2.061595	.200	.974961
.250	2.132745	.250	1.014107
.300	2.192562	.300	1.052469
.400	2.321354	.400	1.131206
.500	2.456008	.500	1.207274
.600	2.593952	.600	1.280013
.700	2.764171	.700	1.366781
.750	2.878238	.750	1.411182
.800	2.995514	.800	1.467537
.900	3.389536	.900	1.624358
.950	3.781165	.950	1.765298
.975	4.158304	.975	1.896196

The ratio of the largest to smallest shape parameter estimate is $20.85/7.45 = 2.80$. From the distribution of w above we see that the probability of a greater ratio due to chance exceeds 0.25. There is therefore no reason for concluding that the shape parameters differ.

Accepting that the 13 samples come from Weibull populations having the same shape parameter we may address the question of whether the scale parameters differ. For $p = 0.632$, the following function is pivotal:

$$t(r, n, k, 0.632) = \bar{\beta} \cdot \ln\{\hat{\eta}_{(\max)} / \hat{\eta}_{(\min)}\}.$$

Using a family error rate of 0.10, we may proclaim as differing in their scale parameters any ordered material pairs (i) and (j) $(i > j)$ for which:

$$\bar{\beta} \times \ln(\hat{\eta}_{(i)} / \hat{\eta}_{(j)}) > 1.62.$$

The 90th percentile of the distribution of $t(10, 10, 13, 0.632) = 1.62$. The smallest ratio for which a difference may be proclaimed, that is, the LSR may thus be computed as:

$$LSR = \exp\left(1.62/\bar{\bar{\beta}}\right).$$

For the values in Table 8.8, $\bar{\bar{\beta}} = 12.82$, so that LSR = 1.13. Thus, any two materials for which their $\hat{\eta}$ ratio exceeds 1.13 may be pronounced different. Material B is thus shown to have a significantly smaller scale parameter than all the others, while the scale parameter for material G is significantly larger than all the others. There are many other interior clusters. I and K do not differ from each other but do differ from all other materials. Other clusters exhibit overlap. For instance, there is no difference among L, F, J, and D, or among F, J, D, and H, but L and H may be said to differ. If the purpose of the testing was to find the material with the largest scale parameter, the test has been successful. If the purpose was to find a definitive ranking of the materials, additional testing is needed.

8.9   INFERENCE ON $P(Y < X)$

The probability that one Weibull random variable Y is less than another independent Weibull random variable X for the case where both X and Y have the same shape parameter was discussed in Section 4.3 where the following expression was cited:

$$P[Y < X] = \frac{1}{1 + \left(\dfrac{\eta_y}{\eta_x}\right)^{\beta}} = \frac{1}{1 + \rho} \tag{8.17}$$

where

$$\rho \equiv \left(\frac{\eta_y}{\eta_x}\right)^{\beta} \tag{8.18}$$

We now consider the interval estimation of this probability based on random samples drawn from the distributions of X and Y. For convenience we assume that both samples are of the same size n and, if type II censored, have the same number of failures r. The material in this section is adapted from McCool (1991).

As noted in Chapter 4, the problem of estimating and of drawing inference about the probability that a random variable Y is less than an independent random variable X arises in a reliability context. When Y represents the random value of a stress that a device will be subjected to in service and X represents the strength that varies from item to item in the population of devices, then the reliability R, that is, the probability that a randomly selected device functions successfully, is equal to $P[Y < X]$. We will accordingly use R as the symbol for $P[Y < X]$ although it only represents reliability in the stress/strength context. The same problem also arises in the context of statistical tolerancing where Y represents, say, the diameter of a shaft and X the diameter of a bearing that is to be mounted on the shaft. The probability that the bearing fits without interference is then $P[Y < X]$ and could be broadly interpreted as the reliability of the fit.

Apart from these two application areas it is suggested by Wolfe and Hogg (1971) and by Enis and Geisser (1971) that in some cases the difference between two random variables is more naturally characterized by $P[Y < X]$ than by the more usual difference of their means (Enis and Geisser, 1971; Wolfe and Hogg, 1971). A case in point cited by Wolfe and Hogg arises in biometry wherein Y represents a patient's remaining years of life if treated with drug A, and X represents the patient's remaining years when treated with drug B. If the choice of drug is left to the patient, his or her deliberations will center on whether $P[Y < X]$ is less than or greater than 0.5.

Because R is monotonic in ρ, inference on ρ is equivalent to inference on R. We hereafter confine attention to the parameter ρ. Large values of ρ correspond to small values of R.

8.9.1   ML Estimation

Given random samples of size n type II censored at the r-th order statistic and drawn from the distributions of X and Y, the ML estimate of the common shape parameter β is the solution of Equation 8.3 for the case that $k = 2$. Dropping the subscript and specializing the notation to the situation we now consider gives the equation:

$$\frac{2}{\hat{\beta}} + \frac{1}{r}\left(\sum_1^r \ln x_i + \sum_1^r \ln y_i \right) - \frac{\sum_1^n x_i^{\hat{\beta}} \ln x_i}{\sum_1^n x_i^{\hat{\beta}}} - \frac{\sum_1^n y_i^{\hat{\beta}} \ln y_i}{\sum_1^n y_i^{\hat{\beta}}} = 0. \qquad (8.19)$$

This estimate pools the data from both samples and is therefore more precise than the estimator obtained using just one of the samples. The observations are assumed to be indexed so that x_i and y_i denote uncensored observations for $i \le r$, and $x_i = x_r$ and $y_i = y_r$ for $i > r$.

As discussed in Section 8.1, it has been shown that

$$v = \hat{\beta} / \beta \qquad (8.20)$$

is a pivotal function whose distribution depends only on n and r.

The ML estimates of η_x and η_y are

$$\hat{\eta}_x = \left(\frac{\sum_1^n x_i}{r} \right)^{1/\hat{\beta}} \tag{8.21}$$

and,

$$\hat{\eta}_y = \left(\frac{\sum_1^n y_i}{r} \right)^{1/\hat{\beta}}. \tag{8.22}$$

Schafer and Sheffield show that when the n and r are the same for both groups,

$$T = u_y - u_x. \tag{8.23}$$

is a pivotal function where

$$u_x = \hat{\beta} \ln \left(\frac{\hat{\eta}_x}{\eta_x} \right). \tag{8.24}$$

and

$$u_y = \hat{\beta} \ln \left(\frac{\hat{\eta}_y}{\eta_y} \right). \tag{8.25}$$

T is symmetrically distributed about its mean $E(T) = 0$. Schafer and Sheffield (1976) give percentage points of T determined by Monte Carlo sampling for uncensored samples of size $n = 5(1)20(4)40(10)100$.

The ML estimate of ρ is

$$\hat{\rho} = \left(\frac{\hat{\eta}_y}{\hat{\eta}_x} \right)^{\hat{\beta}} \tag{8.26}$$

Taking logarithms, adding, and subtracting $\hat{\beta} \ln(n_y)$ and $\hat{\beta} \ln(n_x)$ using Equations 8.24, 8.25, and 8.20 and rearranging gives:

$$\ln \hat{\rho} = u_y - u_x + \left(\frac{\hat{\beta}}{\beta} \right) \ln \left(\frac{\eta_y}{\eta_x} \right)^{\beta} = T + v \ln(\rho). \tag{8.27}$$

The expected value of $\ln \hat{\rho}$ is:

$$E(\ln \hat{\rho}) = E(T) + E(v) \ln \varrho \tag{8.28}$$

Values of $E(v)$ obtained by Monte Carlo sampling as discussed further below are listed in Table 8.9 for 28 combinations of n and r. For small samples, $\ln\hat{\rho}$ is a highly biased estimator of $\ln\hat{\rho}$, that is, $E(v) > 1.0$. For $n = r = 30$, the bias is inconsequential.

8.9.2 Normal Approximation

The variance of $\ln\hat{\rho}$ is:

$$var(\ln\hat{\rho}) = \sigma_T^2 + (\ln\rho)^2\sigma_v^2 + 2\ln\hat{\rho}\times cov(T, v). \tag{8.29}$$

where

$$cov(T, v) = E(Tv) - E(T)\times E(v) = E(Tv) = E(u_x v) - E(u_y v).$$

Since u_x and u_y are identically distributed as long as the sample sizes for X and Y are the same, $E(T) = 0$ and $E(u_x v) = E(u_y v)$. Thus,

$$Cov(T, v) = 0.$$

Assuming approximate normality for $\ln\hat{\rho}$, two-sided $100\ (1 - \alpha)\%$ probability limits for $\ln\hat{\rho}$ may be computed for fixed ρ as,

$$E(v)\ln\hat{\rho} \pm z_{\frac{\alpha}{2}}[\sigma_T^2 + (\ln\rho)^2\ \sigma_v^2]^{\frac{1}{2}} \tag{8.30}$$

where z_p is the upper p-th percentage point of the standard normal distribution. Values of σ_T and σ_v obtained by Monte Carlo sampling are given in Table 8.9. Approximate $100(1 - \alpha)\%$ confidence intervals for ρ may be obtained by equating the observed value of $\ln\hat{\rho}$ to the upper and lower ends of the above interval and solving for ρ. The upper and lower limits for $\ln\rho$ are thus found from the following expression:

$$\frac{E(v)\ln\hat{\rho} \pm z_{\alpha/2}\sqrt{\sigma_v^2\left[(\ln\hat{\rho})^2 - z_{\frac{\alpha}{2}}^2\sigma_T^2\right] + (E(v)\times\sigma_T)^2}}{E^2(v) - z_{\alpha/2}^2\sigma_v^2}. \tag{8.31}$$

Values of σ_T and σ_v have been obtained by Monte Carlo sampling. The values listed are the averages of two simulations of 10,000 cases each.

Table 8.9   Values of $E(v)$, σ_v, and σ_T for various r and n

n	r	$E(v)$	σ_v	σ_T
5	3	1.6490	1.1651	1.8818
5	5	1.2782	0.4142	0.9090
10	5	1.5006	0.5213	0.9705
10	10	1.1150	0.2188	0.5196
15	5	1.3900	0.5313	0.9943
15	10	1.1420	0.2541	0.5355
15	15	1.0732	0.1655	0.4020
20	5	1.3988	0.5485	0.9900
20	10	1.1528	0.2676	0.5450
20	15	1.1210	0.2261	0.4131
20	20	1.0545	0.1378	0.3398
25	5	1.4079	0.5622	0.7821
25	10	1.1570	0.2701	0.5458
25	15	1.0941	0.1950	0.4110
25	20	1.0628	0.1524	0.3456
25	25	1.0421	0.1211	0.2988
30	5	1.4028	0.5644	1.0046
30	10	1.1634	0.2794	0.5434
30	15	1.0943	0.1972	0.4135
30	20	1.0673	0.1600	0.3466
30	25	1.0488	0.1316	0.3005
30	30	1.0343	0.1072	0.2708
40	10	1.1645	0.2833	0.5472
40	20	1.0714	0.1663	0.3437
40	40	1.0268	0.0921	0.2313
50	25	1.0556	0.1439	0.3027
50	50	1.0271	0.0813	0.2058
100	100	1.0103	0.0569	0.1444

Example

An uncensored sample of size $n = 30$ was drawn from the population $W(1, 1.5)$ to represent the random variable X. Another uncensored sample was drawn from $W(0.630, 1.5)$ to represent Y. The true value of ρ was:

$$\rho = \left(\frac{\eta_y}{\eta_x}\right)^{\beta} = \left(\frac{0.630}{1.0}\right)^{1.5} = 0.5.$$

And so the true value of $R = 1/(1 + \rho) = 0.67$.

The common shape parameter was estimated by solving Equation 8.19 to be $\hat{\beta} = 1.57$. The individual scale parameters were estimated to be:

$$\hat{\eta}_y = 0.6511$$

and

$$\hat{\eta}_y = 0.8442.$$

This gives the ML estimate of ρ:

$$\hat{\rho} = \left(\frac{\hat{\eta}_y}{\hat{\eta}_x}\right)^{\hat{\beta}} = \left(\frac{0.6511}{0.8442}\right)^{1.57} = 0.665.$$

The corresponding ML estimate of R is computed as $1/(1 + 0.665) = 0.601$.
From Table 8.9 for $n = r = 30$, $E(v) = 1.0343$, $\sigma_v = 0.1072$, and $\sigma_T = 0.2708$.
For 90% limits use $z_{0.05} = 1.645$. The limits on $\ln\rho$ computed as the two roots of Equation 8.31 are:

$$-0.849 < \ln\rho < 0.187.$$

So the limits on ρ are:

$$0.428 < \rho < 1.205.$$

Finally, using $R = \dfrac{1}{1+\rho}$ the 90% limits for R become

$$0.453 < R < 0.70,$$

which includes the true population value of $R = 0.67$.

8.9.3 An Exact Simulation Solution

It is possible to compute exact confidence limits for ρ and hence for R, but the method is cumbersome. In McCool (1991), simulation was used to compute percentage points of $\ln\hat{\rho}$ for various values of ρ using

$$\ln\hat{\rho} = u_y - u_x + v\ln\rho.$$

The simulation was carried out for $\rho = 0.001$, 0.01, 0.02, 0.05, 0.1(0.1)1.0, 2.0, 5.0, 10.0, 20.0.
For $\rho = 1.0$ the distribution of $\ln\hat{\rho}$ reduces to the distribution of $T = u_y - u_x$. The program was run for 21 combinations of n and r. The same values of n and r were used for the samples of X and Y, although the method is generally applicable.

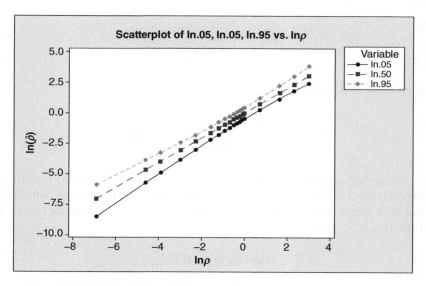

Figure 8.9 The 5th, 50th, and 95th percentiles of $\ln\hat{\rho}$ versus $\ln\rho$ for $n = r = 30$.

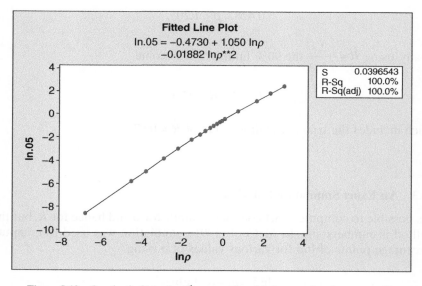

Figure 8.10 Quadratic fit to the 5th, percentile of $\ln\hat{\rho}$ versus $\ln\rho$ for $n = r = 30$.

Figure 8.9 shows the 5th, 50th, and 95th percentiles of $\ln\hat{\rho}$ plotted against $\ln\rho$ for $n = r = 30$.

The 5th and 95th percentiles are well fitted by quadratic functions as shown in Figures 8.10 and 8.11.

The median is well fitted by a straight line (Fig. 8.12):

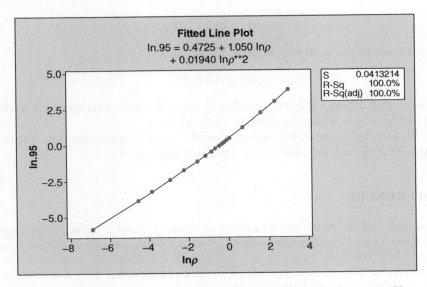

Figure 8.11 Quadratic fit to the 95th, percentile of $\ln\hat{\rho}$ versus $\ln\rho$ for $n = r = 30$.

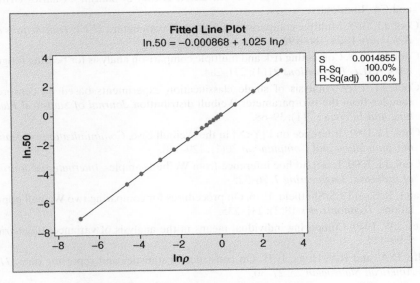

Figure 8.12 Linear fit to the median of $\ln\hat{\rho}$ versus $\ln\rho$ for $n = r = 30$.

8.9.4 Confidence Intervals

Alternatively the 5th and 95th percentiles may be plotted against ρ (log scales recommended) and the interpolation carried out graphically.

For the example, the following interval is obtained by using the equations fitted to the foregoing graphs:

$$-0.852 < \ln \rho < 0.062,$$

which leads to the following interval for R:

$$0.484 < R < 0.701.$$

This interval includes the true population value $R = 0.67$ corresponding to the populations from which the samples were taken.

This interval is somewhat tighter than the normal approximation, but the upper limit of both intervals are in excellent agreement.

REFERENCES

Brown, P.F. and J.R. Potts. 1977. Evaluation of Powder Processed Turbine Engine Ball Bearings Air Force Aero-Propulsion Laboratory Wright Patterson Air Force Base, AFAPL-TR-77-26.

Enis, P. and S. Geisser. 1971. Estimation of the probability that Y<X. *Journal of the Acoustical Society of America* 66: 162–168.

Kendall, M.G. and A. Stuart. 1961. *The Advanced Theory of Statistics*. Charles Griffin and Co., London.

McCool, J.I. 1975. Multiple comparisons for Weibull parameters. *IEEE Transactions on Reliability* R-24: 186–192.

McCool, J.I. 1978. Competing risk and multiple comparison analysis for bearing fatigue tests. *ASLE Transactions* 21(4): 271–284.

McCool, J.I. 1979. Analysis of single classification experiments based on censored samples from the two-parameter Weibull distribution. *Journal of Statistical Planning and Inference* 3(1): 39–68.

McCool, J.I. 1991. Inference on P[Y<X] in the Weibull case. *Communications in Statistics. Simulation and Computation* 20(1): 129–148.

McCool, J.I. 2000. Exact ad hoc inference from Weibull samples. *International Journal of Industrial Engineering* 7: 26–32.

Schafer, R.E. and T.S. Sheffield. 1976. On procedures for comparing two Weibull populations. *Technometrics* 18(2): 231–235.

Tukey, J.W. 1949. Comparing individual means in the analysis of variance. *Biometrics* 5: 99–114.

Wolfe, D.A. and R.V. Hogg. 1971. On constructing statistics and reporting data. *The American Statistician* 25: 27–30.

EXERCISES

1. A set of $k = 3$ samples of size $n = 10$, type II censored at $r = 5$ were analyzed individually. The largest estimated shape parameter was 2.72 and the smallest was 1.96. Test the hypothesis of a common shape parameter using an experiment-wise error rate $\alpha = 0.10$.

2. Following acceptance of the plausibility of a common shape parameter for the case described in exercise 1, the raw ML estimate of the common shape parameter was computed to be $\hat{\beta}_1 = 1.68$. Compute a median unbiased estimate and 90% confidence limits for the population value of the common shape parameter.

3. If when the data described in problem 1 are combined into a single sample of size 30, the maximum likelihood estimate of the shape parameter is computed to be $\hat{\beta}_0 = 1.10$, does the use of the SPR test with $\alpha = 0.05$ indicate that the scale parameters differ?

4. Continuing with the same case, if the largest and smallest of the three scale parameter estimates among the three samples are $\hat{\eta}_{max} = 44.0$ and $\hat{\eta}_{max} = 11.0$, using the appropriate critical value of t_1 with $\alpha = 0.05$, do the corresponding population scale parameters differ?

5. Using Table 8.5 with $k = 5$ samples and a common sample size $n = 15$, how many failures r must be observed in each sample so that the precision measure R for shape parameter estimation is less than 2.0?

6. Four uncensored samples of size $n = r = 10$ were tested at four different temperatures. The ML estimates of the shape and scale parameters were reported but the individual data values were not. The estimates are as follows:

| $\hat{\beta}$ | 4.606 | 3.538 | 2.857 | 4.156 |
| $\hat{\eta}$ | 893.8 | 471.9 | 408.4 | 403.9 |

Use the Multi-Weibull software to test for the equality of the shape parameters and then the scale parameters.

7. Samples of size $n = 20$, type II censored with $r = 10$ were drawn from each of two Weibull populations assumed to share a common shape parameter. Denoting the random variables as X and Y we assume $X \sim W(\eta_x, \beta)$ and $Y \sim W(\eta_y, \beta)$. The ML estimate of the common shape parameter was computed to be $\hat{\beta} = 1.72$. The ML scale parameter estimates using the common shape parameter were $\hat{\eta}_x = 84.1$ and $\hat{\eta}_x = 36.6$. Compute 90% confidence limits on $R = P[X < Y]$ using Equation 8.31 and the appropriate values listed in Table 8.9.

CHAPTER 9

Weibull Regression

In Chapter 8 we considered a multiple sample model in which the issue was whether the scale parameters were unaffected by the levels of an external factor or whether they varied in some unspecified way with the level of the external factor. There was no systematic pattern proposed for the manner in which they differed if indeed they were not the same. If the factor were a quantitative one such as temperature or voltage, the scale parameters might vary systematically with the level of that factor and, if so, it would be of interest to quantify that relationship for purposes of interpolating or extrapolating to factor levels not directly tested. One application of this approach is accelerated testing wherein tests are run at levels of a factor such as load or temperature that are more "severe" than met under "use" conditions in order to hasten test completion. A fitted relationship between the factor level and one or more parameters of the lifetime distribution is then used to predict the life distribution under generally milder "use" conditions. Nelson (1990) gives a broad survey of the many models and analyses related to accelerated testing. Our discussion is limited to a model often called the power law model. Exact inference for the power law model was considered by McCool and is the basis for the discussion that follows (McCool, 1980, 1986).

9.1 THE POWER LAW MODEL

Under the power law model, the variable that we generically call stress and designate by the symbol s is presumed to have a multiplicative effect on the Weibull scale parameter but no effect on the shape parameter. The cumulative distribution of life conditional on s may be expressed as:

$$F(x \mid s) = 1 - \exp\left(-\frac{x}{\eta(s)}\right)^{\beta}. \tag{9.1}$$

Using the Weibull Distribution: Reliability, Modeling, and Inference, First Edition. John I. McCool.
© 2012 John Wiley & Sons, Inc. Published 2012 by John Wiley & Sons, Inc.

The effect of stress on the scale parameter is modeled as:

$$\eta(s) = \eta_0 s^{-\gamma}. \tag{9.2}$$

The parameter η_0 is the scale parameter when $s = 1$. The constant γ is known as the stress-life exponent. When γ is positive, the scale parameter diminishes with s. This is the usual case when s is an accelerating factor such as load and temperature. If s has a beneficial effect such that larger values of s increase the scale parameter, then γ will be negative. The model is thus applicable more broadly than just as an accelerated test model. The variable s could be regarded as a covariate that is measured on each of the test elements and which has either a beneficial or a detrimental effect as described by the power law model. On a log scale the power law model becomes a linear one:

$$\ln \eta(s) = \ln \eta_0 - \gamma \ln s \tag{9.3}$$

The stress-life exponent is the slope and $\ln \eta_0$ the intercept in the linear relationship between the dependent variable $\ln \eta(s)$ and the independent variable $\ln s$. This is the form of a simple linear regression model and hence the term Weibull regression.

The power law model for the scale parameter implies that the $100p$-th percentile at stress s may be expressed as:

$$x_p(s) = (k_p)^{\frac{1}{\beta}} \eta(s) = (k_p)^{\frac{1}{\beta}} \eta_0 s^{-\gamma}. \tag{9.4}$$

The context we consider is where life tests are conducted at k levels of the stress variable, designated $s_1, s_2 \ldots s_k$. The tests at each level are type II censored. At stress level s_i the sample size is n_i and the number of failed items is r_i.

Given the results of these tests, our objectives are:

1. To verify that the shape parameter does not vary with stress level. This is implemented by the same technique introduced in Chapter 8 based on the ratio of the largest to the smallest of the maximum likelihood (ML) shape parameter estimates computed at each stress level.
2. To form ML estimates of the parameters β, η_0, and γ and any percentile x_p at a specified stress s.
3. To calculate confidence intervals for β and γ.
4. To calculate confidence intervals for $x_p(s)$ both for s values at which testing was performed and for values outside this range.
5. To compute median unbiased estimates of β, $x_p(s)$, and γ.
6. To test the hypothesis that the scale parameter varies in accordance with a power law model against the alternative that the scale parameters do not have a monotonic relationship with the levels of s.

9.2   ML ESTIMATION

The likelihood function for the Weibull regression model is the same as for the model of Chapter 8, except that η_i is replaced by:

$$\eta_i = \eta_o s_i^{-\gamma}. \tag{9.5}$$

Setting the derivative of the log likelihood function with respect to η_o equal to 0 and solving for η_o results in the solution:

$$\hat{\eta}_0 = \left(\frac{1}{R} \sum_{i=1}^{k} \sum_{j=1}^{n_i} \left[\frac{x_{ij}}{s_i^{-\hat{\gamma}}} \right]^{\hat{\beta}} \right)^{1/\hat{\beta}} \tag{9.6}$$

where,

$$R \equiv \sum_{i=1}^{k} r_i. \tag{9.7}$$

Setting to zero the derivatives of the log likelihood function with respect to γ and β, and substituting the expression above for η_o leads, after considerable simplification, to the pair of nonlinear equations given below. The ML estimates of β and γ are the simultaneous solution of this pair of equations:

$$\sum_{i=1}^{k} r_i \ln s_i - \frac{R \sum_{i=1}^{k} s_i^{\hat{\gamma} \hat{\beta}} \ln s_i \sum_{j=1}^{n_i} [x_{ij}]^{\hat{\beta}}}{\sum_{i=1}^{k} s_i^{\hat{\gamma} \hat{\beta}} \sum_{j=1}^{n_i} [x_{ij}]^{\hat{\beta}}} = 0 \tag{9.8}$$

and

$$\frac{1}{\hat{\beta}} + \frac{1}{R} \sum_{i=1}^{k} \sum_{j=1}^{\bar{n}} \ln x_{ij} - \frac{\sum_{i=1}^{k} s_i^{\hat{\gamma} \hat{\beta}} \sum_{j=1}^{n_i} x_{ij}^{\hat{\beta}} \ln x_{ij}}{\sum_{i=1}^{k} s_i^{\hat{\gamma} \hat{\beta}} \sum_{j=1}^{n_i} [x_{ij}]^{\hat{\beta}}} = 0. \tag{9.9}$$

After solving these two equations simultaneously for $\hat{\beta}$ and $\hat{\gamma}$ they are substituted into Equation 9.6 to compute $\hat{\eta}_0$. This pair of equations has the property that they are unchanged if the scale of the s variable is changed. So, for example, the solutions will remain the same whether s is expressed in the metric or English system of units. An advantage stemming from this fact is that stresses in their actual units could be rescaled by dividing by the smallest one, for instance, so that the smallest stress is 1.0 in the transformed system. For example, if $k = 2$ and $s_1 = 1000$ and $s_2 = 1200$ psi, they could be expressed as $s_1 = 1$ and $s_2 = 1.2$ without affecting the solution. The value of $\hat{\eta}_0$ will vary with the choice of units for stress but the estimate of γ will not. The equations

EXAMPLE   **279**

are also invariant with respect to the units in which the random variable is expressed, that is, hours, weeks, and so on, although again, $\hat{\eta}_0$ will vary with those units.

9.3   EXAMPLE

The three columns of data in Table 9.1 are sorted, simulated, uncensored samples of size 10 at each of three stresses, $s_1 = 1, s_2 = 1.1$, and $s_3 = 1.2$. At $s = 1$, the data were drawn from the Weibull population $W(56.46,1.3)$ for which it will be recalled that the 10th percentile $x_{0.10}$ is 10.0. For the other stresses the scale parameter was computed using a power law model with $\gamma = 9.0$. The $x_{0.10}$ values at s_2 and s_3 are 4.24 and 1.94, respectively, computed using the relation:

$$x_{0.10}(s) = (x_{0.10})_0 \, s^{-9}.$$

The term $(x_{0.10})_0$ refers to the population 10th percentile value at $s = 1$. Also shown in the table are the estimated values of the shape and scale parameters and the 10th percentile based on just the 10 lives at each stress.

To test whether the shape parameters are consistent with the common shape parameter assumption we form the ratio:

$$\frac{\hat{\beta}_{max}}{\hat{\beta}_{min}} = \frac{1.31}{1.03} = 1.27.$$

Table 9.1   Simulated Sample of Power Law Data; $n = r = 10$, $\eta_0 = 56.46$, $\gamma = 9$, $\beta = 1.3$

	$s = 1$	$s = 1.1$	$s = 1.2$
	0.213	4.41	2.36
	7.09	6.70	3.14
	24.4	14.8	4.79
	26.9	19.6	6.53
	29.8	23.6	7.38
	34.8	27.1	8.15
	44.3	32.0	9.95
	56.3	33.0	11.1
	96.4	64.5	19.9
	97.1	90.1	49.9
$\hat{x}_{0.10}$	4.78	6.14	1.68
$\hat{\eta}$	42.16	34.37	12.88
$\hat{\beta}$	1.03	1.31	1.11

From *ASLE Transactions* 29(1), copyright 1986.
Reprinted with permission of STLE.

From Table 8.1 in Chapter 8 the upper 5% critical value for $n = r = 10$ and $k = 3$ is 2.23 so we do not reject the common shape parameter hypothesis.

The 10th percentile estimate at $s=1.1$ is seen to be larger than at $s = 1.0$. The estimated scale parameter is smaller at $s = 1.1$, but the greater shape parameter estimate at that stress resulted in a larger estimated 10th percentile.

The Mathcad module given at the end of this chapter solves for the common shape parameter and the stress-life exponent and then the estimated scale parameters and 10th percentiles at each stress level. The 10th percentiles computed using the estimated common shape parameter are now monotonically decreasing with stress: 6.591, 3.536, and 2.003. The common shape parameter was estimated to be 1.121 and the stress-life exponent was estimated to be 6.533. Discussion of the other numerical results in the Mathcad module is taken up in Section 9.8.

9.4 PIVOTAL FUNCTIONS

A simulated value of the Weibull variable x_{ij} appearing in the two simultaneous expressions for $\hat{\beta}$ and $\hat{\gamma}$ can be computed in terms of a uniform random variable u as:

$$x_{ij} = \eta_i (\ln u_{ij})^{1/\beta} = \eta_0 s_i^{-\gamma} (\ln u_{ij})^{1/\beta}. \tag{9.10}$$

The u_{ij} are identically distributed uniform random variables over the interval $(0,1)$ irrespective of the subscripts i and j. Substituting this expression in Equations 9.8 and 9.9 reveals, after considerable algebraic manipulation, that the variables cluster into two groups, namely

$$q = \frac{\hat{\beta}}{\beta} \tag{9.11}$$

and

$$w^* = (\hat{\gamma} - \gamma)\hat{\beta}. \tag{9.12}$$

The solution of the two equations for q and w^* depends on the stress levels and the u_{ij} but not on any of the Weibull parameters. Thus, q and w^* are pivotal functions whose distributions can be found by Monte Carlo sampling. Additionally one can show that the following function is also pivotal:

$$u^*(r, n, k, p, s) = \hat{\beta} \ln \left[\frac{\hat{x}_p(s)}{x_p(s)} \right] \tag{9.13}$$

Where $x_p(s)$ is the true value of the $100p$-th percentile at some arbitrary stress s and $\hat{x}_p(s)$ is its ML estimate. The stress does not have to be a stress at which the life tests were conducted.

9.5 CONFIDENCE INTERVALS

With the distributions of q, w^*, and u^* determined by Monte Carlo sampling for specified n, r, k, p, and s, one may bias correct and compute confidence intervals for β and $x_p(s)$ as before. The precision measures R and $R_{0.50}^{\beta}$ are also computed as before. For the stress-life exponent γ we write the probability statement:

$$\text{Prob}[w_{0.05}^* < (\hat{\gamma} - \gamma)\hat{\beta} < w_{0.95}^*] = 0.90. \tag{9.14}$$

Solving the inequalities for γ results in the 90% confidence interval:

$$\hat{\gamma} - \frac{w_{0.95}^*}{\hat{\beta}} < \gamma < \hat{\gamma} - \frac{w_{0.05}^*}{\hat{\beta}}. \tag{9.15}$$

A median unbiased estimate of γ is:

$$\hat{\gamma}' = \hat{\gamma} - \frac{w_{0.50}^*}{\hat{\beta}}. \tag{9.16}$$

The precision measure for the stress-life exponent can be based on the difference L between the upper and lower confidence intervals. For consistency we choose a 90% confidence interval for this purpose, giving:

$$L = \frac{w_{0.95}^* - w_{0.05}^*}{\hat{\beta}}. \tag{9.17}$$

L will vary from sample to sample since it depends on the shape parameter estimate. The median value of L will occur when the shape parameter estimate is equal to its median which is $\beta q_{0.50}$. Thus a reasonable precision measure for γ estimation is:

$$\beta L_{0.50} = \frac{w_{0.95}^* - w_{0.05}^*}{q_{0.50}} \tag{9.18}$$

Where $L_{0.50}$ denotes the median value of L.

9.6 TESTING THE POWER LAW MODEL

As shown in Chapter 8, the common shape parameter estimate $\hat{\beta}_1$ is unaffected by differences in the scale parameters. On the other hand, the shape parameter estimate under the power law assumption tends to decrease when the power law model fails to hold. Intuitively the reason is that when data are constrained to follow a power law and they, in fact, do not, it requires a small shape parameter estimate to "explain" the large apparent scatter relative to the fitted relationship. (Recall that scatter increases as the shape parameter decreases.) A reasonable test of the power law relationship therefore is the ratio $\dfrac{\hat{\beta}_1}{\hat{\beta}}$ of the unconstrained shape parameter estimate to the value estimated under the power law model. If the data do not follow the power law, this ratio will tend to increase. An upper percentage point of the distribution of $\dfrac{\hat{\beta}_1}{\hat{\beta}}$ obtained when the power law holds is the critical value for such a test. The test is inapplicable if $k = 2$ since a power law will always fit when there are only two stresses.

An alternative to this shape parameter test is a likelihood ratio test. The numerator of the likelihood ratio is the likelihood function evaluated using the estimated parameters of the power law model. The denominator is the likelihood function evaluated under the model of Chapter 8 where there is no constraint on the scale parameters. In both cases the likelihood is computed assuming that the shape parameter is the same at all stress levels. When the power law model holds, the numerator and denominator should be close in value. When it does not hold, the ratio λ will be small and $-2\ln\lambda$ will be large. The numerator has three parameters, η_0, β, and γ. The denominator has the k values of η_i and β. For large enough samples, $-2\ln\lambda$ is approximately chi-square distributed with $k + 1 - 3 = k - 2$ degrees of freedom when the power law model holds. The computation of the shape parameter ratio and the likelihood ratio tests are carried out for the example data in the Mathcad model given at the end of this chapter. Our Monte Carlo computations discussed presently did not include computation of the likelihood ratio so the correction factor needed to produce greater agreement between $-2\ln\lambda$ and the chi-square distribution is not known for this application. Therefore, there is a distinct possibility that the uncorrected likelihood ratio may be misleading.

9.7   MONTE CARLO RESULTS

Simulations were conducted with $k = 2$ and 3 stress levels and uncensored samples of size 5, 10, 15, or 20 at each stress. In performing the simulations it is convenient to take $\beta = 1$, $\eta_0 = 1$, and $\gamma = 0$. For $k = 2$ the stress levels were taken as $s_1 = 1$ and $s_2 = 1.2$. For $k = 3$ the stress levels chosen were $s_1 = 1$, $s_2 = 1.1$, and $s_3 = 1.2$. The results are shown in Tables 9.2 and 9.3, respectively. For each pivotal function five percentage points are displayed, namely $\alpha = 0.05, 0.10, 0.50, 0.90$, and 0.95. For the $k = 2$ case these percentage points are displayed for $\dfrac{\hat{\beta}_{max}}{\hat{\beta}_{min}}$, q, w^*, and u^* for seven values of s and with $p = 0.10$. The seven stresses include the two stresses at which the life tests are run and five others ranging from 0.5 to 0.9 in increments of 0.1. (This range could

Table 9.2 Selected Percentage Points of q, w^*, and u^* at Seven Stress Levels for Various Uncensored Sample Sizes, $p = 0.10$ and $k = 2$

n	r	α	$\dfrac{\hat{\beta}_{max}}{\hat{\beta}_{min}}$	$q = \dfrac{\hat{\beta}}{\beta}$	$w^* = \hat{\beta}(\hat{\gamma}-\gamma)$	$u^* = \ln[\hat{x}_p(s_i)/x_p(s_i)]$						
						$s_1 = 1.0$	$s_2 = 1.2$	$s_3 = 0.5$	$s_4 = 0.6$	$s_5 = 0.7$	$s_6 = 0.8$	$s_7 = 0.9$
5	5	0.05	1.041	0.7773	-7.804	-0.9715	-0.9659	-5.462	-4.102	-2.994	-2.080	-1.402
5	5	0.10	1.081	0.8513	-5.773	-0.7115	-0.6995	-3.960	-2.9818	-2.1927	-1.5482	-1.0213
5	5	0.50	1.4897	1.1945	0.0063	0.3875	0.3839	0.3625	0.3662	0.3780	0.3700	0.3836
5	5	0.90	2.7708	1.8085	5.9944	2.1083	2.0903	5.6100	4.5375	3.6457	2.9662	2.4507
5	5	0.95	3.3892	2.0557	8.0413	2.7499	2.7814	7.6222	6.1781	4.9786	4.0547	3.2901
10	10	0.05	1.0265	0.8130	-4.6974	-0.7161	-0.7237	-3.4091	-2.5862	-109.339	-1.4019	-0.9769
10	10	0.10	1.0535	0.8614	-3.5384	-0.5488	-0.5564	-2.5984	-1.9721	-1.4601	-1.0506	-0.7527
10	10	0.50	1.3008	1.0869	-0.0473	0.1587	0.1661	0.1376	0.1271	0.1317	0.1333	0.1472
10	10	0.90	1.8732	1.4056	3.5025	1.1100	1.1200	3.1742	2.5532	2.0688	1.6433	1.3312
10	10	0.95	2.1341	1.5155	4.6450	1.4443	1.4336	4.1621	3.3630	2.6899	2.1355	1.7189
15	15	0.05	1.0185	0.8361	-3.6863	-0.6146	-0.6088	-2.7589	-2.1186	-1.5897	-1.1637	-0.8285
15	15	0.10	1.0392	0.8799	-2.8395	-0.4726	-0.4653	-2.1274	-1.6325	-1.2190	-0.8873	-0.6421
15	15	0.50	1.2337	1.0560	0.0106	0.1081	0.1117	0.1054	0.1001	0.1029	0.0995	0.1010
15	15	0.90	1.6486	1.2893	2.7878	0.8275	0.8268	2.4582	1.9663	1.5697	1.2248	0.9781
15	15	0.95	1.8165	1.3694	3.6751	1.0595	1.0668	3.1614	2.5497	2.0308	1.5992	1.2766
20	20	0.05	1.0169	0.8547	-3.1063	-0.5513	-0.5319	-2.3749	-1.8127	-1.3552	-0.9945	-0.7161
20	20	0.10	1.0347	0.8927	-2.3793	-0.4077	-0.4083	-1.7902	-1.3804	-1.0385	-0.7612	-0.5471
20	20	0.50	1.1938	1.0400	0.0129	0.0723	0.0689	0.0755	0.0745	0.0734	0.0769	0.0781
20	20	0.90	1.5218	1.2335	2.3890	0.6742	0.6673	2.0787	1.6576	1.3127	1.0313	0.8152
20	20	0.95	1.6509	1.2996	3.1080	0.8737	0.8784	2.6450	2.1133	1.6786	1.3193	1.0439

From *ASLE Transactions* 29(1), copyright 1986. Reprinted with permission of STLE.

Table 9.3 Selected Percentage Points of q, w^*, and u^* at Seven Stress Levels for Various Uncensored Sample Sizes, $p = 0.10$ and $k = 3$

n	r	α	$\hat{\beta}_{max}/\hat{\beta}_{min}$	$q = \hat{\beta}/\beta$	$\hat{\beta}_1/\beta$	$w^*\hat{\beta}(\hat{\gamma}-\gamma)$	$u^* = \ln[\hat{x}_p(s_i)/x_p(s_i)]$						
							$s_1 = 1.0$	$s_2 = 1.1$	$s_3 = 1.2$	$s_4 = 0.6$	$s_5 = 0.7$	$s_6 = 0.8$	$s_7 = 0.9$
5	5	0.05	1.2071	0.7952	0.9975	−7.0713	−0.8843	−0.7203	−0.8827	−3.9332	−2.9006	−2.0320	−1.3337
5	5	0.10	1.3033	0.8536	0.9996	−5.4627	−0.6561	−0.5443	−0.6594	−2.9666	−2.1830	−1.5386	−1.0142
5	5	0.50	1.9537	1.1181	1.0268	0.0427	0.2278	0.2342	0.2327	0.2371	0.2380	0.2295	0.2290
5	5	0.90	3.5866	1.5156	1.1732	5.3414	1.4439	1.3005	1.4213	3.7956	3.0141	2.3428	1.8197
5	5	0.95	4.5141	1.6738	1.2485	7.1454	1.8901	1.7392	1.8913	4.9303	3.9107	3.0604	2.3840
10	10	0.05	1.1328	0.8349	0.9985	−4.6243	−0.6542	−0.5514	−0.6403	−2.6562	−1.9830	−1.4224	−0.9549
10	10	0.10	1.1844	0.8785	0.9996	−3.4778	−0.4999	−0.4142	−0.4886	−2.0123	−1.4860	−1.0555	−0.7251
10	10	0.50	1.5347	1.0543	1.0409	−0.0057	0.1023	0.1091	0.1089	0.0818	0.0869	0.0907	0.0955
10	10	0.90	2.2266	1.2891	1.0697	3.4271	0.8486	0.7637	0.8434	2.3337	1.8408	1.4164	1.0695
10	10	0.95	2.5022	1.3675	1.1005	4.4144	1.0934	0.9894	1.0930	2.9980	2.3692	1.8231	1.3962
15	15	0.05	1.0968	0.8598	0.9989	−3.5061	−0.5527	−0.4622	−0.5395	−2.0700	−1.5557	−0.8657	−0.6012
15	15	0.10	1.1417	0.8946	0.9997	−2.7407	−0.4193	−0.3545	−0.4118	−1.6088	−1.2034	−0.8657	−0.6012
15	15	0.50	1.4074	1.0364	1.0066	0.0161	0.0694	0.0701	0.0708	0.0688	0.0747	0.0678	0.0681
15	15	0.90	1.8741	1.2127	1.0441	2.7406	0.6532	0.5701	0.6282	1.8237	1.4247	1.1060	0.8343
15	15	0.95	2.0790	1.2768	1.0630	3.5200	0.8366	0.7553	0.8321	2.3185	1.8193	1.4038	1.0654
20	20	0.05	1.0848	0.8731	0.9991	−2.9759	−0.4832	−0.4096	−0.4746	−1.7597	−1.3249	−0.9670	−0.6884
20	20	0.10	1.1226	0.9024	0.9997	−2.3106	−0.3723	−0.3153	−0.3687	−1.3814	−1.0406	−0.7588	−0.5281
20	20	0.50	1.3288	1.0258	1.0050	0.0016	0.0514	0.0505	0.0542	0.0490	0.0532	0.0512	0.0491
20	20	0.90	1.7046	1.1781	1.0329	2.3629	0.5395	0.4804	0.5247	1.5615	1.2152	0.9278	0.6992
20	20	0.95	1.8268	1.2239	1.0468	3.0590	0.6930	0.6160	0.6798	2.0126	1.5772	1.2022	0.9073

From *ASLE Transactions* 29(1), copyright 1986. Reprinted with permission of STLE.

conveniently be written in shorthand notation as 0.5(0.1)0.9.) Upper percentage points of $\dfrac{\hat{\beta}_{max}}{\hat{\beta}_{min}}$ may be used to verify that the shape parameters are the same at every level. The same quantities are given in Table 9.3 for the $k = 3$ case, except that this table also contains percentage points of $\dfrac{\hat{\beta}_1}{\hat{\beta}}$ for testing the validity of the power law model and the stress levels beyond the three at which testing is conducted are 0.6(0.1)0.9.

9.8 EXAMPLE CONCLUDED

We have already established that for the data of Table 9.1 the common shape parameter test holds. The shape parameter estimate $\hat{\beta}_0$ which results when all the data points are combined into a single group of size 30 is 1.001. The estimate $\hat{\beta}_1$ is 1.140 so $\dfrac{\hat{\beta}_1}{\hat{\beta}_0} = 1.140$ and exceeds the 90th percentile 1.116 given in Table 8.3a and is just slightly less than the 95th percentile . Thus, we reject the hypothesis that there is no difference among the scale parameters. The next question is whether the scale parameters vary in accordance with a power law model. The power law constrained estimate of the shape parameter is $\hat{\beta} = 1.121$.

To test the power law model, we form the ratio $\dfrac{\hat{\beta}_1}{\hat{\beta}} = 1.018$. This is well below the 90th percentile of 1.0697 given in Table 9.3. The scale parameter differences are thus consistent with a power law model. This is confirmed by the likelihood ratio test. The test statistic, $-2\ln\lambda$ is computed to be 1.08. Under the chi-square distribution with $k - 1 = 1$ degrees of freedom, there is a 30% chance of exceeding 1.08.

Having accepted the power law model, we can compute confidence intervals on β, γ, and $x_p(s)$.

Using $q_{0.050} = 0.8349$ and $q_{0.95} = 1.3675$ from Table 9.3, a 90% confidence interval for the shape parameter is:

$$0.820 = \frac{1.121}{1.3675} < \beta < \frac{1.121}{0.8349} = 1.34.$$

The interval just barely contains the true value of $\beta = 1.3$. A median unbiased estimate is computed as:

$$\hat{\beta}' = \frac{1.121}{1.0543} = 1.06.$$

The validity of the interval for the shape parameter depends on the validity of the power law model. The interval computed in Chapter 8 is valid no matter how the scale parameters vary and is a viable alternate choice. In the former case three parameters are estimated from the data, namely η_0, β, and γ. In the latter case $k + 1$ parameters are estimated, namely k values of η_i, and β. For $k = 2$ the number of parameters estimated is the same in both cases. Also when $k = 2$ the power law model must hold because with only two sets of data, a power law model can always be fit. For $k > 2$, the number of estimated parameters is greater when the power law is not assumed and hence the confidence limits for the shape parameter may not be as tight as under the power law model provided that the power law model is true. For $k = 3$ or 4 it may be preferable to use the methodology of Chapter 8 even though the power law model cannot be formally rejected.

A 90% interval for the stress-life exponent is calculated using $w_{0.05}^* = -4.62$ and $w_{0.95}^* = 4.41$

$$2.56 = 6.53 - \frac{4.41}{1.121} < \gamma < 6.53 - \frac{(-4.62)}{1.121} = 10.65.$$

This is a wide interval. It does contain the true value $\gamma = 9.0$.

A median unbiased estimate of γ is:

$$\hat{\gamma}' = 6.53 - \frac{-0.0473}{1.121} = 6.57.$$

A confidence interval for $x_{0.10}$ at $s = 1.0$ is calculated using $u_{0.05}^* = -0.6542$ and $u_{0.95}^* = 1.0934$.

$$2.49 = 6.59 \exp\left[-\frac{1.093}{1.121}\right] < x_{0.10}(s = 1) < 6.59 \exp\left[\frac{0.652}{1.121}\right] = 11.8.$$

A median unbiased estimate of the 10th percentile at $s = 1$ is:

$$\hat{x}_{0.10}' = 6.59 \exp\left[-\frac{0.1023}{1.121}\right] = 6.02.$$

In the same way, confidence intervals and median unbiased estimates may be computed at each of the other stresses listed in Table 9.3. The results are summarized in Table 9.4 below. The last column shows the true value of $x_{0.10}$ computed from the population power law model used to generate the simulated data:

The true value of $x_{0.10}$ is within the confidence limits for all seven stresses. Figure 9.1 is a plot of the logarithms of the median unbiased estimates and the 90% confidence limits as a function of the stress s. The loss in precision due to extrapolation outside the stress range where the data were obtained is evident.

Table 9.4   90% Confidence Intervals for $x_{0.10}$ Computed at Seven Stress Levels under the Estimated Power Law Model

Stress	$\hat{x}'_{0.10}$	Lower Confidence Limit	Upper Confidence Limit	True $x_{0.10}$
0.60	169.1	12.8	1980	992.290
0.70	61.8	8.18	397	247.809
0.80	25.8	5.57	100.7	74.506
0.90	12.0	3.80	30.9	25.812
1.00	6.02	2.49	11.8	10.000
1.10	3.23	1.46	5.79	4.241
1.20	1.83	0.754	3.54	1.938

From *ASLE Transactions* 29(1), copyright 1986. Reprinted with permission of STLE.

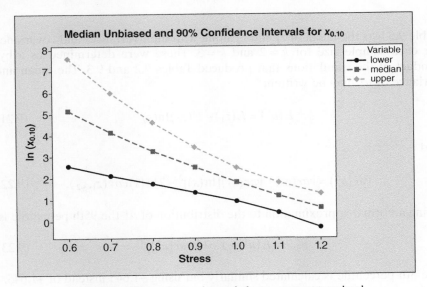

Figure 9.1   90% confidence intervals for $x_{0.10}$ versus stress level.

9.9   APPROXIMATING u^* AT OTHER STRESS LEVELS

For any stress s the quantity $u^* = \hat{\beta}\ln\left(\dfrac{\hat{x}_{0.10}(s)}{x_{0.10}(s)}\right)$ may be expressed as:

$$u^* = \hat{\beta}\ln\left(\frac{\hat{x}_{0.10}(s)}{x_{0.10}(s)}\right) = \hat{\beta}\ln\left(\frac{\hat{x}_{0.10}(s=1)}{x_{0.10}(s=1)}\right) - \hat{\beta}(\hat{\gamma}-\gamma)\ln(s). \tag{9.19}$$

The first term is u^* at $s = 1$ and the second term is the product of w^* and $\ln(s)$. For notational simplicity we write:

$$u^* = z_1 - z_2 \ln(s). \tag{9.20}$$

Table 9.5   Means, Variances and Covariance of z_1 and z_2 for Various Values of n, r, and k

n	r	k	$E(z_1)$	$E(z_2)$	$Var(z_1)$	$Var(z_2)$	$Cov(z_1,z_2)$
5	5	2	0.5789	0.0405	1.414	24.14	1.121
10	10	2	0.2371	−0.0322	0.4516	8.198	0.7156
15	15	2	0.1497	−0.0103	0.2664	5.031	0.4577
20	20	2	0.1111	0.0104	0.1863	3.509	0.3169
5	5	3	0.3340	0.0030	0.7520	19.27	1.769
10	10	3	0.1467	0.0338	0.2869	7.518	0.6800
15	15	3	0.0971	0.0065	0.1787	4.632	0.4360
20	20	3	0.0716	0.0082	0.1274	3.386	0.3193

From *ASLE Transactions* 29(1), copyright 1986. Reprinted with permission of STLE.

Table 9.5 lists the mean of z_1 and z_2 along with their variances and covariance for each sample size for $k = 2$ and $k = 3$. These were determined as a by-product of the simulations that produced Tables 9.2 and 9.3. The mean and variance of u^* may be written:

$$E(u^*) = E(z_1) - E(z_2)\ln(s) \tag{9.21}$$

and

$$var(u^*) = var(z_1) + var(z_2)[\ln(s)]^2 - 2\ln(s)cov(z_1, z_2). \tag{9.22}$$

Using a normal approximation to the distribution of u^* the 95th percentile is:

$$u^*_{0.95} = E(u^*) + 1.645[var(u^*)]^{1/2}. \tag{9.23}$$

The 5th percentile is calculated similarly but using −1.645 instead of +1.645.

As an example and to verify the accuracy of the approximation, we will approximate the percentage points at $s = 0.6$ for $k = 3$ and $n = r = 20$.

The mean is:

$$E(u^*) = 0.0716 - 0.0082 \times \ln(0.6) = 0.076.$$

The variance is:

$$var(u^*) = 0.1274 + 3.386 \times [\ln(0.6)]^2 - 2 \times \ln(0.6)(0.3193) = 1.337.$$

The 5th percentile is approximated by −1.826. The 95th percentile is approximated by 1.978. These approximations compare to the values from direct simulation −1.76 and 2.01. Since the actual limits are wider than the approximate limits, the approximation is not conservative in this case.

9.10   PRECISION

Tables 9.6 and 9.7 list for $k = 2$ and $k = 3$, respectively, the precision measures R and $\beta L_{0.50}$ at each of the four sample sizes along with the values of $R^{\beta}_{0.50}$ at each of the seven stresses. Also shown for comparison are the values of R and $R^{\beta}_{0.50}$ that would result if all kn items were tested at the single stress $s = 1$. For $k = 2$, the values of $R^{\beta}_{0.50}$ differ slightly at the two stress levels probably due to sampling error. These values exceed the value of $R^{\beta}_{0.50}$ corresponding to a single sample. It is conjectured that at $s = 1.1$, midway between the two stresses used in the life tests, the precision could be nearly comparable to the single sample precision. This is so for the $k = 3$ case where the precision at $s = 1.1$ compares favorably to the single sample values.

The uncertainty in estimation of γ is greatly affected by the closeness of the stress levels. For $k = 2$ and $n = r = 5$, an evaluation at $s_1 = 1$ and $s_2 = 2$ gave a value of $\beta L_{0.50}$ of 3.49 compared with the value 13.3 for $s_1 = 1$ and $s_2 = 1.2$. As in ordinary regression, the precision with which the slope is estimated improves with the distance between the largest and smallest values of the independent variable. The hazard is that separating the stress levels by too large a margin may introduce a different failure mode than that under study and invalidate the stress-life model.

9.11   STRESS LEVELS IN DIFFERENT PROPORTIONS THAN TABULATED

If with $k = 2$ the stresses are not in the proportion $1:1.2$ corresponding to the tabled values, it is possible to transform them to equivalent stresses that are in the required proportions. For example, if $s_1 = 1$ and $s_2 = 2$, one may find a transformation:

$$\sigma = s^{\delta}. \tag{9.24}$$

Such that $\sigma_1 = 1$ and $\sigma_2 = 1.2$. The defining relation is:

$$1.2 = 2^{\delta}.$$

So that,

$$\delta \ln 2 = \ln 1.2; \ \delta = \frac{\ln 1.2}{\ln 2} = 0.263.$$

The estimated power law in the transformed stress is:

$$\eta = \eta_0 \sigma^{-\tilde{\gamma}} = \eta_0 (s^{0.263})^{\tilde{\gamma}}.$$

Table 9.6 Precision Measures for Estimated Parameters $k = 2$

		$R^\beta_{0.50}$									Single Sample $2n, 2r$	
n	r	$s_1 = 1$	$s_2 = 1.2$	$s_3 = 0.5$	$s_4 = 0.6$	$s_5 = 0.7$	$s_6 = 0.8$	$s_7 = 0.9$	R	$\beta L_{0.50}$	$R^\beta_{0.50}$	R
5	5	22.5	23.0	5.72E4	5.47E3	792	170	50.8	2.64	13.3	15.3	2.49
10	10	7.30	7.28	1.06E3	238	70.4	25.9	11.9	1.86	8.60	6.13	1.82
15	15	4.88	4.89	272	83.2	30.8	13.7	7.34	1.64	6.97	4.22	1.62
20	20	3.86	3.88	125	43.6	18.5	9.25	5.43	1.52	5.98	3.47	1.51

From *ASLE Transactions* 29(1), copyright 1986. Reprinted with permission of STLE.

Table 9.7 Precision Measures for Estimated Parameters $k = 3$

		$R^\beta_{0.50}$									Single Sample $3n, 3r$	
n	r	$s_1 = 1$	$s_2 = 1.1$	$s_3 = 1.2$	$s_4 = 0.6$	$s_5 = 0.7$	$s_6 = 0.8$	$s_7 = 0.9$	R	$\beta L_{0.50}$	$R^\beta_{0.50}$	R
5	5	11.9	9.00	11.9	2.76E3	440	94.8	27.7	2.10	12.7	8.41	2.03
10	10	5.03	4.15	4.96	186	55.9	20.1	8.79	1.63	8.57	4.22	1.62
15	15	3.82	3.24	3.76	69.0	26.0	11.6	5.96	1.48	6.78		1.48
20	20	3.15	2.72	3.08	39.5	16.9	8.29	4.74	1.40	5.88		1.40

From *ASLE Transactions* 29(1), copyright 1986. Reprinted with permission of STLE.

Here $\tilde{\gamma}$ is the estimated stress-life exponent that results from using $s_2 = 1.2$. The actual estimated stress-life exponent is then $0.263\tilde{\gamma}$ or, in general, $\hat{\gamma} = \delta\tilde{\gamma}$. Using $s_2 = 1.2$ instead of an actual value s only affects the stress-life exponent and not the shape parameter estimate. When life tests are conducted at $k = 3$ stresses they can be mapped to 1, 1.1, and 1.2 only if they are spaced so that the same exponent transforms the two normalized larger stresses to 1.1 and 1.2. In planning the tests it should be possible to select s_2 so that a common exponent transforms the stresses to the proportions 1, 1.1, and 1.2. For example, suppose we wish to use $k = 3$ and have selected the largest and smallest stresses for testing to be 2000 and 1000. These are transformed to $s_1 = 1$ and $s_3 = 2$ by dividing by 1000. We also know that raising 2 to the power 0.263 will transform it to 1.2. We now seek the intermediate stress s_2 so that the same exponent will transform it to $s_2 = 1.1$. That is, we seek the stress s_2 such that:

$$s_2^{0.263} = 1.1.$$

The solution is $s_2 = 1.436$ or, in original units, $s_2 = 1436$. This choice, only modestly different from the midpoint choice of $s_2 = 1500$, will allow the use of the tables of simulation results computed for $k = 3$.

9.12 DISCUSSION

The Weibull regression or power law model has been shown to be amenable to exact inference with readily calculated measures of precision wherefrom required sample sizes may be determined. Tabular values are limited as to sample sizes and stress levels. A software program comparable to the multiweibull model discussed in Chapter 8 could be developed for the Weibull regression model to extend its applicability to a wider range of sample sizes, censoring and stress levels.

9.13 THE DISK OPERATING SYSTEM (DOS) PROGRAM REGEST

The DOS program REGEST will compute the ML estimates of the parameters β and γ under the power law model described in this chapter as well as the estimates of five quantiles of the life distribution at each stress. The initial screen shown as Figure 9.2 asks the user for the number of sample groups and then asks in turn for the stress level for each group and the sample size and number of failures for the group. It then asks for the data pertinent to that group (not shown). When the data for all groups are entered, the program asks for initial guesses for the shape parameter and for the stress-life exponent. Figure 9.3 shows the computed results and several quantiles of the life distribution at each stress level. The data used in this example are given in Table

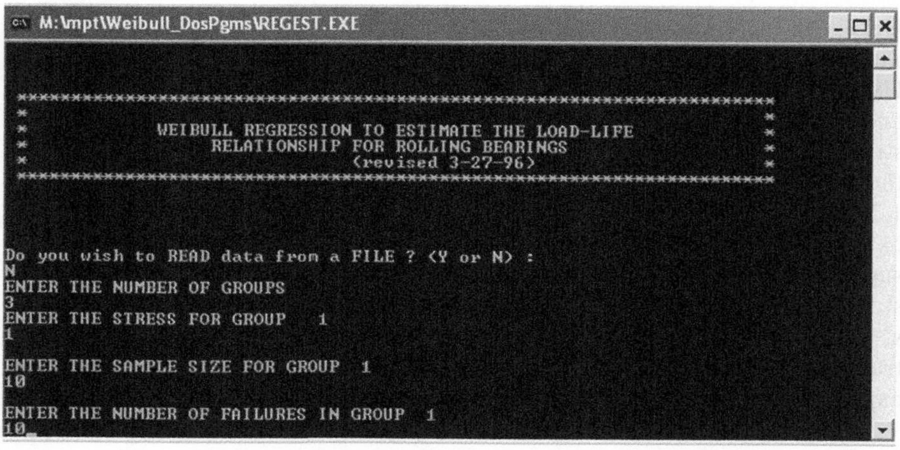

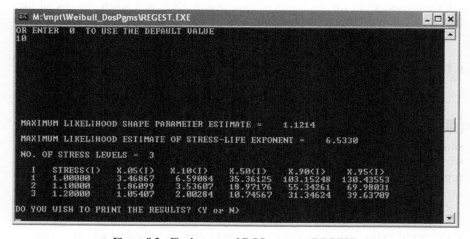

Figure 9.2 Initial screen for DOS program REGEST.

Figure 9.3 Final screen of DOS program REGEST.

9.1. The same data are used in the Mathcad module that follows and the results are seen to be in agreement.

Mathcad Module for Weibull Regression

This Mathcad module is designed to estimate the Weibull parameters and the exponent of a power function relationship between the Weibull scale parameter and a quantitative independent variable such as stress. It also computes the test statistics for assessing whether a power function relation is acceptable.

Set the number of levels k, the sample size at each level, n, and the number of failures at each stress level r. $\mathbf{n}$ and $\mathbf{r}$ are vectors of dimension k.

$$k := 3 \quad \mathbf{n} := \begin{pmatrix} 10 \\ 10 \\ 10 \end{pmatrix} \quad \mathbf{r} := \mathbf{n} \quad \mathbf{r} = \begin{pmatrix} 10 \\ 10 \\ 10 \end{pmatrix}.$$

$$\mathbf{s} := \begin{pmatrix} 1 \\ 1.1 \\ 1.2 \end{pmatrix} \qquad \text{the } s \text{ vector gives the stress levels at which testing was}$$
performed.

The vector y contains the data in column form. Column 1 contains the data at the first value of s, column 2 the second, and so on.

$$\mathbf{y} := \begin{pmatrix} 0.213 & 4.41 & 2.36 \\ 7.09 & 6.70 & 3.14 \\ 24.4 & 14.8 & 4.79 \\ 26.9 & 19.6 & 6.53 \\ 29.8 & 23.6 & 7.38 \\ 34.8 & 27.1 & 8.15 \\ 44.3 & 32.0 & 9.95 \\ 56.3 & 33.0 & 11.1 \\ 96.4 & 64.5 & 19.9 \\ 97.1 & 90.1 & 49.9 \end{pmatrix}$$

Transpose the data into the matrix x so that each row corresponds to a value of s.

$$\mathbf{x} := \mathbf{y}^{\mathrm{T}}$$

x =		1	2	3	4	5	6	7	8	9
	1	0.213	7.09	24.4	26.9	29.8	34.8	44.3	56.3	96.4
	2	4.41	6.7	14.8	19.6	23.6	27.1	32	33	64.5
	3	2.36	3.14	4.79	6.53	7.38	8.15	9.95	11.1	19.9

Guess values for the exponent and shape parameter

$$\beta := 1 \quad \gamma := 10$$

Define R and X

$$R := \sum_{i=1}^{k} r_i \quad X := \sum_{i=1}^{k} \sum_{j=1}^{r_i} \ln(x_{i,j}) \quad X = 83.104.$$

Now solve for the power law model parameters

$$\eta = \eta_0 \times s^{-\gamma}.$$

Given

$$\frac{1}{\beta} + \frac{X}{R} - \frac{\sum_{i=1}^{k}\left[(s_i)^{\beta \times \gamma} \times \sum_{j=1}^{n_i}\left[(x_{i,j})^{\beta} \times \ln(x_{i,j})\right]\right]}{\sum_{i=1}^{k}\left[(s_i)^{\beta \times \gamma} \times \sum_{j=1}^{n_i}(x_{i,j})^{\beta}\right]} = 0.$$

$$\sum_{i=1}^{k}(r_i \times \ln(s_i)) - \frac{R \times \sum_{i=1}^{k}\left[(s_i)^{\beta \times \gamma} \times \ln(s_i) \times \sum_{j=1}^{n_i}(x_{i,j})^{\beta}\right]}{\sum_{i=1}^{k}\left[(s_i)^{\beta \times \gamma} \times \sum_{j=1}^{n_i}(x_{i,j})^{\beta}\right]} = 0$$

$$\beta > 0$$

$$b := \text{Find}(\beta, \gamma)$$

$$\beta := b_1 \quad \gamma := b_2$$

$$\beta = 1.121 \quad \gamma = 6.533.$$

Compute the scale parameter at $s = 1.0$

$$\eta 0 := \left[\frac{\sum_{i=1}^{k}\left[(s_i)^{\beta \cdot \gamma} \cdot \sum_{j=1}^{n_i}(x_{i,j})^{\beta}\right]}{R}\right]^{\frac{1}{\beta}}$$

$$\eta 0 = 49.031.$$

Compute estimated scale parameters and 10th percentile at each s value

$$i := 1 \ldots k$$

$$\eta_i = \eta 0 \cdot (s_i)^{-\gamma}$$

$$x10 := 0.10536^{\frac{1}{\beta}} \cdot \eta_i.$$

Compute the log likelihood under the power law model

$$LL2 := [R \cdot \ln(\beta) + (\beta - 1) \cdot (X)] - \left[\beta \cdot \left[\sum_{i=1}^{k}(r_i \cdot \ln(\eta_i))\right]\right] - \sum_{i=1}^{k}\sum_{j=1}^{n_i}\left(\frac{x_{i,j}}{\eta_i}\right)^{\beta}$$

$$LL2 = -127.084$$

Store scale parameter estimates under the power law model

$$i := 1 \ldots k$$

$$u_i := \eta_i$$

$S_i =$		$i =$		$\eta_i =$		$x10_i =$
1		1		49.031		6.591
1.1		2		26.306		3.536
1.2		3		14.9		2.003

Consider the case with no stress-related constraint placed on scale parameter values.

initial guess $\beta_1 := b_1$

Given

$i =$		$s_i =$		$u_i =$		$\eta_i =$
1		1		49.031		43.454
2		1.1		26.306		32.864
3		1.2		14.9		13.07

$$\frac{1}{\beta_1} + \frac{X}{R} - \sum_{i=1}^{k} \frac{\sum_{j=1}^{k}\left[r_i \cdot (x_{i,j})^{\beta_1} \cdot \ln(x_{i,j}) \right]}{R \cdot \sum_{j=1}^{n_i} (x_{i,j})^{\beta_1}} = 0$$

$$\beta_1 > 0$$

$$\beta_1 := \text{Find}(\beta_1)$$

$$\beta_1 = 1.14.$$

Compute unconstrained scale parameter estimates

$$i := 1 \ldots k$$

$$\eta_i := \left[\frac{\sum_{j=1}^{n_i} (x_{i,j})^{\beta_1}}{r_i} \right]^{\frac{1}{\beta_1}}$$

Compare constrained and unconstrained estimates

$$i := 1 \ldots k.$$

The u_i are scale parameter estimates under the power law model.

The η_i are scale parameter estimates with no relation among levels. Compare shape parameter estimates:

$$\frac{\beta}{\beta_1} = 0.983.$$

Compute log likelihood with no constraint on scale parameters

$$LL0 := [R \cdot \ln(\beta_1) + (\beta_1 - 1) \cdot (X)] - \left[\beta_1 \cdot \left[\sum_{i=1}^{k} (r_i \cdot \ln(\eta_i)) \right] \right] - \sum_{i=1}^{k} \sum_{j=1}^{n_i} \left(\frac{x_{i,j}}{\eta_i} \right)^{\beta_1}$$

$$LL0 = -126.544$$

$$T := 2 \cdot (LL0 - LL2)$$

$$T = 1.08.$$

T = $-2\ln\lambda$ is approximately $\chi 2$ distributed with $(k + 1) - 3 = k - 2 = 1$ degrees of freedom. It tests whether the power relation is reasonable.

$$p = \text{Prob}\,[T > 1.08] = 1 - pchisq(T, k - 2)$$

$$p := 1 - pchisq(T, k - 2)$$

$$p = 0.299.$$

REFERENCES

McCool, J.I. 1980. Confidence limits for Weibull regression with censored data. *IEEE Transactions on Reliability* 29: 145–150.

McCool, J.I. 1986. Using Weibull regression to estimate the load-life relationship for rolling bearings. *ASLE Transactions* 29: 91–101.

Nelson, W. 1990. *Accelerated Testing: Statistical Models, Test Plans and Data Analyses.* John Wiley and Sons, Inc., New York.

EXERCISES

1. Three uncensored samples of size $n = 5$ were simulated at three stresses using $\eta_0 = 100$, $\gamma = 3$, and $\beta = 2$. The data are tabled below:

Stress (psi)	Lifetimes (hours)				
1000	22.8686124	42.5531329	100.800217	117.730573	165.318143
1100	46.2908985	74.5227136	117.288484	123.728826	132.190278
1200	45.5254146	66.8071674	67.2936268	70.4535347	81.6412457

Use the DOS program REGEST to estimate the regression model parameters. Use the DOS program WEIBEST to compute the common shape parameter estimate with no power law assumption. Use the tabulated percentage points in (9.3) to test for the validity of the power law model. Compute 90% confidence limits for $x_{0.10}$ at stress $s = 1000$, the Weibull shape parameter, and the stress-life exponent.

CHAPTER 10

The Three-Parameter Weibull Distribution

10.1 THE MODEL

The introduction of a third parameter extends the flexibility and usefulness of the Weibull distribution as a model for random phenomena.

Under the three-parameter version of the Weibull distribution the cumulative distribution function (CDF) is written as:

$$F(x) = 1 - \exp\left[-\left(\frac{x-\gamma}{\eta}\right)^{\beta}\right]; x > \gamma. \tag{10.1}$$

The parameter γ is alternately known as the location parameter, the threshold parameter, or, when the random variable x represents a lifetime, the guarantee time. If indeed a random lifetime followed the three-parameter Weibull distribution it would be of great benefit provided that γ were sufficiently large because then the reliability function $R(x)$ could be claimed to be 1.0 for any lives x that were less than γ.

The three-parameter Weibull has been applied to model the distribution of the diameter of trees, (Bailey and Dell, 1973), the fatigue life of roller chains (Sun et al., 1993), fretting fatigue life of an aluminum alloy (Poon and Hoeppner, 1979), and the volume of oil spills in the Gulf of Mexico (Harper et al., 2008). It competes with the two-parameter Weibull as a model for fracture strength of brittle materials. (Alqam et al., 2002; Lu et al., 2002).

The expression for the probability density function $f(x)$ is the derivative of $F(x)$ in Equation 10.1.

$$f(x) = \frac{dF(x)}{dx} = \frac{\beta}{\eta}\left[\frac{x-\gamma}{\eta}\right]^{\beta-1} \exp\left[-\left(\frac{x-\gamma}{\eta}\right)^{\beta}\right]; x > \gamma. \tag{10.2}$$

Its appearance is identical to the ordinary two-parameter Weibull density, except it starts at $x = \gamma$ rather than $x = 0$. The p-th quantile is expressed as:

$$x_p = (-\ln(1-p))^{1/\beta}\eta + \gamma = (k_p)^{1/\beta}\eta + \gamma. \qquad (10.3)$$

This is the expression for the p-th quantile of the two-parameter distribution augmented by the location parameter γ.

It is sometimes convenient to write the Weibull CDF with a percentile as scale parameter. Solving Equation 10.3 for η and substituting in Equation 10.1 gives:

$$F(x) = 1 - \exp - k_p\left[\frac{x-\gamma}{x_p-\gamma}\right]^{\beta}; x > \gamma. \qquad (10.4)$$

The population mean or expected value of a three-parameter Weibull distribution is:

$$E(x) = \gamma + \eta\Gamma\left(1 + \frac{1}{\beta}\right). \qquad (10.5)$$

Just as for the percentiles, the mean is the same as in the two-parameter case but with the addition of the location parameter. The variance is exactly the same as for the two-parameter Weibull. The additive constant of the location parameter does not affect the variance.

Inference for the three-parameter Weibull model is not so well developed as for the two-parameter case.

In particular, estimating the parameters by the method of maximum likelihood (ML) is complicated by the fact that the likelihood equations may have no solution; that is, the likelihood function may have no stationary point. In other cases the solution of the likelihood equations may converge to a saddle point and not to the maximum of the likelihood function. (Lawless, 2002; Rockette et al., 1974).

Exact interval estimation based on the ML estimates has not been addressed in the literature.

A simple method is given here for testing the two-parameter Weibull distribution against the three parameter alternative in type II censored samples (McCool, 1998). It is based on the ratio of (i) the ML estimate of the shape parameter calculated with the observations artificially censored at the r_1-th ordered value to (ii) the corresponding estimate using all r ($>r_1$) observed values in the type II censored sample of size n ($\geq r$). Some critical values for conducting the test are given for $10 < n < 100$ and various values of r and r_1. The values of r_1 given in the table were chosen to maximize the power of the test. Software is described for generating the critical values for any choice of sample sizes.

The acceptance region for the test can be translated into a lower confidence interval estimate for the location parameter. Two numerical examples illustrate the calculations.

10.2   ESTIMATION AND INFERENCE FOR THE WEIBULL LOCATION PARAMETER

The procedure uses two ML estimates of the shape parameter: one based on the complete sample and one based on just the early-order statistics. The idea behind the procedure is readily grasped in the context of graphical estimation.

For the two-parameter Weibull distribution we have established that

$$y(x) \equiv \ln \ln \left[\frac{1}{1-F(x)} \right] = \beta \ln x - \beta \ln \eta. \qquad (10.6)$$

$y(x)$ is a linear function of $\ln x$ having slope β and intercept $-\beta\ln\eta$ and is the basis for the linearization of the Weibull CDF used in the construction of Weibull probability paper. For the three-parameter case:

$$y(x) \equiv \ln \ln \left[\frac{1}{1-F(x)} \right] = \beta \ln (x-\gamma) - \beta \ln \eta; \ x > \gamma. \qquad (10.7)$$

The slope of a plot of $y(x)$ against $\ln x$ in the three-parameter case is:

$$\frac{dy(x)}{dx} = \frac{\beta}{x-\gamma}. \qquad (10.8)$$

The slope is infinite at $x = \gamma$ and decreases monotonically with x thereafter to an asymptote of β.

If a data sample is drawn from a two-parameter Weibull distribution, the estimated function $\hat{y}(x)$, plotted using estimated values of $F(x)$ at each ordered failure time, will tend to plot against the failure time logarithms as a straight line with a slope β. If $\gamma > 0$, that is, the population is a three-parameter Weibull distribution, the estimated CDF will tend to be a concave function of $\ln x$ approaching a constant slope β for large x values.

Figure 10.1 shows how a plot of $\hat{y}(x)$ versus $\ln x$ might appear for a sample drawn from a three-parameter Weibull distribution.

If these data were regarded as a two-parameter Weibull sample a graphical estimate of the shape parameter $\hat{\beta}_A$ could be found as the slope of the straight line that best fits the complete data sample.

If only a subset comprising the smallest ordered values were used in graphically estimating the shape parameter, the estimate $\hat{\beta}_L$ would be obtained. For three-parameter Weibull data $\hat{\beta}_L$ will tend to exceed $\hat{\beta}_A$. On the other hand,

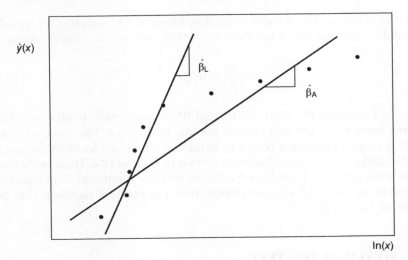

Figure 10.1 Weibull plot of three-parameter Weibull data. From Journal of Quality Technology, copyright 1998, American Society for Quality. Reprinted with permission. No further distribution allowed without permission.

when the sample is drawn from a two-parameter Weibull population ($\gamma = 0$), $\hat{\beta}_A$ and $\hat{\beta}_L$ will be comparable in magnitude.

10.3 TESTING THE TWO- VERSUS THREE-PARAMETER WEIBULL DISTRIBUTION

Rather than graphical estimation we consider ML estimation of β due to its precision and freedom from subjectivity. As we have seen in Section 5.6.3 for a sample of size n, type II censored at the r-th ordered observation x_r, the ML estimate of β for a two-parameter Weibull distribution is the solution of the nonlinear equation:

$$\frac{1}{\beta} + \frac{1}{r}\sum_{i=1}^{r} \ln x_i - \frac{\sum_{i=1}^{r} x_i^{\beta} \ln x_i + (n-r)x_r^{\beta}}{\sum_{i=1}^{n} x_i^{\beta}} = 0. \tag{10.9}$$

Denoting the solution of Equation 10.9 as $\hat{\beta}_A$ and the solution with censoring at $r_1 < r$, as $\hat{\beta}_L$, the distribution of $w = \dfrac{\hat{\beta}_L}{\hat{\beta}_A}$ will depend only upon n, r_1, and r when the underlying distribution is indeed of the two-parameter Weibull form ($\gamma = 0$). When the underlying distribution is the three-parameter Weibull the mean value of $\hat{\beta}_L$ will tend to exceed the mean value of $\hat{\beta}_A$.

With the percentiles of w determined by Monte Carlo sampling for specified r_1, r, and n, one may reject the hypothesis that $\gamma = 0$ at the $100\alpha\%$ level if

$$\frac{\hat{\beta}_L}{\hat{\beta}_A} > w_{1-\alpha}. \tag{10.10}$$

Table 10.1 contains the 50-th, 90-th, and 95-th percentage points of w for n ranging from 10 to 100 and various choices of r_1 and r. The value of r_1 was selected to give maximum power in detecting a nonzero location parameter for specified n and r. This is discussed further in Section 10.4. These percentage points were obtained from 10,000 realizations of w computed from simulated two-parameter Weibull samples drawn from a population having a 10th percentile of 1.0 and $\beta = 1.0$.

10.4 POWER OF THE TEST

The ordered members x_i of a three-parameter Weibull sample are expressible in terms of the corresponding ordered members y_i of a sample from the standard exponential distribution as

$$x_i = \eta y_i^\beta + \gamma. \tag{10.11}$$

Substituting Equation 10.11 into Equation 10.9 and simplifying reveals that $\hat{\beta}$ depends on γ through the ratio γ/η. The probability of accepting $\gamma > 0$ therefore depends on the true value of γ relative to the scale parameter or relative to any percentile of the two-parameter distribution which the data follow when γ is subtracted from each observation.

To examine the power of the test, the non-null distribution of w was found for various n, r_1, and r using 10,000 samples from the three-parameter Weibull distribution having a location parameter $\gamma = 10$, which is equal to the 10th percentile of the two-parameter distribution used to generate the null distribution of w. The probability P_a of accepting the location parameter as zero in a 10% significance level test was then found by interpolation in the non-null distribution of w. These acceptance probabilities are listed in Table 10.2.

The power, $(1 - P_a)$, is seen to increase with r for fixed n and r_1 and with n for fixed r and r_1. For a given value of n and r, the best choice of r_1 appears to be 5 for $n \leq 30$, 7 for $n = 40$ to 60, and 9 for $n = 80$ to 100.

10.5 INTERVAL ESTIMATION

Given that a random variable x is drawn from a three-parameter Weibull population having location parameter γ, the transformed variable $y = x - \gamma$

Table 10.1 Percentage Points of w for Various n, r, and r_1

n	r_1	r	$w_{0.50}$	$w_{0.90}$	$w_{0.95}$
10	5	6	0.988	1.488	1.789
10	5	7	1.035	1.730	2.132
10	5	8	1.077	1.902	2.352
10	5	9	1.116	2.022	2.517
10	5	10	1.138	2.126	2.683
15	5	10	1.141	2.094	2.579
15	5	15	1.223	2.408	3.055
20	5	6	0.990	1.498	1.759
20	5	10	1.137	2.073	2.582
20	5	12	1.172	2.198	2.784
20	5	15	1.210	2.345	2.974
20	5	18	1.238	2.417	3.135
20	5	20	1.254	2.491	3.188
25	5	10	1.146	2.118	2.622
25	5	14	1.199	2.321	2.924
25	5	15	1.211	2.238	2.850
25	5	18	1.237	2.460	3.087
25	5	20	1.250	2.515	3.148
25	5	25	1.278	2.540	3.278
30	5	6	0.990	1.467	1.734
30	5	10	1.139	2.079	2.602
30	5	15	1.213	2.340	2.915
30	5	20	1.256	2.457	3.119
30	5	25	1.278	2.544	3.224
30	5	30	1.294	2.600	3.279
40	7	15	1.098	1.755	2.074
40	7	20	1.136	1.888	2.240
40	7	25	1.157	1.937	2.299
40	7	30	1.172	1.984	2.364
40	7	40	1.198	2.039	2.430
50	7	25	1.152	1.941	2.292
50	7	30	1.165	1.995	2.342
50	7	40	1.182	2.049	2.420
50	7	50	1.191	2.070	2.466
60	7	30	1.167	2.014	2.406
60	7	40	1.183	2.062	2.470
60	7	50	1.199	2.097	2.527
60	7	60	1.203	2.128	2.564
80	9	40	1.122	1.771	2.054
80	9	50	1.133	1.793	2.091
80	9	60	1.140	1.824	2.108
80	9	70	1.146	1.844	2.121
80	9	80	1.150	1.850	2.143
100	9	50	1.132	1.787	2.087
100	9	60	1.140	1.809	2.101
100	9	70	1.147	1.827	2.129
100	9	80	1.151	1.837	2.152
100	9	90	1.152	1.840	2.146
100	9	100	1.155	1.846	2.173

Table 10.2   **Probability of Accepting $\gamma = 0$ when $\gamma = x_{0.10}$**

n	r_1	r	P_a
10	2	10	0.822
10	5	10	0.788
10	6	10	0.797
10	7	10	0.809
10	8	10	0.833
20	5	10	0.740
20	6	10	0.750
20	7	10	0.770
20	8	10	0.800
20	9	10	0.830
20	5	15	0.680
20	6	15	0.700
20	7	15	0.710
20	9	15	0.740
20	10	15	0.760
20	5	20	0.630
20	7	20	0.650
20	9	20	0.680
20	10	20	0.700
20	15	20	0.760
30	2	20	0.703
30	5	20	0.555
30	7	20	0.582
30	8	20	0.616
30	10	20	0.649
30	2	30	0.674
30	5	30	0.477
30	7	30	0.498
30	9	30	0.536
30	10	30	0.547
40	2	20	0.662
40	5	20	0.512
40	7	20	0.507
40	8	20	0.556
40	10	20	0.594
40	5	40	0.360
40	7	40	0.340
40	9	40	0.370
40	11	40	0.390
40	13	40	0.430
50	5	50	0.250
50	7	50	0.240
50	8	50	0.240
50	9	50	0.260
50	10	50	0.270

Table 10.2 *(Continued)*

n	r_1	r	P_a
60	5	60	0.180
60	7	60	0.170
60	9	60	0.170
60	10	60	0.180
60	11	60	0.190
80	5	80	0.100
80	7	80	0.072
80	9	80	0.067
80	10	80	0.071
80	11	80	0.072
100	9	100	0.021
100	10	100	0.023
100	11	100	0.027

will follow a two-parameter Weibull distribution with the same scale and shape parameters as the three-parameter distribution. Thus, if the true location parameter γ is subtracted from the observed data prior to calculating $\hat{\beta}_A$ and $\hat{\beta}_L$ from Equation 10.9, the resulting ratio denoted,

$$w(\gamma) = \frac{\hat{\beta}_L}{\hat{\beta}_A}, \tag{10.12}$$

will follow the null distribution of w determined by Monte Carlo sampling from a two-parameter Weibull population for given values of n, r_1, and r.

We may thus write the $100(1 - \alpha)\%$ probability statement:

$$\mathrm{Prob}[w(\gamma) < w_{1-\alpha}] = 1 - \alpha. \tag{10.13}$$

We also need the fact that if an arbitrary amount λ $(\lambda < x_1)$ is subtracted from each observation in a given sample prior to calculating $\hat{\beta}_A$ and $\hat{\beta}_L$, $w(\lambda) = \frac{\hat{\beta}_L}{\hat{\beta}_A}$ will be a decreasing function of λ.

Accordingly, we may invert the inequality of Equation 10.12 to give a $100(1 - \alpha)\%$ lower confidence limit for γ, that is,

$$\gamma > w^{-1}(w_{1-\alpha}) = \gamma_\alpha. \tag{10.14}$$

The notation adopted here is to specify the γ value, which will be exceeded with 90% confidence as $\gamma_{0.10}$. Figure 10.2 shows schematically a typical data-

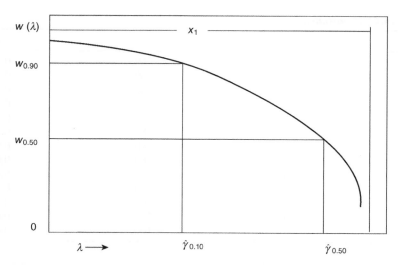

Figure 10.2 w(λ) as a function of λ. From the *Journal of Quality Technology*, copyright 1998, American Society for Quality. Reprinted with permission. No further distribution allowed without permission.

dependent function $w(\lambda)$ plotted over the range of physically meaningful values of the location parameter, that is, $(0, x_1)$. The 90% lower confidence limit for the location parameter value is shown as the value of λ for which the function $w(\lambda)$ is equal to $w_{0.90}$. Correspondingly the median unbiased estimate of γ is the λ value for which $w(\lambda)$ is equal to $w_{0.50}$. It is clear from the graph that unless $w(0)$ exceeds $w_{1-\alpha}$, $\hat{\gamma}_\alpha$ will be negative and hence uninformative since we presuppose that γ is positive or 0. Accordingly, for negative solutions $\hat{\gamma}_\alpha$ is taken to be zero.

In practice the inversion indicated by Figure 10.2 is readily accomplished using a golden section search procedure. This technique is implemented in the DOS program LOCEST illustrated in the next section using fatigue data on rolling chains obtained at a load of 5.5 kgf and given in Sun et al. (1993).

As an example we have modified the sample of 10 observations first introduced in Chapter 5 and drawn from the two parameter Weibull distribution $W(56.46,1.3)$ by adding 50 to each observation. The sample thus became a random sample from a three-parameter distribution with $\gamma = 50$. Then the value of the ratio $w = \hat{\beta}_L / \hat{\beta}_A$ was computed with $r = 10$ and $r_1 = 5$ after subtracting the quantity λ from each observation as λ varied over the range from 0 to 64. The resultant plot is shown as Figure 10.3.

The value of $w_{0.50} = 1.13$ intersects the $w(\lambda)$ curve at a value close to 50. The median value was calculated using the software described next in Section 10.6

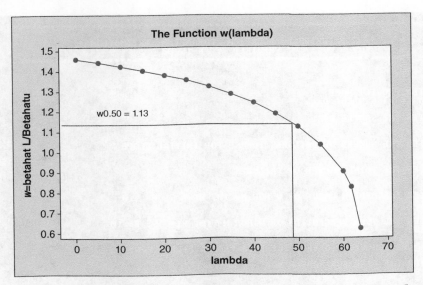

Figure 10.3 $w(\lambda)$ for Weibull sample of size 10 with $\eta = 56.46$, $\beta = 1.3$, $\gamma = 50$; $r_1 = 5$.

to be 49.46. Since the value of $w(0) = 1.459$ is less than $w_{0.90} = 2.126$ the lower 90% limit is taken to be zero.

10.6 INPUT AND OUTPUT SCREENS OF LOCEST.exe

The input screen shown in Figure 10.4 asks the user to input a short title and then the values of n, r_1, and r. It then asks for the median and 95th quantiles $w_{0.50}$ and $w_{0.95}$ associated with those values of n, r_1, and r. The file input and output capabilities are useful if repeated runs are envisioned.

The output screen associated with this data set is shown as Figure 10.5. The lower 95% confidence limit for γ is shown to be 0. This means that $w(0)$ was less than $w_{0.95}$. The median estimate is 0.706. If $w_{0.90}$ is input instead of $w_{0.95}$ the value shown on the output as the lower 95% confidence limit will be the lower 90% limit. Similarly if any other percentage point of w_{1-p} is input the corresponding output will be $\hat{\gamma}_p$.

Shown at the bottom of the screen are the Weibull parameters estimated after subtracting $\hat{\gamma}_{0.50}$ from each data point.

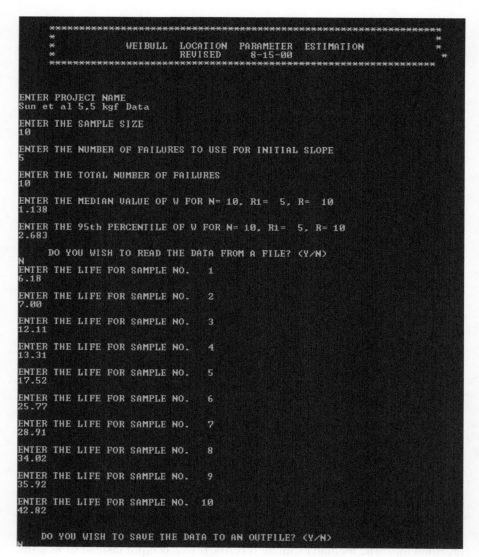

```
**********************************************************************
  **                                                              **
  **          WEIBULL  LOCATION  PARAMETER  ESTIMATION            **
  **                 REVISED      8-15-00                         **
  **                                                              **
**********************************************************************


ENTER PROJECT NAME
Sun et al 5.5 kgf Data

ENTER THE SAMPLE SIZE
10

ENTER THE NUMBER OF FAILURES TO USE FOR INITIAL SLOPE
5

ENTER THE TOTAL NUMBER OF FAILURES
10

ENTER THE MEDIAN VALUE OF W FOR N= 10, R1=  5, R=  10
1.138

ENTER THE 95th PERCENTILE OF W FOR N= 10, R1=  5, R= 10
2.683

      DO YOU WISH TO READ THE DATA FROM A FILE? <Y/N>
N
ENTER THE LIFE FOR SAMPLE NO.    1
6.18

ENTER THE LIFE FOR SAMPLE NO.    2
7.00

ENTER THE LIFE FOR SAMPLE NO.    3
12.11

ENTER THE LIFE FOR SAMPLE NO.    4
13.31

ENTER THE LIFE FOR SAMPLE NO.    5
17.52

ENTER THE LIFE FOR SAMPLE NO.    6
25.77

ENTER THE LIFE FOR SAMPLE NO.    7
28.91

ENTER THE LIFE FOR SAMPLE NO.    8
34.02

ENTER THE LIFE FOR SAMPLE NO.    9
35.92

ENTER THE LIFE FOR SAMPLE NO.    10
42.82

      DO YOU WISH TO SAVE THE DATA TO AN OUTFILE? <Y/N>
N
```

Figure 10.4 Input screen for LOCEST.exe.

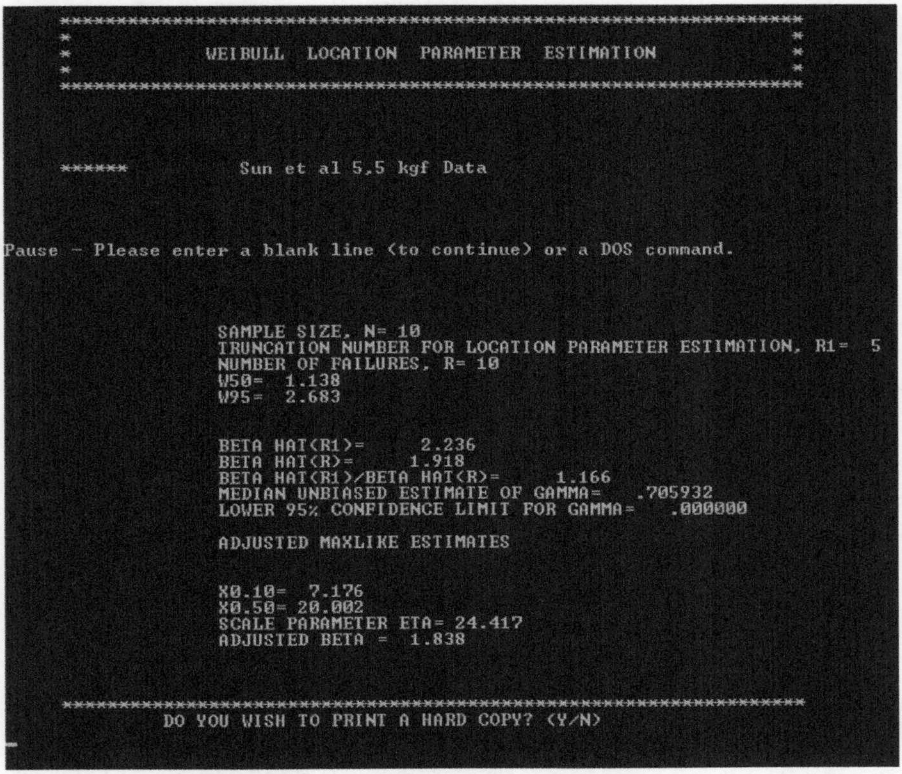

Figure 10.5 Output screen for LOCEST.exe.

10.7 THE PROGRAM LocationPivotal.exe

A powerful program capable of generating percentage points of w for any choice of n, r_1, and r was developed by Christopher Garrell as a student project assignment. It can be used to evaluate the power of the hypothesis test on γ. It is dubbed LocationPivotal.exe. It may be downloaded from the author's website along with a copy of Christopher's project report.

The input screen is shown as Figure 10.6.

In addition to the values of the sample size characteristics n, r_1, and r, the user can select the Weibull population from which to develop the distribution of w. By varying the location parameter one may determine the probability of accepting $\gamma = 0$ as a function of the true value of γ.

Figure 10.7 shows the output screen corresponding to the input in Figure 10.6. Sixteen percentage points of w are computed. (Only 11 are visible in the screen shot.) Under Tools the user may indicate which input and output

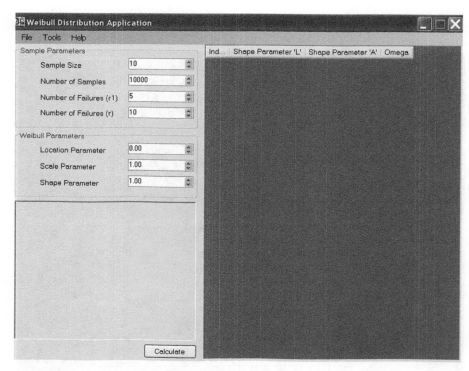

Figure 10.6 Input screen for locationpivotal.

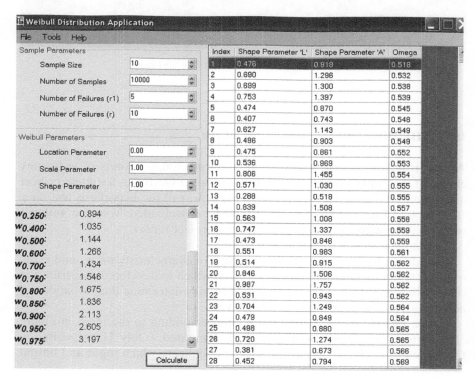

Figure 10.7 Output screen for locationpivotal.

Table 10.3 Simulated Sample Size of 20; $\eta = 100$, $\beta = 1.5$, and $\gamma = 100$

118.6	126.3	134.7	138.4	148.8
150.3	150.9	151.2	155.6	177.5
180.1	197.9	204.6	238.6	263.8
281.9	291.3	297.4	314.2	336.7

variables are of interest. These may be written to a file using *Save As* under the *File* menu. The percentage points of w displayed are consistent with those given in Table 10.1 for $n = r = 10$, $r_1 = 5$.

10.8 SIMULATED EXAMPLE

As a second example, and to illustrate the methodology on data for which the distributional assumptions are known to apply, a sample of size 20 was drawn using simulation from the Weibull distribution having $\eta = 100$, $\beta = 1.5$, and $\gamma = 100$. Three-parameter data are easily simulated by first drawing a two-parameter sample and then adding the location parameter to each observation.

The sorted sample is given in the Table 10.3.

The shape parameter estimates obtained by censoring the data at $r_1 = 5$ and $r = 20$ are $\hat{\beta}(5) = 31.97$ and $\hat{\beta}(20) = 9.542$. Using $w_{0.50} = 1.254$ in LOCEST gives the median unbiased estimate of the location parameter:

$$\hat{\gamma}_{0.50} = 110.34.$$

A lower 90% confidence interval for γ is computed by inputting $w_{0.90} = 2.491$ from Table 10.2 and results in:

$$\hat{\gamma}_{0.10} = 50.72.$$

Adjusting the data by subtracting $\hat{\gamma}_{0.50}$ from every observation and re-estimating η and β gives $\hat{\beta} = 1.34$ and $\hat{\eta} = 100.6$. These "adjusted" ML estimates and $\hat{\gamma}_{0.50}$ are in good accord with the true population values.

REFERENCES

Alqam, M., R. Bennett, and A. Zureick. 2002. Three-parameter vs. two-parameter Weibull distribution for pultruded composite material properties. *Composite Structures* 58(4): 497–503.

Bailey, R. and T. Dell. 1973. Quantifying diameter distributions with the Weibull function. *Forest Science* 19(2): 97–104.

Harper, W.V., T.R. James, and T.G. Eschenbach. 2008. Maximum likelihood estimation methodology comparison for the three parameter Weibull distribution with applications to offshore oil spills in the Gulf of Mexico. *Joint Statistical Meetings* Denver, CO: American Statistical Association 7.

Lawless, J.F. 2002. *Statistical Models and Methods for Lifetime Data.* Wiley, New York, NY.

Lu, C., R. Danzer, and F.D. Fischer. 2002. Fracture statistics of brittle materials: Weibull or normal distribution. *Physical Review E* 65(6): 67102.

McCool, J. 1998. Inference on the Weibull location parameter. *Journal of Quality Technology* 30: 119–126.

Poon, C. and D.W. Hoeppner. 1979. The effect of environment on the mechanism of fretting fatigue. *Wear* 52: 175–191.

Rockette, H., C. Antle, and L.A. Klimko. 1974. Maximum likelihood estimation with the Weibull model. *Journal of the American Statistical Association* 69(345): 246–249.

Sun, Y.S., M. Xie, T.N. Goh, and H.L. Ong. 1993. Development and applications of a three parameter Weibull distribution with load-dependent location and scale parameters. *Reliability Engineering and System Safety* 40: 133–137.

EXERCISES

1. An uncensored sample size of 20 when artificially censored at $r_1 = 5$ resulted in a raw ML shape parameter estimate of 7.25. The ML shape parameter estimate for the complete sample was 2.25. Test the hypothesis that the location parameter is 0.

2. Generate a sample of size 10 from $W(100,1.5)$. Add 100 to each observation in the sample. Use $r_1 = 5$ and the appropriate values in Table 10.1 as input to LOCEST and determine the median and 90% lower confidence limit for the location parameter.

3. Use the LocationPivotal software to estimate the probability that $w > 2.683$ for $n = 10, r_1 = 5$ when the sample actually is drawn from the three-parameter Weibull population with $\beta = 1.5$, $\eta = 100$, and $\gamma = 100$.

CHAPTER 11

Factorial Experiments with Weibull Response

Experiments conducted at all combinations of the levels of two or more factors are called factorial experiments. Factorial experiments have been shown to be more efficient in exploring the effects of external factors on a response variable than non-factorial arrangements of factor levels. In this chapter we present a methodology for the analysis of Weibull-distributed data obtained at all combinations of the levels of two factors. The response variable is assumed to follow the two-parameter Weibull distribution with a shape parameter that, although unknown, does not vary with the factor levels. The purpose of the analysis is (1) to compute interval estimates of the common shape parameter and (2) to assess whether either factor has an effect on the Weibull scale parameter and hence on any percentile of the distribution.

11.1 INTRODUCTION

Zelen (1960) considered the two-factor classification for censored observations drawn from the one- and two-parameter exponential distributions. He derived likelihood ratio tests for the significance of main and interaction effects when those effects are multiplicative with the exponential scale parameter. He evaluated three alternative approximations for the small sample distribution of the likelihood ratio.

In work that was actually done subsequent to that reported in his 1960 article, Zelen (1959) showed by means of Monte Carlo sampling experiments that the analysis he developed for the exponential distribution is not robust against Weibull alternatives He showed that a more robust approach is to treat the logarithmic transform of the scale parameter estimate formed using the data in the (i,j)-th cell, as a normal variate, and to use it as the response in an unreplicated factorial experiment. A drawback of this approach is that for the

---

Using the Weibull Distribution: Reliability, Modeling, and Inference, First Edition. John I. McCool.

2×2 design there are then no degrees of freedom left for error. Similarly, for most other designs of practical size, the error degrees of freedom would be too few for reasonable power.

An alternative approach, not discussed by Zelen, is to divide the sample in each cell of the layout into two or more subsamples to give replication. This could be done unambiguously for uncensored samples but becomes problematic when censoring is present. With type II censoring, for example, if one divides a sample of size n into two groups each containing half of the r failures, it is possible that neither group would qualify as being a type II censored subsample. Apart from the problem of allocating censored units to subgroups it is not clear how many subsamples should be formed to provide optimum overall experimental precision.

In what follows we consider the two-way layout with Weibull response for the case where the shape parameter, though unknown, is common from cell to cell. This situation is more complex than the exponential situation inasmuch as data from *all* cells are used in the estimation of the shape parameter. This chapter is based on a dissertation and elaborations contained in three subsequent publications: McCool (1993); McCool (1996a); McCool (1996b); and McCool and Baran (1999).

11.2   THE MULTIPLICATIVE MODEL

In the present chapter we examine the combined effect of two external factors such as load and speed on the distribution of a Weibull random variable.

Each factor occurs at a number of levels; for example, the factor temperature might be set at 100, 150, and 200°C. Data is taken at each combination of the factor levels, so that, for example, if one factor has three and the second factor has four levels, $3 \times 4 = 12$ tests are performed.

We will discuss a methodology for analyzing such experiments to determine whether neither, one, or both factors have a significant influence on the observed random variable.

If factor A has a levels and factor B has b levels, the cumulative distribution function (CDF) of the Weibull distribution at the conditions corresponding to level i of factor A and level j of factor B is taken to be:

$$F(x) = 1 - \exp\left[-\left(\frac{x}{\eta_{ij}}\right)^{\beta}\right]. \tag{11.1}$$

The $100p$-th percentile x_p life at that combination of conditions is thus:

$$(x_p)_{ij} = \eta_{ij}(k_p)^{1/\beta}. \tag{11.2}$$

One may think of the levels of factor A as forming the rows and the levels of factor B the columns of a two-way table or layout.

We further express the scale parameter of η_{ij} in terms of a multiplicative row effect a_i due to the i-th level of factor A, a column effect b_j due to the j-th level of factor B, and an interaction effect c_{ij} due to the particular synergy of the factor levels in row i and column j, that is,

$$\eta_{ij} = a_i b_j c_{ij} \eta;\ (i = 1 \cdots a,\ j = 1, \cdots b). \tag{11.3}$$

η is a base level scale parameter value. Introducing the additional constraints:

$$\prod_{i=1}^{a} a_i = 1 \tag{11.4}$$

$$\prod_{j=1}^{b} b_j = 1 \tag{11.5}$$

and

$$\prod_{i=1}^{a} c_{ij} = \prod_{j=1}^{b} c_{ij} = 1. \tag{11.6}$$

serves to define η as the geometric mean of the cell scale parameter values taken over all the cells; that is:

$$\eta = \left[\prod_{i=1}^{a} \prod_{j=1}^{b} \eta_{ij} \right]^{1/ab}. \tag{11.7}$$

Thus, for example, given the following 2×2 table of scale parameter values:

$\eta_{11} = 2$	$\eta_{12} = 4$
$\eta_{21} = 3$	$\eta_{22} = 6$

We have,

$$\eta = (2x4x3x6)^{1/4} = \sqrt{12}.$$

From the constraints:

$$a_2 = \frac{1}{a_1}$$

$$b_2 = \frac{1}{b_1}$$

$$c_{11} = \frac{1}{c_{12}} = c_{22} = \frac{1}{c_{21}}.$$

Equating the numerical values of η_{ij} to their multiplicative expressions gives:

$$\eta_{11} = a_1 b_1 c_{11} \sqrt{12} = 2$$

$$\eta_{12} = \frac{a_1}{b_1} \frac{1}{c_{11}} \sqrt{12} = 4$$

$$\eta_{21} = \frac{b_1}{a_1 c_{11}} \sqrt{12} = 3$$

$$\eta_{22} = \frac{c_{11}}{a_1 b_1} \sqrt{12} = 6.$$

The solutions are:

$$a_1 = \sqrt{2/3}; \quad a_2 = \sqrt{3/2}$$
$$b_1 = \sqrt{1/2}; \quad b_2 = \sqrt{2/1}$$

and

$$c_{11} = c_{12} = c_{21} = c_{22} = 1.0.$$

In this instance, multiplicative row and column factors accounted for all the η_{ij} values. The cell specific c_{ij} values were all unity. For the values in the table above it may be said that "interaction" is absent.

In general, if c_{ij} is unity for all i and j, we say that interaction does not occur. When interaction is absent, the data may be "explained" by the simpler "reduced" model wherein:

$$\eta_{ij} = a_i b_j \eta \ (i = 1 \cdots a, \ j = 1 \cdots b). \tag{11.8}$$

Similarly if, in addition, factor B has no effect, η_{ij} may be written,

$$\eta_{ij} = \eta_i = a_i \eta \ (i = 1 \cdots a) \tag{11.9}$$

while if factor A has no effect,

$$\eta_{ij} = \eta_j = b_j \eta. \tag{11.10}$$

Finally, if there are no row, column, or interaction effects the model simply reduces to:

$$\eta_{ij} = \eta. \tag{11.11}$$

11.3 DATA

A test is presumed to be conducted for each combination of factor levels. For simplicity we take the sample size n to be the same for each cell in the a × b array. The n observations in each cell may be type II censored at the r-th ordered value. If so, r is also presumed to be the same for each cell.

The total sample size is thus abn and the total number of number of uncensored observations is abr. When the observed lifetimes within each cell are sorted from low to high the k-th ordered value is denoted $x_{ij(k)}$.

11.4 ESTIMATION

The method of maximum likelihood was applied to estimate the shape parameter and effects under each of the five models described by Equations 11.3 and 11.8–11.11. The corresponding shape parameter estimates are denoted $\hat{\beta}_1$ to $\hat{\beta}_5$, respectively. For model number 2, $\hat{\beta}_2$ and the estimates of the effects a_i and b_j must generally be found by the simultaneous solution of $a + b + 1$ nonlinear equations. However, for the special case of the 2×2 layout ($a = b = 2$) $\hat{\beta}_2$ maybe solved separately and then the effect estimates computed. In what follows we restrict consideration of model 2 to the case where $a = b = 2$. The estimates of the scale parameters η_{ij} are obtained by multiplying the base level scale parameter by the relevant effect estimates.

Table 11.1 lists the equations for estimating $\hat{\beta}_1$ to $\hat{\beta}_5$. The auxiliary quantities used in this table are defined as follows:

$$v_{ij} = \sum_1^n x_{ij(k)}^{\hat{\beta}} \tag{11.12}$$

$$v_{i.} = \left[\prod_{j=1}^b v_{ij} \right]^{1/b} \tag{11.13}$$

$$v_{j.} = \left[\prod_{i=1}^a v_{ij} \right]^{1/a} \tag{11.14}$$

$$v_{..} = \left[\prod_{i=1}^a \prod_{j=1}^b v_{ij} \right]^{1/ab} \tag{11.15}$$

Table 11.1 Maximum Likelihood (ML) Estimation Equations for Factorial Experiments under Various Models

Model	ML Shape Parameter Found by Solving	Eq. for $\hat{\eta}_{ij}^{\hat{\beta}}$
$\eta_{ij} = a_i b_j c_{ij} \eta$	$\dfrac{1}{\hat{\beta}_1} + \dfrac{s_{..}}{abr} - \dfrac{1}{ab} \displaystyle\sum_{i=1}^{a} \sum_{j=1}^{b} \dfrac{T_{ij}}{v_{ij}} = 0$	v_{ij}/r
$\eta_{ij} = a_i b_i \eta (a = b = 2)$	$\dfrac{1}{\hat{\beta}_2} + \dfrac{s_{..}}{abr} - \dfrac{1}{k} \displaystyle\sum_{i=1}^{a} \sum_{j=1}^{b} \dfrac{T_{ij}}{v_{i.} v_{.j}} = 0$ $\left(k \equiv \displaystyle\sum_{i=1}^{a} \sum_{j=1}^{b} \dfrac{v_{ij}}{v_{i.} v_{.j}} \right)$	$k v_{i.} v_{.j}/abr$
$\eta_{ij} = a_i \eta$	$\dfrac{1}{\hat{\beta}_3} + \dfrac{s_{..}}{abr} - a^{-1} \displaystyle\sum_{i=1}^{a} \left(\dfrac{\displaystyle\sum_{j=1}^{b} T_{ij}}{\displaystyle\sum_{j=1}^{b} v_{ij}} \right) = 0$	$\displaystyle\sum_{j=1}^{b} v_{ij} / br$
$\eta_{ij} = b_j \eta$	$\dfrac{1}{\hat{\beta}_4} + \dfrac{s_{..}}{abr} - b^{-1} \displaystyle\sum_{j=1}^{b} \left(\dfrac{\displaystyle\sum_{i=1}^{a} T_{ij}}{\displaystyle\sum_{i=1}^{a} v_{ij}} \right) = 0$	$\displaystyle\sum_{i=1}^{a} v_{ij} / ar$
$\eta_{ij} = \eta$	$\dfrac{1}{\hat{\beta}_5} + \dfrac{s_{..}}{abr} - \displaystyle\sum_{i=1}^{a} \sum_{j=1}^{b} T_{ij} \bigg/ \displaystyle\sum_{i=1}^{a} \sum_{j=1}^{b} v_{ij} = 0$	$\displaystyle\sum_{i=1}^{a} \sum_{j=1}^{b} v_{ij} / abr$

$$T_{ij} = \sum_{k=1}^{n} x_{ij(k)}^{\hat{\beta}} \ln x_{ij(k)} \tag{11.16}$$

$$s_{ij} = \sum_{k=1}^{r} \ln x_{ij(k)} \tag{11.17}$$

$$s_{..} = \sum_{i=1}^{a} \sum_{j=1}^{b} s_{ij}. \tag{11.18}$$

Also listed in Table 11.1 is the expression for the maximum likelihood estimate of the cell scale parameters raised to a power equal to the shape parameter estimate appropriate to that model. Thus, for example, to estimate η_{ij} under the last model of Table 11.1 one computes:

$$\hat{\eta}_{ij} = \sum_{i=1}^{a} \sum_{j=1}^{b} v_{ij} / br. \tag{11.19}$$

Except for model 2, the shape parameter estimates are special cases of the shape parameter estimate applicable to k groups of Weibull data under the assumption that the shape parameter is the same for all groups. This methodology was presented in Chapter 8.

$\hat{\beta}_1$ is the estimate obtained when each of the cells in the data array is taken as a group. In this case $k = ab$. $\hat{\beta}_3$ results when each row is taken as a group, that is, the columns are collapsed. In this case $k = a$. Correspondingly, $\hat{\beta}_4$ is obtained by collapsing rows and treating the data in each column as a group. This gives $k = b$. $\hat{\beta}_5$ results when all of the data in the array are treated as a single large group, that is, $k = 1.0$.

11.5 TEST FOR THE APPROPRIATE MODEL

The full model given by Equation 11.3 is the least restrictive. Under this model η_{ij} is estimated using only the data in the cell i–j along with the common shape parameter estimate $\hat{\beta}_1$ obtained using all of the data in the entire array. Succeeding models are successively more restrictive with the last model $\eta_{ij} = \eta$ representing the case where the entire sample of abn items come from a single Weibull population. When a restrictive model is inappropriate, a consequence is that the shape parameter estimated under that model will tend to be smaller than it is when the appropriate model is used.

It has been shown that one may use the ratio of shape parameter estimates as the basis for a test of whether a more restrictive model is tenable. For example, if interaction is absent, $\hat{\beta}_1 / \hat{\beta}_2$ should be about unity. If interaction is present, however, $\hat{\beta}_2$ will be relatively smaller than $\hat{\beta}_1$ and the ratio $\hat{\beta}_1 / \hat{\beta}_2$ will therefore tend to be larger than unity.

In the language of hypothesis testing, our so-called null hypothesis would be that interaction is absent, that is,

$$H_0 : c_{ij} = 1.0 \quad (\text{all } i, j)$$

This hypothesis would be rejected with significance level $\alpha = 0.05$ if

$$\frac{\hat{\beta}_1}{\hat{\beta}_2} > \left[\frac{\hat{\beta}_1}{\hat{\beta}_2}\right]_{0.95}. \tag{11.20}$$

$(\hat{\beta}_1 / \hat{\beta}_2)_{0.95}$ represents the 95th percentile of the distribution of the ratio $(\hat{\beta}_1 / \hat{\beta}_2)$ applicable when the hypothesis is true. It serves as a measure of the relative rarity of larger $\hat{\beta}_1 / \hat{\beta}_2$ ratios. Only 5% of the time will a larger value be encountered due to chance alone when interaction is absent. If we encounter a larger value we proclaim that interaction exists and tolerate a 5% risk that our proclamation is wrong. If the hypothesis of no interaction were accepted one could proceed to test $\hat{\beta}_1 / \hat{\beta}_3$ to see if the column effects are also negligible.

11.6 MONTE CARLO RESULTS

Monte Carlo simulation was used to produce 10,000 simulated factorial experiments of various sizes in which the data in all cells were drawn from a common Weibull population. This was done for values of sample size n and censoring number r ranging from $n = r = 2$ to $n = r = 10$ and for 2×2, 2×3, and 3×3 layouts.

For each simulated experiment the values $\hat{\beta}_1 - \hat{\beta}_5$ were computed using the equations in Table 11.1. (For the 2×3 and 3×3 experiments, $\hat{\beta}_2$ was not calculated.) The ratios of $\hat{\beta}_1$ to the other values of $\hat{\beta}$ were calculated for each experiment and sorted from low to high to determine the percentiles.

Tables 11.2–11.4 list the upper 90%, 95%, and 99% points of these ratios along with the 5%, 10%, 50%, 90%, and 95% points of

$$v = \frac{\hat{\beta}_1}{\beta}. \tag{11.21}$$

These latter values are used for setting confidence limits on the shape parameter β as in Chapter 8. For 90% confidence limits one uses:

$$\frac{\hat{\beta}_1}{v_{0.95}} < \beta < \frac{\hat{\beta}_1}{v_{0.05}}. \tag{11.22}$$

11.7 THE DOS PROGRAM TWOWAY

A computer program named "TWOWAY" was written in the Basic language to analyze a factorial experiment using the methodology described here. For a 2×2 experiment the program computes the roots of the nonlinear equations listed in Table 11.1 corresponding to each of the five hypotheses on the scale parameters of the cells. The analysis employs a bisection technique to isolate the value of the roots within 0.00001. Having found $\hat{\beta}_k$ corresponding to the hypothesis H_k, the program proceeds to compute the ML estimates of the scale parameter η_{ij} applicable to each cell of the table using the appropriate formula in Table 11.1. For other than 2×2 experiments TWOWAY omits the computation of $\hat{\beta}_2$ and the scale parameter estimates associated with H_2.

11.8 ILLUSTRATION OF THE INFLUENCE OF FACTOR EFFECTS ON THE SHAPE PARAMETER ESTIMATES

To illustrate the analysis of a 2×2 Weibull factorial experiment and to demonstrate the discriminating power of the tests, a hypothetical 2×2 experiment is considered for which no row, column or interaction effects are present. The

Table 11.2 Selected Percentage Points of the Distributions of $\hat{\beta}_1/\beta$ and $\hat{\beta}_1/\hat{\beta}_k$ ($k = 2$–5) for Various n and r; $a = b = 2$

n	r	$\hat{\beta}_1/\beta$					$\hat{\beta}_1/\hat{\beta}_2$			$\hat{\beta}_1/\hat{\beta}_3$ and $\hat{\beta}_1/\hat{\beta}_4$			$\hat{\beta}_1/\hat{\beta}_5$		
		0.05	0.10	0.50	0.90	0.95	0.90	0.95	0.99	0.90	0.95	0.99	0.90	0.95	0.99
2	2	0.944	1.069	1.764	3.284	4.019	1.742	2.060	3.160	2.066	2.439	3.729	2.417	2.889	4.538
3	3	0.870	0.960	1.357	2.068	2.362	1.307	1.437	1.762	1.437	1.599	1.942	1.601	1.782	2.220
4	2	0.861	0.979	1.610	3.015	3.690	1.741	2.063	3.132	2.078	2.440	3.701	2.420	2.879	4.412
4	4	0.857	0.929	1.238	1.705	1.903	1.184	1.261	1.461	1.285	1.365	1.576	1.378	1.481	1.714
5	3	0.755	0.832	1.181	1.801	2.049	1.290	1.417	1.726	1.438	1.575	1.912	1.575	1.745	2.169
5	5	0.859	0.920	1.179	1.549	1.683	1.131	1.186	1.305	1.209	1.270	1.417	1.277	1.346	1.522
6	3	0.735	0.810	1.151	1.755	1.997	1.288	1.413	1.722	1.437	1.574	1.906	1.568	1.740	2.165
6	6	0.857	0.913	1.140	1.455	1.560	1.100	1.141	1.236	1.161	1.208	1.322	1.214	1.267	1.396
7	3	0.722	0.796	1.131	1.727	1.967	1.288	1.412	1.721	1.435	1.571	1.904	1.566	1.737	2.153
7	7	0.862	0.912	1.114	1.392	1.489	1.082	1.120	1.212	1.134	1.176	1.272	1.180	1.227	1.339
8	4	0.689	0.745	0.997	1.380	1.538	1.171	1.235	1.421	1.263	1.339	1.529	1.348	1.441	1.655
8	8	0.867	0.911	1.099	1.345	1.433	1.071	1.099	1.164	1.112	1.144	1.219	1.149	1.189	1.273
9	4	0.678	0.733	0.980	1.359	1.515	1.170	1.235	1.418	1.262	1.337	1.527	1.347	1.440	1.655
9	9	0.872	0.911	1.088	1.317	1.391	1.061	1.089	1.149	1.099	1.130	1.199	1.131	1.165	1.243
10	5	0.663	0.710	0.914	1.207	1.318	1.118	1.163	1.272	1.188	1.242	1.371	1.251	1.313	1.462
10	10	0.873	0.913	1.076	1.285	1.353	1.053	1.073	1.125	1.085	1.112	1.169	1.115	1.144	1.215

From McCool, The analysis of a two way factorial design with Weibull response. *Communications in Statistics-Simulation and Computation* 25(1): 263–286. Reprinted by permission of Taylor and Francis (http://www.tandfonline.com).

Table 11.3   Selected Percentage Points of the Distributions of $\hat{\beta}_1/\beta$ and $\hat{\beta}_1/\hat{\beta}_k$ ($k = 3$–5) for Various n and r; $a = 2, b = 3$

n	r	$\hat{\beta}_1/\beta$					$\beta/\hat{\beta}_3$			$\hat{\beta}_1/\hat{\beta}_4$			$\hat{\beta}_1/\hat{\beta}_5$		
		0.05	0.10	0.50	0.90	0.95	0.90	0.95	0.99	0.90	0.95	0.99	0.90	0.95	0.99
2	2	1.028	1.144	1.736	2.846	3.303	2.115	2.400	3.247	1.911	2.149	2.862	2.284	2.606	3.492
3	3	0.936	1.007	1.353	1.869	2.074	1.475	1.589	1.842	1.391	1.494	1.747	1.561	1.687	1.988
4	2	0.940	1.044	1.586	2.599	3.019	2.110	2.397	3.215	1.920	2.149	2.844	2.275	2.596	3.462
4	4	0.910	0.973	1.229	1.593	1.723	1.309	1.381	1.527	1.250	1.309	1.451	1.361	1.437	1.606
5	3	0.814	0.878	1.177	1.626	1.807	1.456	1.563	1.793	1.376	1.475	1.712	1.537	1.657	1.930
5	5	0.906	0.955	1.169	1.465	1.569	1.224	1.272	1.395	1.185	1.229	1.332	1.262	1.314	1.449
6	3	0.793	0.856	1.147	1.584	1.762	1.454	1.560	1.792	1.376	1.472	1.706	1.535	1.652	1.927
6	6	0.899	0.946	1.136	1.383	1.463	1.176	1.215	1.292	1.144	1.181	1.258	1.205	1.249	1.343
7	3	0.779	0.841	1.127	1.558	1.734	1.454	1.558	1.786	1.374	1.469	1.701	1.535	1.651	1.923
7	7	0.900	0.942	1.110	1.335	1.407	1.144	1.179	1.254	1.118	1.149	1.215	1.170	1.208	1.293
8	4	0.733	0.784	.991	1.287	1.397	1.286	1.352	1.485	1.231	1.288	1.417	1.334	1.399	1.553
8	8	0.903	0.940	1.097	1.288	1.350	1.122	1.150	1.213	1.099	1.125	1.183	1.144	1.174	1.242
9	4	0.721	0.771	.975	1.267	1.376	1.284	1.350	1.481	1.230	1.287	1.414	1.332	1.398	1.549
9	9	0.904	0.940	1.085	1.263	1.325	1.108	1.131	1.182	1.087	1.109	1.155	1.127	1.152	1.205
10	5	0.701	0.739	.909	1.141	1.227	1.202	1.246	1.354	1.168	1.208	1.298	1.238	1.286	1.406
10	10	0.906	.941	1.074	1.239	1.294	1.093	1.114	1.163	1.076	1.095	1.138	1.111	1.133	1.183

From McCool. The analysis of a two way factorial design with Weibull response. *Communications in Statistics-Simulation and Computation* 25(1): 263–286. Reprinted by permission of Taylor and Francis (http://www.tandfonline.com).

Table 11.4 Selected Percentage Points of the Distributions of $\hat{\beta}_1/\beta$ and $\hat{\beta}_1/\beta_k$ ($k = 3\text{--}5$) for Various n and r; $a = 3$ and $b = 3$

n	r	$\hat{\beta}_1/\beta$					$\hat{\beta}_1/\hat{\beta}_3$ and $\hat{\beta}_1/\hat{\beta}_4$			$\hat{\beta}_1/\hat{\beta}_5$		
		0.05	0.10	0.50	0.90	0.95	0.90	0.95	0.99	0.90	0.95	0.99
2	2	1.118	1.224	1.717	2.514	2.829	1.953	2.170	2.670	2.160	2.401	3.004
3	3	0.991	1.057	1.343	1.753	1.888	1.432	1.513	1.713	1.526	1.627	1.845
4	2	1.023	1.118	1.571	203,02	2.595	1.953	2.165	2.682	2.155	2.393	2.977
4	4	0.955	1.008	1.226	1.512	1.611	1.275	1.328	1.155	1.338	1.398	1.529
5	3	0.963	0.919	1.169	1.529	1.648	1.416	1.496	1.676	1.506	1.599	1.796
5	5	0.945	0.988	1.167	1.392	1.465	1.202	1.239	1.320	1.247	1.290	1.378
6	3	0.841	0.895	1.139	1.489	1.606	1.414	1.493	1.672	1.504	1.595	1.794
6	6	0.938	0.976	1.132	1.324	1.392	1.159	1.188	1.245	1.196	1.226	1.297
7	3	0.827	0.879	1.119	1.464	1.579	1.413	1.492	1.671	1.503	1.593	1.793
7	7	0.935	0.970	1.109	1.283	1.340	1.130	1.153	1.209	1.160	1.185	1.251
8	4	0.769	0.811	0.989	1.222	1.303	1.256	1.303	1.415	1.313	1.367	1.490
8	8	0.932	0.962	1.094	1.249	1.301	1.111	1.131	1.174	1.137	1.160	1.208
9	4	0.756	0.798	0.973	1.202	1.284	1.255	1.302	1.414	1.312	1.367	1.487
9	9	0.932	0.959	1.081	1.221	1.268	1.097	1.114	1.151	1.119	1.139	1.181
10	5	0.734	0.768	0.907	1.086	1.144	1.183	1.217	1.289	1.226	1.263	1.345
10	10	0.932	0.960	1.027	1.202	1.244	1.086	1.102	1.135	1.106	1.124	1.158

From McCool. The analysis of a two way factorial design with Weibull response. *Communications in Statistics-Simulation and Computation* 25(1): 263–286. Reprinted by permission of Taylor and Francis (http://www.tandfonline.com).

example data are analyzed and then successively reanalyzed after the introduction of various effects. The data are shown below:

1.0	1.0
2.0	2.0
3.0	3.0
4.0	4.0
1.0	1.0
2.0	2.0
3.0	3.0
4.0	4.0

Since each cell contains exactly the same four values, the comparisons to
be made will be free of the random contribution of cell-to-cell variability.

An initial screen for TWOWAY asks for the number of rows and columns
and the common value of the sample size n and censoring number r for each
cell. The following screen, shown as Figure 11.1, then appears for the user to
input the data values in each of the cells of the two-way array.

After all the data have been entered, the following screen (Figure 11.2)
appears asking for a range of values within which all of the shape parameter
estimates will lie. If the range that the user enters is not sufficiently wide, the

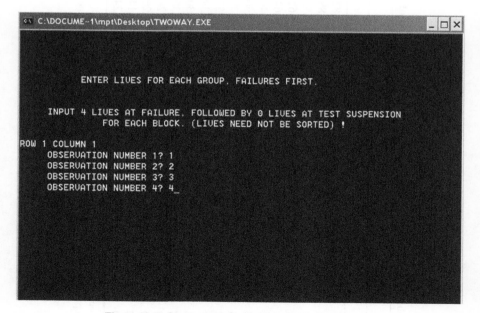

Figure 11.1 Input screen for the data in row 1 column 1.

Figure 11.2   This screen appears when all data have been entered.

program will ask for a revised range. When the factor effects are large, a larger range will be required.

The output screen shown in Figure 11.3 gives the five-shape parameter estimates and the ML estimates of the scale parameters under each model.

The analysis yields the following estimates of the shape parameter:

$$\hat{\beta}_1 = \hat{\beta}_2 = \hat{\beta}_3 = \hat{\beta}_4 = \hat{\beta}_5 = 2.4532.$$

All of the estimates are seen to be equal when the data in each cell are identical.

A multiplicative row effect is now introduced by multiplying the data in row 1 by 1/2 and the data in row 2 by 2.0. The resultant data are tabled below:

0.5	0.5
1.0	1.0
1.5	1.5
2.0	2.0
2.0	2.0
4.0	4.0
6.0	6.0
8.0	8.0

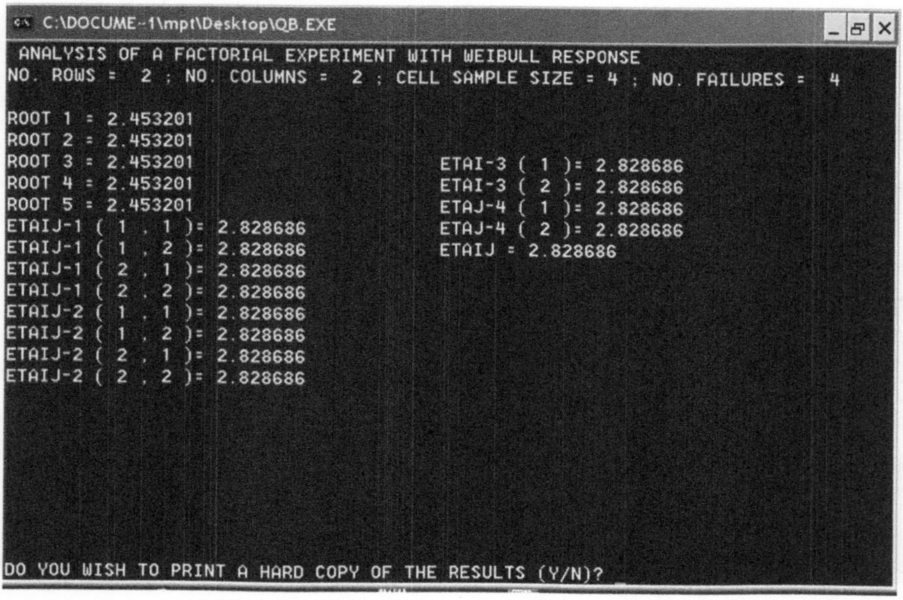

```
cs  C:\DOCUME~1\mpt\Desktop\QB.EXE                                    _ ⊟ ×
ANALYSIS OF A FACTORIAL EXPERIMENT WITH WEIBULL RESPONSE
NO. ROWS =  2 ; NO. COLUMNS =  2 ; CELL SAMPLE SIZE = 4 ; NO. FAILURES =  4

ROOT 1 = 2.453201
ROOT 2 = 2.453201
ROOT 3 = 2.453201                        ETAI-3 ( 1 )= 2.828686
ROOT 4 = 2.453201                        ETAI-3 ( 2 )= 2.828686
ROOT 5 = 2.453201                        ETAJ-4 ( 1 )= 2.828686
ETAIJ-1 ( 1 , 1 )= 2.828686             ETAJ-4 ( 2 )= 2.828686
ETAIJ-1 ( 1 , 2 )= 2.828686             ETAIJ = 2.828686
ETAIJ-1 ( 2 , 1 )= 2.828686
ETAIJ-1 ( 2 , 2 )= 2.828686
ETAIJ-2 ( 1 , 1 )= 2.828686
ETAIJ-2 ( 1 , 2 )= 2.828686
ETAIJ-2 ( 2 , 1 )= 2.828686
ETAIJ-2 ( 2 , 2 )= 2.828686






DO YOU WISH TO PRINT A HARD COPY OF THE RESULTS (Y/N)?
```

Figure 11.3 The results screen for TWOWAY.

The shape parameter estimates for these data are:

$$\hat{\beta}_1 = \hat{\beta}_2 = \hat{\beta}_3 = 2.4532$$

$$\hat{\beta}_4 = \hat{\beta}_5 = 1.2939.$$

The multiplicative row effect has left $\hat{\beta}_1$ to $\hat{\beta}_3$ unchanged since they are unaffected by multiplicative factors applied to the rows. $\hat{\beta}_4 = \hat{\beta}_5$ and both reflect the row effect to the same degree. The ratio $\hat{\beta}_1 / \hat{\beta}_4 = 1.896$ is well beyond the 99-th percentile of the null distribution as may be seen from the percentage points listed in Table 11.2.

Adding a column effect by multiplying column 1 by 1/4 and column 2 by 4 transforms the data to:

0.125	2.0
0.250	4.0
0.375	6.0
0.500	8.0
0.5	8.0
1.0	16.0
1.5	24.0
2.0	32.0

The β estimates now become:

$$\hat{\beta}_1 = \hat{\beta}_2 = 2.4532$$

$$\hat{\beta}_3 = 0.77832 \quad \hat{\beta}_4 = 1.2939 \quad \hat{\beta}_5 = 0.6744.$$

It is seen that $\hat{\beta}_2$ is unchanged since this estimator allows for the presence of both row and column effects. $\hat{\beta}_3$ is smaller than $\hat{\beta}_4$, reflecting the fact that the column effect that was introduced was twice as large as the row effect. $\hat{\beta}_5$ is diminished by both the row and column multipliers and thus is smaller than both $\hat{\beta}_3$ and $\hat{\beta}_4$. The ratio $\hat{\beta}_1 / \hat{\beta}_3 = 3.151$ is highly significant. The ratio $\hat{\beta}_1 / \hat{\beta}_4 = 1.896$ remains the same as before, indicating that the introduction of a column effect has not altered the significance of the row effect.

Finally, transforming the original data by the factors $c_{11} = c_{22} = \frac{1}{2}$ and $c_{12} = c_{21} = 2$ results in the following data representing a pure interaction effect.

0.5	2.0
1.0	4.0
1.5	6.0
2.0	8.0
2.0	0.5
4.0	1.0
6.0	1.5
8.0	2.0

The estimates are:

$$\hat{\beta}_1 = 2.4532$$

$$\hat{\beta}_2 = \hat{\beta}_3 = \hat{\beta}_4 = \hat{\beta}_5 = 1.2939.$$

It is seen that $\hat{\beta}_1$ is unchanged and that each of the other β estimates is affected in the same amount. The individual tests are all highly significant, but a significant interaction effect makes row and column effect tests immaterial inasmuch as each cell must be separately interpreted when interaction exists.

11.9 NUMERICAL EXAMPLES

To illustrate the analysis on data which are known to conform to its inherent assumptions, a 2×2 array was developed using simulation. Two uncensored samples of size n = 5 were generated from a Weibull population having $\eta = 2$ and $\beta = 2$. These data were used to form the first row of the 2×2 layout. Two further samples of size n = 5 were drawn from the Weibull population having $\eta = 1/2$ and $\beta = 2$ to form the second row. The sorted data are tabled below:

0.6297	1.021
0.7960	1.107
0.9468	1.502
2.208	1.945
2.147	2.727
0.1117	0.1999
0.2361	0.3451
0.3038	0.6332
0.3310	0.7275
0.6333	0.7447

In terms of the model, these data represent the case where the base scale parameter $\eta = 1$, there is a row effect $a_1 = 2$ and $a_2 = 1/2$ but no column or interaction effect ($b_1 = b_2 = c_{11} = 1$).

The computed estimates of the shape parameter are listed below:

$$\hat{\beta}_1 = 2.358$$

$$\hat{\beta}_2 = 2.325$$

$$\hat{\beta}_3 = 2.222$$

$$\beta_4 = 1.362$$

$$\beta_5 = 1.354$$

The ratios $\hat{\beta}_1 / \hat{\beta}_k$ are tabled below for $k = 2$–5, along with the associated P values as estimated from the tabular Monte Carlo distributions of the various ratios as determined under the null hypothesis.

Effect	Shape Parameter Ratio	p
Interaction	$\hat{\beta}_1 / \hat{\beta}_2 = 1.014$	0.59
Column	$\hat{\beta}_1 / \hat{\beta}_3 = 1.061$	0.50
Row	$\hat{\beta}_1 / \hat{\beta}_4 = 1.731$	<0.01
All	$\hat{\beta}_1 / \hat{\beta}_5 = 1.741$	<0.01

It is seen that the analysis has correctly detected the row effect and, also, correctly failed to show a significant column or interaction effect. The ratio for "ALL" will react to all significant effects. The fact that its magnitude is close to the magnitude of the ratio for the row effect further reflects the fact that only the row effect is real. The estimated parameters, assuming only row effects are meaningful, are:

$$\hat{a}_1 = 1 / \hat{a}_2 = 1.847$$

$$\hat{\eta} = 0.9082$$

$$\hat{\beta} = 2.358.$$

A 90% confidence interval for β may be estimated from Equation 11.22 using the percentage points of $v = \hat{\beta}_1 / \beta$ listed in Table 11.2.

$$1.40 = 2.358 / 1.683 < \beta < 2.358 / 0.859 = 2.75.$$

It is noted that this interval includes the true value $\beta = 2.0$. A median unbiased estimate of the shape parameter is computed using the median value, $v_{0.50}$, of $v = \hat{\beta}_1 / \beta$ listed in Table 11.2 as:

$$\hat{\beta}_1' = 2.358 / 1.179 = 2.0.$$

It is seen that in an unusual coincidence, the median unbiased estimate is exactly equal to the true value of β.

It is of interest to analyze these same data by means of the two-way analysis of variance (ANOVA) after first applying a logarithmic transformation. As discussed in Section 3.3, this is a reasonable approximate approach inasmuch as the transformed data will satisfy the additivity and constant variance assumptions of the analysis of variance if the untransformed data follow the multiplicative Weibull model.

After transforming the data by taking natural logarithms, the computed ANOVA table is as follows:

Source	Degrees of Freedom	SS	MS	F	p
Rows	1	8.635	8.635	28.2	0.0
Columns	1	0.813	0.813	2.65	0.12
Interaction	1	0.079	0.079	0.288	0.62
Error	16	4.90	0.306		

It is seen that the analysis correctly assesses the significance of the the row effect and the absence of interaction. The p value for the column effect is small enough, however, to mislead many experimenters into accepting that the column effect is real. The shape parameter ratio test, on the other hand, gave no indication of a column effect. The cell and marginal means of the transformed data are shown below and seem to suggest a spurious column effect.

0.162	0.440	0.301
−1.28	−0.748	−1.01
−0.558	−0.155	

An example of a 2×2 factorial experiment with rolling contact fatigue life as the response variable is given in McCool (1996b).

As a second example of the methodology we will analyze a 2×2 experiment in which the shear strengths of 10 specimens of a polymer material used in dental restorations were measured at all four combinations of the levels of two factors: (1) the presence and absence of silanation, a treatment designed to bond the silica filler to the polymer matrix. and (2) whether or not the specimens had been soaked to saturation in a 50:50 mixture of ethanol and water. The raw data and Weibull plots are reported in an article by McCool and Baran (1999). The ML estimates of the shape parameters for the four individual samples are displayed below:

	Unsoaked	Soaked
Silanated	25.4	17.9
Unsilanated	17.1	18.5

To test whether the data are consistent with a common shape parameter assumption we compute:

$$w(10, 10, 4) = \frac{\hat{\beta}_{max}}{\hat{\beta}_{min}} = \frac{25.4}{17.1} = 1.49.$$

Running the Multi-Weibull software described in Chapter 8, the 30th percentile of $w(10,10,4) = 1.50$. There is thus no reason to reject the common shape parameter hypothesis ($p \approx 0.70$).

The five estimates of the shape parameter obtained under the five models were computed to be as follows:

$$\hat{\beta}_1 = 19.018$$
$$\hat{\beta}_2 = 6.421$$
$$\hat{\beta}_3 = 2.327$$
$$\hat{\beta}_4 = 6.419$$
$$\hat{\beta}_5 = 2.111.$$

The various ratios and the upper 0.90, 0.95, and 0.99 points of their null distribution are tabled below:

Ratio of Shape Parameter Estimates	Upper Percentage Points		
	0.90	0.95	0.99
$\hat{\beta}_1 / \hat{\beta}_2 = 2.96$	1.053	1.073	1.1125
$\hat{\beta}_1 / \hat{\beta}_3 = 8.17$	1.085	1.112	1.1171
$\hat{\beta}_1 / \hat{\beta}_4 = 2.96$	1.085	1.112	1.1171
$\hat{\beta}_1 / \hat{\beta}_5 = 9.013$	1.115	1.114	1.215

$\hat{\beta}_1 / \hat{\beta}_2$ is significant, indicating that the interaction effect is real. When the data exhibit a significant interaction effect the other tests become irrelevant. Each cell of the matrix must be separately estimated subject to the commonality of the shape parameter. The shape parameter used to test for a column effect, $\hat{\beta}_3$, is reduced in magnitude both by the interaction effect and by the column effect, if there is one. Likewise the row effect shape parameter is reduced in magnitude both by the interaction effect and by the actual row effect, if any. From the table above the shape parameter ratio for testing the row effect is numerically about the same (2.96) as the ratio for the interaction, indicating that the row effect itself is not large. Adopting model 1 as the most appropriate characterization of the data, the estimated scale parameters are calculated to be:

	Unsoaked	Soaked
Silanated	42.8	21.84
Unsilanated	62.1	16.1

The interaction is evident in this tabulation. When silanated specimens are soaked, the strength is reduced by roughly half. When unsilanated specimens are soaked, the strength is reduced by to about a fourth of its unsoaked value.

REFERENCES

McCool, J.I. 1993. The Analysis of a Two Way Layout with Two Parameter Weibull Response. Department of Statistics. Philadelphia PA, Temple University. Ph.D: 143.

McCool, J.I. 1996a. The analysis of a two way factorial design with Weibull response. *Communications in Statistics-Simulation and Computation* 25(1): 263–286.

McCool, J.I. 1996b. The analysis of two factor rolling contact fatigue test results. *Tribology Transactions* 39: 59–66.

McCool, J.I. and G.R. Baran. 1999. The analysis of 2×2 factorial fracture experiments with brittle materials. *Journal of Material Science* 34: 3181–3188.

Zelen, M. 1959. Factorial experiments in life testing. *Technometrics* 1(3): 269–288.

Zelen, M., ed. 1960. *Analysis of Two Factor Classifications with Respect to Life Tests. Contributions to Statistics*. Stanford University Press, Palo Alto, CA.

EXERCISES

1. A 2×2 experiment was conducted with two factors A and B. For each combination of factor levels $n = 5$ uncensored life tests were run. The results are shown in the table below. Use the TWOWAY program to analyze the data and test for a row, column, and interaction effect.

	A1	A2
	13.95	22.41
	89.72	135.40
B1	103.28	149.38
	104.65	178.08
	118.09	219.12
	25.40	21.97
	50.76	29.61
B2	80.59	41.53
	162.03	191.90
	163.49	239.63

2. In a 2×2 experiment with $\eta_0 = 100$ and $a_1 = 2$ and $b_2 = 3$, compute the scale parameters for each combination of factor levels assuming there is no interaction.

Index

WILEY SERIES IN PROBABILITY AND STATISTICS
ESTABLISHED BY WALTER A. SHEWHART AND SAMUEL S. WILKS

Editors: *David J. Balding, Noel A. C. Cressie, Garrett M. Fitzmaurice, Harvey Goldstein, Iain M. Johnstone, Geert Molenberghs, David W. Scott, Adrian F. M. Smith, Ruey S. Tsay, Sanford Weisberg*
Editors Emeriti: *Vic Barnett, J. Stuart Hunter, Joseph B. Kadane, Jozef L. Teugels*

The *Wiley Series in Probability and Statistics* is well established and authoritative. It covers many topics of current research interest in both pure and applied statistics and probability theory. Written by leading statisticians and institutions, the titles span both state-of-the-art developments in the field and classical methods.

Reflecting the wide range of current research in statistics, the series encompasses applied, methodological and theoretical statistics, ranging from applications and new techniques made possible by advances in computerized practice to rigorous treatment of theoretical approaches.

This series provides essential and invaluable reading for all statisticians, whether in academia, industry, government, or research.

† ABRAHAM and LEDOLTER · Statistical Methods for Forecasting
AGRESTI · Analysis of Ordinal Categorical Data, *Second Edition*
AGRESTI · An Introduction to Categorical Data Analysis, *Second Edition*
AGRESTI · Categorical Data Analysis, *Second Edition*
ALTMAN, GILL, and McDONALD · Numerical Issues in Statistical Computing for the Social Scientist
AMARATUNGA and CABRERA · Exploration and Analysis of DNA Microarray and Protein Array Data
ANDĚL · Mathematics of Chance
ANDERSON · An Introduction to Multivariate Statistical Analysis, *Third Edition*
* ANDERSON · The Statistical Analysis of Time Series
ANDERSON, AUQUIER, HAUCK, OAKES, VANDAELE, and WEISBERG · Statistical Methods for Comparative Studies
ANDERSON and LOYNES · The Teaching of Practical Statistics
ARMITAGE and DAVID (editors) · Advances in Biometry
ARNOLD, BALAKRISHNAN, and NAGARAJA · Records
* ARTHANARI and DODGE · Mathematical Programming in Statistics
* BAILEY · The Elements of Stochastic Processes with Applications to the Natural Sciences
BAJORSKI · Statistics for Imaging, Optics, and Photonics
BALAKRISHNAN and KOUTRAS · Runs and Scans with Applications
BALAKRISHNAN and NG · Precedence-Type Tests and Applications
BARNETT · Comparative Statistical Inference, *Third Edition*
BARNETT · Environmental Statistics
BARNETT and LEWIS · Outliers in Statistical Data, *Third Edition*
BARTHOLOMEW, KNOTT, and MOUSTAKI · Latent Variable Models and Factor Analysis: A Unified Approach, *Third Edition*
BARTOSZYNSKI and NIEWIADOMSKA-BUGAJ · Probability and Statistical Inference, *Second Edition*
BASILEVSKY · Statistical Factor Analysis and Related Methods: Theory and Applications

*Now available in a lower priced paperback edition in the Wiley Classics Library.
†Now available in a lower priced paperback edition in the Wiley–Interscience Paperback Series.

---

*Now available in a lower priced paperback edition in the Wiley Classics Library.

†Now available in a lower priced paperback edition in the Wiley–Interscience Paperback Series.

---

*Now available in a lower priced paperback edition in the Wiley Classics Library.

†Now available in a lower priced paperback edition in the Wiley–Interscience Paperback Series.

*Now available in a lower priced paperback edition in the Wiley Classics Library.

†Now available in a lower priced paperback edition in the Wiley–Interscience Paperback Series.

*Now available in a lower priced paperback edition in the Wiley Classics Library.

†Now available in a lower priced paperback edition in the Wiley–Interscience Paperback Series.

*Now available in a lower priced paperback edition in the Wiley Classics Library.

†Now available in a lower priced paperback edition in the Wiley–Interscience Paperback Series.

*Now available in a lower priced paperback edition in the Wiley Classics Library.
†Now available in a lower priced paperback edition in the Wiley–Interscience Paperback
 Series.

*Now available in a lower priced paperback edition in the Wiley Classics Library.

†Now available in a lower priced paperback edition in the Wiley–Interscience Paperback Series.